T0296796

Online Learning and Adaptive Filters

Learn to solve the unprecedented challenges facing online learning and adaptive signal processing in this concise, intuitive text. The ever-increasing amount of data generated every day requires new strategies to tackle issues such as combining data from a large number of sensors; improving spectral usage, utilizing multiple antennas with adaptive capabilities; or learning from signals placed on graphs, generating unstructured data. Solutions to all of these and more are described in a condensed and unified way, enabling us to expose valuable information from data and signals quickly and economically. The up-to-date techniques explained here can be implemented in simple electronic hardware or as part of multipurpose systems. Also featuring alternative explanations for online learning, including newly developed methods and data selection, and several easily implemented algorithms, this one-of-a-kind book is an ideal resource for graduate students, researchers, and professionals in online learning and adaptive filtering.

Paulo S. R. Diniz is a professor at the Universidade Federal do Rio de Janeiro and a life fellow of the Institute of Electrical and Electronics Engineers and fellow of the European Association for Signal Processing. He is a coauthor of a Cambridge University Press textbook *Digital Signal Processing.* He is also a member of the National Academy of Engineering and the Brazilian Academy of Science.

Marcello L. R. de Campos is a professor at the Universidade Federal do Rio de Janeiro.

Wallace A. Martins is a research scientist at the University of Luxembourg.

Markus V. S. Lima is an associate professor at the Universidade Federal do Rio de Janeiro.

José A. Apolinário Jr. is an associate professor at the Instituto Militar de Engenharia.

Online Learning and Adaptive Filters

PAULO S. R. DINIZ
Universidade Federal do Rio de Janeiro

MARCELLO L. R. DE CAMPOS
Universidade Federal do Rio de Janeiro

WALLACE A. MARTINS
University of Luxembourg

MARKUS V. S. LIMA
Universidade Federal do Rio de Janeiro

JOSÉ A. APOLINÁRIO JR.
Instituto Militar de Engenharia

Shaftesbury Road, Cambridge CB2 8EA, United Kingdom

One Liberty Plaza, 20th Floor, New York, NY 10006, USA

477 Williamstown Road, Port Melbourne, VIC 3207, Australia

314–321, 3rd Floor, Plot 3, Splendor Forum, Jasola District Centre,
New Delhi – 110025, India

103 Penang Road, #05–06/07, Visioncrest Commercial, Singapore 238467

Cambridge University Press is part of Cambridge University Press & Assessment,
a department of the University of Cambridge.

We share the University's mission to contribute to society through the pursuit of
education, learning and research at the highest international levels of excellence.

www.cambridge.org
Information on this title: www.cambridge.org/9781108842129

DOI: 10.1017/9781108896139

First published 2023

A catalogue record for this publication is available from the British Library.

Library of Congress Cataloging-in-Publication Data
Names: Diniz, Paulo Sergio Ramirez, 1956– author.
Title: Online learning and adaptive filters / Paulo S.R. Diniz,
Universidade Federal do Rio de Janeiro, Marcello L.R. de Campos,
Universidade Federal do Rio de Janeiro, Wallace A. Martins, Universidade
Federal do Rio de Janeiro, Markus V. S. Lima, Universidade Federal do
Rio de Janeiro, Joco A. Apolinário Jr., Military Institute of Engineering.
Description: Cambridge, United Kingdom ; New York, NY, USA : Cambridge
University Press, [2023] | Includes bibliographical references and index.
Identifiers: LCCN 2022027011 | ISBN 9781108842129 (hardback) | ISBN
9781108896139 (ebook)
Subjects: LCSH: Adaptive signal processing – Mathematics. | Machine
learning – Mathematics. | Signal processing – Digital
techniques – Mathematics. | Digital filters (Mathematics)
Classification: LCC TK5102.9 .D634 2023 | DDC 621.382/2–dc23/eng/20220826
LC record available at https://lccn.loc.gov/2022027011

ISBN 978-1-108-84212-9 Hardback

Contents

Preface

The current trend of pervasive acquisition, transmission, and storage of data has brought about many technological challenges. Although the field of *Digital Signal Processing* has been developed for nearly half a century, there are many issues to be addressed originating from the ever-increasing amount of data generated every day. In particular, the field of *Adaptive Signal Processing*, despite being quite mature, requires new strategies to deal with a large number of data-acquiring sensors, which in turn gives rise to unprecedented challenges. Some of them are related to issues such as:

- How to combine data from a large number of sensors?
- How to improve spectral usage utilizing multiple antennas with adaptive capabilities?
- How to learn from signals placed on graphs, generating unstructured data?
- How to exploit sparsity in the estimation parameters, particularly in high-dimensional problems?
- How to minimize the amount of information exchange among the nodes of network elements, in order to save transmission power, bandwidth, and storage capability?
- How to detect and replace, to a certain extent, a faulty node in a sensor network?
- How to deal with the increased nonlinearities originating from faulty sensors and overloaded power amplifiers?

By applying adaptive filtering techniques appropriately, one can surpass many of these challenges. However, no textbook includes more recent tools to answer the questions above within a single bound.

The topics covered in this book are the subject of research of many groups with interest in supervised and unsupervised learning algorithms, including our group. In our particular case, we address algorithms that can be potentially implemented in embedded systems meant to provide online solutions. The philosophy is to present many selected algorithms straightforwardly to allow their prompt implementation.

Chapter 1 provides a brief description of the classical adaptive filtering algorithms starting with the definition of the actual objective function each algorithm minimizes. It also includes a summary of their expected performance according to available results from the open literature. In particular, the chapter briefly presents the least mean squares (LMS), recursive least squares (RLS), affine projection (AP), normalized LMS (NLMS), and set-membership AP algorithms.

Chapter 2 presents several strategies to exploit sparsity in the parameters being estimated in order to obtain better estimates, accelerate convergence, or reduce the computational complexity. It is worth mentioning that sparse systems received significant attention in the last two decades as they appear in many challenging problems like acoustic echo paths, underwater communication channels, 5G wireless communication channels, and medical images, among many others. In Chapter 2, the two main approaches employed to exploit sparsity in adaptive filtering are addressed. The first approach forces the parameters to be sparse, explicitly employing the application of a *sparsity-promoting penalty function* (or regularization). Within this section, this chapter covers adaptive filtering algorithms, employing the l_1-norm regularization and the smoothed l_0-norm regularization. While the l_1 norm leads to convex optimization problems, the addition of a regularization that approximates the l_0 norm leads to a nonconvex optimization problem, thus requiring deeper study. However, this drawback is compensated by the many advantages of the smoothed l_0 norm over the l_1-norm regularization. Indeed, by employing the former regularization, the resulting algorithms are capable of converging faster, obtaining better estimates, taking advantage of systems presenting lower degrees of sparsity, and not harming the estimates when applying these algorithms to estimate dispersive systems (i.e., systems which are not sparse). The second approach, on the other hand, herein called the *proportionate approach*, relies on the idea of updating each coefficient with a step size proportional to its own magnitude. In this way, large coefficients (in magnitude) update much faster than small ones. Finally, after discussing the proportionate approach, some of the most important algorithms of this class are presented, including the proportionate NLMS (PNLMS), the improved PNLMS (IPNLMS), the μ-law PNLMS (MPNLMS), the set-membership PNLMS (SM-PNLMS), as well as their AP generalizations. After explaining and discussing the pros and cons of these two kinds of regularization, some of the most widely used NLMS- and AP-based algorithms employing these regularizations are presented.

Chapter 3 explains the basic concepts of kernel-based methods, a widely used tool in machine learning. For example, in signal processing, the kernel methods can be adapted to perform the online parameter estimation of nonlinear models. The idea is to employ a kernel trick that enables the derivation of learning algorithms for nonlinear adaptive filters that resemble their linear counterparts. The emphasis of the chapter is to introduce the kernel version of classical algorithms such as LMS, RLS, AP, and SM-AP. In particular, we discuss how to keep the dictionary of the kernel finite through a series of model reduction strategies. This way, all discussed kernel algorithms are tailored for online implementation.

Chapter 4 shows how the classical adaptive filtering algorithms can be adapted to distributed learning. In distributed learning, there is a set of adaptive filters placed at nodes utilizing local input and desired signals. These distributed networks of sensor nodes are located at distinct positions, which might improve the reliability and robustness of the parameter estimation in comparison to stand-alone adaptive filters. In distributed adaptive networks, parameter estimation might be obtained in a centralized form or a decentralized form. In the centralized case, the signals from all nodes of the network are processed in a single fusion center, whereas in the decentralized case, they are processed locally, followed by a proper combination of partial estimates to result in a consensus parameter estimate. The main drawbacks of the centralized configuration are its data communication and computational costs, particularly in networks with a large number of nodes. On the other hand, the decentralized estimators require fewer data to feed the estimators and improve on robustness. Chapter 4 provides a discussion on equilibrium and consensus using arguments drawn from the pari-mutuel betting system. The expert opinion pool is the concept to induce improved estimation and data modeling, utilizing DeGroots algorithm and Markov chains as tools to probate equilibrium at consensus. Chapter 4 also introduces the distributed versions of the LMS and RLS adaptive filtering algorithms with emphasis on the decentralized parameter estimation case. Then, the chapter discusses a strategy to incorporate a data selection based on the SM adaptive filtering. It concludes with an extension of the concept to decentralized adaptive detection.

Chapter 5 discusses some techniques to derive constrained adaptive algorithms aiming at applications related to beamforming and array processing in general. The chapter briefly introduces the main concepts of array signal processing, emphasizing those related to adaptive beamforming, and discusses how to impose linear constraints on adaptive filtering algorithms to achieve the beamforming effect. Adaptive beamforming, emphasizing the incoming signal impinging from a known direction by means of an adaptive filter, is the primary objective of the array signal processing addressed in this chapter. We start this study with the narrowband beamformer. The constrained LMS, RLS, conjugate gradient, and SM-AP algorithms are introduced along with the generalized sidelobe canceller, and the Householder constrained structures; sparsity-promoting adaptive beamforming algorithms are also addressed in this chapter. After this, Chapter 5 introduces the concepts of frequency-domain and time-domain broadband adaptive beamforming and shows their equivalence. The chapter wraps up with brief discussions and reference suggestions on essential topics related to adaptive beamforming, including the numerical robustness of adaptive beamforming algorithms.

Chapter 6 briefly discusses how to process data that are originated on irregular discrete domains, an emerging area called graph signal processing (GSP). Basically, the type of graph we deal with consists of a network with vertices and weighted edges defining the neighborhood and the connections among the nodes. As such, the graph signals are collected in a vector whose entries represent the values of the signal nodes at a given time. A common issue related to GSP is the sampling problem, given the irregular structure of the data, where some sort of interpolation is possible whenever the graph signals are bandlimited or nearly bandlimited. These interpolations can be performed through the extension of the conventional adaptive filtering to signals distributed on graphs where there is no traditional data structure. The chapter presents the LMS, NLMS, and RLS algorithms for GSP along with their analyses and applications to estimate bandlimited signals defined on graphs. In addition, the chapter presents a general framework for data-selective adaptive algorithms for GSP.

This text aims to present a very concise introduction to the classical adaptive filtering techniques, a subject covered in many good books, and concentrate on explaining the new trends in this field, which were primarily developed in the last two decades and are still the subject of intensive research. Furthermore, many relevant algorithms belonging to the set of newly developed adaptive filtering solutions are described in detail in this book.

Acknowledgments

The support of the Department of Electronics and Computer Engineering of the Polytechnic School (undergraduate school of engineering) of Universidade Federal do Rio de Janeiro (UFRJ) and the Program of Electrical Engineering of Coordenação do Programas de Pos Graduação em Engenharia (COPPE) of UFRJ has been fundamental to complete this work.

We are very grateful to Professors Luiz W. P. Biscainho, Sergio L. Netto, and Eduardo A. B. da Silva, UFRJ, our technical brothers who share the Signals Multimedia and Telecommunications (SMT) Laboratory of UFRJ, with us. They represent much more than merely colleagues, providing the incentive for everything we do. They are inspiring and friendly, making our environment very pleasant to work. Our manager Michelle Nogueira keeps all of us on the right track, taking care of even the impossible. We have no words to thank her for taking care of us, our students, our staff, and the SMT. Professor T. N. Ferreira of the Universidade Federal Fluminense, who is always with us, has provided us with many valuable suggestions.

The long-term collaboration between the SMT and the Department of Electrical Engineering of the Instituto Militar de Engenharia originated the co-authorship of Professor José A. Apolinário Jr. to this book.

The financial support of the Brazilian research councils Conselho Nacional de Desenvolvimento Científico e Tecnológico (CNPq), Coordenação de Aperfeiçoamento de Pessoal de Nível Superior (CAPES), and Fundação de Amparo à Pesquisa do Estado do Rio de Janeiro (FAPERJ) was highly appreciated; without them, we would not have been able to complete this book.

1 Introduction

1.1 Introduction

Adaptive filtering has been a widely utilized signal processing tool, and is the subject of many textbooks including [1–9] just to mention a few. Usually, the classical adaptive filtering algorithms belong to the class of supervised learning algorithms, although there are unsupervised versions of them, such as the blind adaptive filters. In the supervised case, filtering structures tend to be quite simple, allowing the use of more complex learning algorithms. In the latter case, filtering structures are usually nonlinear and can be quite complex so that employing sophisticated learning algorithms might not be possible.

This book aims to describe more recent developments in this field not fully addressed in the classical texts. This chapter summarizes the main equations and features related to the most popular adaptive filtering algorithms. Often, these algorithms serve as a basis to the algorithms introduced in the subsequent chapters. In all cases, we describe the actual objective function utilized by each algorithm so that, from the engineering point of view, one can get a grasp of what the algorithms are minimizing. In addition to that, this Introduction establishes the notation used in this book. However, an experienced researcher in adaptive filtering may choose to skip this chapter without further consequences.

1.2 Data Description

The most common adaptive filtering configuration belongs to the class of supervised learning where an input signal is transformed into an output signal that in turn tries to track the behavior of a reference signal. Figure 1.1 depicts the typical adaptive filtering setup.

In most cases, the signals to be processed by an adaptive filter consist of an input signal denoted by $x(k)$ and a reference or desired signal denoted by $d(k)$. The input signal in the simplest environment is collected in a delay line vector as

$$\mathbf{x}(k) = [x(k)\, x(k-1)\, \cdots\, x(k-N)]^{\mathrm{T}}, \tag{1.1}$$

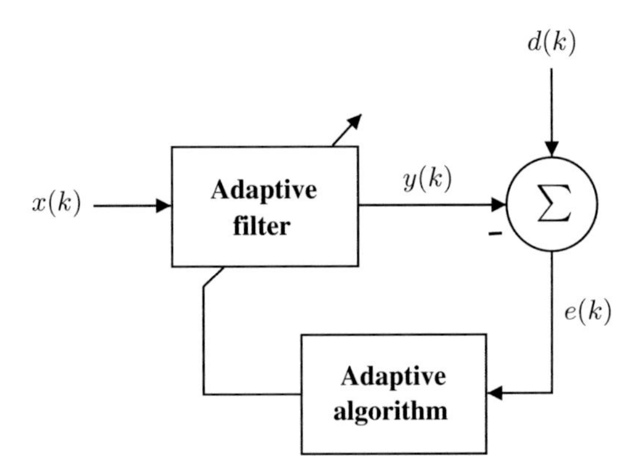

Figure 1.1 Adaptive filtering configuration.

where vector $\mathbf{x}(k) \in \mathbb{C}^{(N+1)\times 1}$, with $\mathbb{C}$ representing the complex numbers. In many situations, $\mathbf{x}(k) \in \mathbb{R}^{(N+1)\times 1}$, where $\mathbb{R}$ represents the real numbers.

The input-signal vector generates the output signal of the adaptive filter through the inner product, also known as the tapped delay line,

$$y(k) = \mathbf{w}^{\mathrm{H}}(k)\mathbf{x}(k), \tag{1.2}$$

where $\mathbf{w}(k) = [w_0(k)\ w_1(k)\ \cdots\ w_N(k)]^{\mathrm{T}}$ represents the adaptive filter coefficients or parameter vector, and $y(k)$ is the adaptive filter output. The latter output signal is compared with the reference signal to form the (*a priori*) error signal as follows:

$$e(k) = d(k) - \mathbf{w}^{\mathrm{H}}(k)\mathbf{x}(k). \tag{1.3}$$

The adaptive filter minimizes a cost function, also known as the objective function, of the error signal represented as

$$\xi(k) = \mathbb{F}[e(k)], \tag{1.4}$$

with $\mathbb{F}[\cdot]$ assuming distinct definitions depending on the algorithm and application.

1.3 The LMS Algorithm

The least mean squares (LMS) algorithm is recognized as the most widely used adaptive filtering algorithm ever, for its implementation simplicity and good performance in many practical situations. The LMS origins are described in the review chapters [10, 11], and its performance and features are addressed in many textbooks [1–9].

The basic concept behind the LMS algorithm comes from the steepest descent approach [1], where the coefficient vector is updated in the opposite direction of the gradient $\nabla_{\mathbf{w}^*}\xi(k)$, where $\xi(k) = \mathbb{E}[|d(k) - \mathbf{w}^{\mathrm{H}}\mathbf{x}(k)|^2]$ is the mean square error (MSE). In fact, the LMS algorithm uses an instantaneous estimate of the gradient at instant k, with the simplest possible option given by

$$\mathbf{w}(k+1) = \mathbf{w}(k) - \mu\nabla_{\mathbf{w}^*}\hat{\xi}(k), \tag{1.5}$$

where $\mu > 0$, known as step size, is responsible for controlling the convergence of the algorithm and $\hat{\xi}(k) = |e(k)|^2$. Another, and more elegant, way to obtain the coefficient update of the LMS algorithm comes from minimizing the following cost function:

$$\begin{aligned}\xi_{\mathrm{LMS}}(k) &= \|\mathbf{w}(k+1) - \mathbf{w}(k)\|^2_{\mathbf{P}(k)} + \mu|\varepsilon(k)|^2 \\ &= \|\mathbf{w}(k+1) - \mathbf{w}(k)\|^2_{\mathbf{P}(k)} + \mu|d(k) - \mathbf{w}^{\mathrm{H}}(k+1)\mathbf{x}(k)|^2,\end{aligned} \tag{1.6}$$

where the weighted quadratic norm is defined as $\|\mathbf{x}\|^2_{\mathbf{P}(k)} = \mathbf{x}^{\mathrm{H}}\mathbf{P}(k)\mathbf{x}$ and $\mathbf{P}(k) = \mathbf{I} - \mu\mathbf{x}(k)\mathbf{x}^{\mathrm{H}}(k)$, with μ being small enough to guarantee that $\mathbf{P}(k)$ is positive definite. It includes two terms: one related to the disturbance on the coefficients $\|\mathbf{w}(k+1) - \mathbf{w}(k)\|^2$ and the other related to the instantaneous squared *a posteriori* error $|\varepsilon(k)|^2$. It is worth mentioning that in Equation (1.6) $\mu \leq \frac{1}{\|\mathbf{x}\|^2}$ to keep the cost function convex.

By computing the gradient of the objective function with respect to $\mathbf{w}(k+1)$, the coefficient update that minimizes the objective function is

$$\mathbf{w}(k+1) = \mathbf{w}(k) + \mu e^*(k)\mathbf{x}(k). \tag{1.7}$$

As can be seen, the coefficient update equation of the LMS algorithm is quite simple and requires low computational costs. There is a vast literature addressing several issues pertaining to the LMS algorithm for which the previously mentioned references [1–9] represent a tiny sample.

1.4 The RLS Algorithm

The standard recursive least squares (RLS) algorithm minimizes the following function [1, 12, 13]:

$$\begin{aligned}\xi^{\mathrm{d}}(k) &= \sum_{i=0}^{k}\lambda^{k-i}|\varepsilon(i)|^2 \\ &= \sum_{i=0}^{k}\lambda^{k-i}\left|d(i) - \mathbf{w}^{\mathrm{H}}(k+1)\mathbf{x}(i)\right|^2,\end{aligned} \tag{1.8}$$

where $\varepsilon(i)$ is defined as the *a posteriori* error measured with the data entries $d(i)$ and $\mathbf{x}(i)$, and λ is the forgetting factor parameter. Another way to express the objective function is

$$\xi^{\mathrm{d}}(k) = |d(k)|^2 - 2\mathrm{Re}\left[d^*(k)\mathbf{w}^{\mathrm{H}}(k+1)\mathbf{x}(k)\right] + \left|\mathbf{w}^{\mathrm{H}}(k+1)\mathbf{x}(k)\right|^2 + \lambda\sum_{i=0}^{k-1}\lambda^{k-1-i}|\varepsilon(i)|^2, \tag{1.9}$$

where $\mathrm{Re}[\cdot]$ means real part of $[\cdot]$ from which a recursive solution can be obtained.

As previously seen for the LMS algorithm, the RLS algorithm also has a *minimum disturbance* cost function that is given as [14]

$$\xi_{\mathrm{RLS}}(k) = \lambda\|\mathbf{w}(k+1) - \mathbf{w}(k)\|^2_{\mathbf{R}_{\mathrm{D}}(k)} + |d(k) - \mathbf{w}^{\mathrm{H}}(k+1)\mathbf{x}(k)|^2, \tag{1.10}$$

where $\mathbf{R}_{\mathrm{D}}(k) = \sum_{i=0}^{k}\lambda^{k-i}\mathbf{x}(i)\mathbf{x}^{\mathrm{H}}(i)$, whose inverse is the matrix $\mathbf{S}_{\mathrm{D}}(k)$ used in the following.

The corresponding update equations of the standard RLS algorithm consist of

$$e(k) = d(k) - \mathbf{w}^{\mathrm{H}}(k)\mathbf{x}(k), \tag{1.11}$$

$$\boldsymbol{\psi}(k) = \mathbf{S}_{\mathrm{D}}(k-1)\mathbf{x}(k), \tag{1.12}$$

$$\mathbf{S}_{\mathrm{D}}(k) = \frac{1}{\lambda}\left[\mathbf{S}_{\mathrm{D}}(k-1) - \frac{\boldsymbol{\psi}(k)\boldsymbol{\psi}^{\mathrm{H}}(k)}{\lambda + \boldsymbol{\psi}^{\mathrm{H}}(k)\mathbf{x}(k)}\right], \text{ and} \tag{1.13}$$

$$\mathbf{w}(k+1) = \mathbf{w}(k) + e^*(k)\mathbf{S}_{\mathrm{D}}(k)\mathbf{x}(k), \tag{1.14}$$

where an initialization is required for the following quantities: $\mathbf{S}_{\mathrm{D}}(-1) = \rho\mathbf{I}$, ρ being the inverse of an estimate of the input signal power times $1-\lambda$, and $\mathbf{x}(-1) = \mathbf{w}(0) = [0\,0\cdots 0]^{\mathrm{T}}$.

In a more general RLS algorithm, employing a time-varying forgetting factor, the objective function we seek to minimize is given by

$$\begin{aligned}\xi^{\mathrm{d}}(k) &= \sum_{i=0}^{k}\lambda^{k-i+1}(i)|\varepsilon(i)|^2 \\ &= \sum_{i=0}^{k}\lambda^{k-i+1}(i)\left|d(i) - \mathbf{w}^{\mathrm{H}}(k+1)\mathbf{x}(i)\right|^2,\end{aligned} \tag{1.15}$$

where $\varepsilon(i)$ is also the *a posteriori* error measured with the comparison between $d(i)$ and $\mathbf{w}^{\mathrm{H}}(k+1)\mathbf{x}(i)$. The cost function can be written in an alternative form as

$$\xi^{\mathrm{d}}(k) = \lambda(k)\left[|d(k)|^2 - 2\mathrm{Re}\left[d^*(k)\mathbf{w}^{\mathrm{H}}(k+1)\mathbf{x}(k)\right] + \left|\mathbf{w}^{\mathrm{H}}(k+1)\mathbf{x}(k)\right|^2\right] + \sum_{i=0}^{k-1}\lambda^{k-i+1}(i)|\varepsilon(i)|^2. \tag{1.16}$$

The basic update equations of the generalized RLS algorithm are given by the following relations:

$$e(k) = d(k) - \mathbf{w}^{\mathrm{H}}(k)\mathbf{x}(k), \tag{1.17}$$

$$\boldsymbol{\psi}(k) = \mathbf{S}_{\mathrm{D}}(k-1)\mathbf{x}(k), \tag{1.18}$$

$$\mathbf{S}_{\mathrm{D}}(k) = \mathbf{S}_{\mathrm{D}}(k-1) - \frac{\lambda(k)\boldsymbol{\psi}(k)\boldsymbol{\psi}^{\mathrm{H}}(k)}{1+\lambda(k)\boldsymbol{\psi}^{\mathrm{H}}(k)\mathbf{x}(k)}, \text{ and} \tag{1.19}$$

$$\mathbf{w}(k+1) = \mathbf{w}(k) + \lambda(k)e^{*}(k)\mathbf{S}_{\mathrm{D}}(k)\mathbf{x}(k). \tag{1.20}$$

In the above relations, the initialization entails choosing $\mathbf{S}_{\mathrm{D}}(-1) = \rho\mathbf{I}$, where ρ can be the inverse of an estimate of the input signal power times $1-\lambda(0)$, and $\mathbf{x}(-1) = \mathbf{w}(0) = [0\,0\cdots 0]^{\mathrm{T}}$.

In the standard RLS algorithm, the value of the forgetting factor is usually chosen as $0 \ll \lambda \leq 1$, whereas in the generalized RLS $0 \ll \lambda(k) \leq 1$ and the choice of $\lambda(k)$ depends on the cost function. The latter case encompasses algorithms like BEACON versions of [17, 18] and of some references therein.

1.5 Affine Projection Algorithms

The affine projection (AP) algorithm assembles and reuses the last $L+1$ input signal vectors in an input-vector matrix according to [1, 15]:

$$\mathbf{X}(k) = [\mathbf{x}(k)\;\mathbf{x}(k-1)\cdots\mathbf{x}(k-L)], \tag{1.21}$$

where $\mathbf{X}(k) \in \mathbb{C}^{(N+1)\times(L+1)}$. At iteration k, we define the desired signal vector and the error vector, respectively, represented by

$$\mathbf{d}(k) = [d(k)\;d(k-1)\;\ldots\;d(k-L)]^{\mathrm{T}} \text{ and} \tag{1.22}$$

$$\mathbf{e}(k) = [e_0(k)\;e_1(k)\;\cdots\;e_L(k)]^{\mathrm{T}}, \tag{1.23}$$

where vectors $\mathbf{e}(k) \in \mathbb{C}^{(L+1)\times 1}$ and $\mathbf{d}(k) \subset \mathbb{C}^{(L+1)\times 1}$ retain information from the $L+1$ last iterations. The entries of the error vector are defined as $e_i(k) = d(k-i) - \mathbf{w}^{\mathrm{H}}(k)\mathbf{x}(k-i)$, for $i \in \{0, 1, \ldots, L\}$.

The AP algorithm presented here minimizes the objective function

$$\xi_{\mathrm{AP}}(k) = \|\mathbf{w}(k+1) - \mathbf{w}(k)\|^2_{\mathbf{P}(k)} + \|\mathbf{d}(k) - \mathbf{X}^{\mathrm{T}}(k)\mathbf{w}^{*}(k+1)\|^2_{\mathbf{A}(k)\mathbf{F}}, \tag{1.24}$$

where

$$\mathbf{A}(k) = \left(\mathbf{X}^{\mathrm{H}}(k)\mathbf{X}(k) + \gamma\mathbf{I}\right)^{-1} \text{ and} \tag{1.25}$$

$$\mathbf{P}(k) = \frac{1}{\mu}\mathbf{I} - \mathbf{X}(k)\mathbf{A}(k)\mathbf{F}\mathbf{X}^{\mathrm{H}}(k), \tag{1.26}$$

where μ represents the step size (or a learning factor) of the adaptive algorithm and γ is the regularization factor introduced to avoid numerical problems when matrix $\left(\mathbf{X}^{\mathrm{H}}(k)\mathbf{X}(k)\right)^{-1}$ becomes ill conditioned. The value of μ is small to keep the cost function convex. Matrix $\mathbf{F}$ might assume different forms depending on the desired characteristics of the adaptive algorithm and the affordable computational complexity [19–21].

The cost function in Equation (1.24) has two terms, where the first one implements the minimum disturbance in the adaptive coefficients weighted by matrix $\mathbf{P}(k)$, and the Euclidean norm of the *a posteriori* error vector properly normalized through matrix $\mathbf{A}(k)$. Matrices $\mathbf{A}(k)$ and $\mathbf{P}(k)$ are positive definite.

The standard AP algorithm utilizes $\mathbf{F} = \mathbf{I}$, leading to the coefficient updating in the form

$$\mathbf{w}(k+1) = \mathbf{w}(k) + \mu\mathbf{X}(k)\left(\mathbf{X}^{\mathrm{H}}(k)\mathbf{X}(k) + \gamma\mathbf{I}\right)^{-1}\mathbf{e}^{*}(k). \tag{1.27}$$

1.6 The Normalized LMS Algorithm

The NLMS algorithm, proposed in 1967 [22, 23] and widely used in practice, came as a response to the main drawbacks of the LMS algorithm, previously introduced in 1960 [24]: the slow convergence for correlated input signals and the need to choose an appropriate step size. One can note that the LMS algorithm updates in the direction of the input vector $\mathbf{x}(k)$. If we assume perfect modeling and no observation error, we could figure out that, given the data pair $d(k)$ and $\mathbf{x}(k)$, the optimal point $\mathbf{w}_{\mathrm{o}}$ would belong to a hyperplane $\mathbf{w}^{\mathrm{H}}\mathbf{x}(k) = d(k)$, which happens to be orthogonal to the input vector $\mathbf{x}(k)$. Therefore, as shown in Figure 1.2, we could choose a step size that leads to a null *a posteriori* error. The variable step size can be easily obtained, and its value corresponds to $\frac{1}{\mathbf{x}^{\mathrm{H}}(k)\mathbf{x}(k)}$. We may also obtain the updating expression of the NLMS algorithm by minimizing the following cost function:

$$\xi_{\mathrm{NLMS}}(k) = \|\mathbf{w}(k+1) - \mathbf{w}(k)\|^2_{\mathbf{P}(k)} \text{ subject to } \mathbf{w}^{\mathrm{H}}(k+1)\mathbf{x}(k) = d(k). \tag{1.28}$$

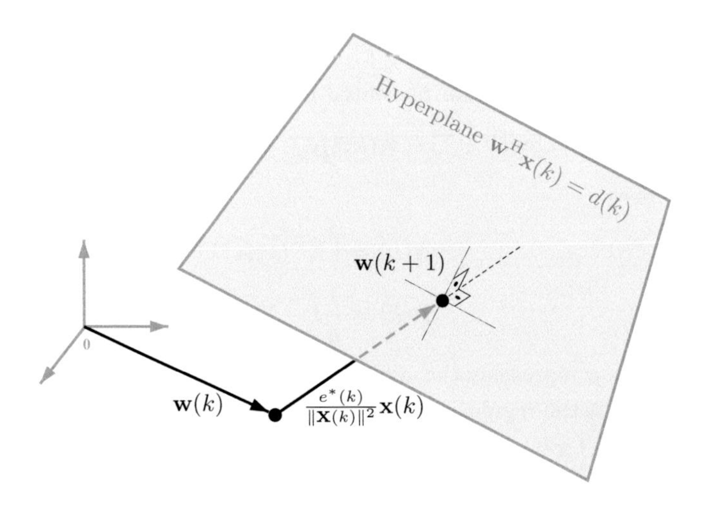

Figure 1.2 Coefficient vector updating of the NLMS algorithm with null *a posteriori* error.

The resulting updating expression of the NLMS algorithm is given as follows, where another step size was added to allow a trade-off between speed of convergence and misadjustment:[1]

$$\mathbf{w}(k+1) = \mathbf{w}(k) + \mu \frac{e^*(k)}{\mathbf{x}^{\mathrm{H}}(k)\mathbf{x}(k) + \gamma}\mathbf{x}(k), \tag{1.29}$$

where γ is a small positive number added to avoid divisions by 0.

Although the NLMS step size can theoretically range from 0 to 2, we recommend its use within $0 < \mu \leq 1$. Since the fastest convergence is obtained with $\mu = 1$, using a larger value would increase misadjustment and decrease convergence speed. Although developed independently and with another approach, the AP algorithm with the particular case where $L = 0$ becomes the NLMS algorithm.

1.7 Set-Membership Affine Projection Algorithms

Set-membership filtering (SMF) is a noteworthy example of connecting set-theoretic estimation with data selection, enabling a reduction of computational burden and, consequently, energy savings. As a bonus, in addition to reducing computational burden, SMF-based algorithms tend to be more robust against noise [16, 29].

The SMF concept appeared in [25] and is suitable for linear adaptive filtering problems. Its criterion aims to estimate the vector $\mathbf{w}$ that leads to an error signal $e = d - \mathbf{w}^{\mathrm{H}}\mathbf{x} \in \mathbb{C}$ whose magnitude is upper bounded by a constant $\overline{\gamma}$ for all possible pairs $\{\mathbf{x}, d\}$. Variable $\overline{\gamma}$ determines how much error is acceptable and is usually chosen based on *a priori* information about the sources of uncertainty. Most of the time, we assume that such uncertainty is caused by an observation error whose variance is σ_n^2, and $\overline{\gamma}$ is chosen as a function of σ_n^2.

Denoting $\mathcal{S}$ as the set comprising all possible pairs $\{\mathbf{x}, d\}$, we can state SMF is interested in finding $\mathbf{w}$ that satisfies $|e| = |d - \mathbf{w}^{\mathrm{H}}\mathbf{x}| \leq \overline{\gamma}$, $\forall \{\mathbf{x}, d\} \in \mathcal{S}$. That is, by defining the *feasibility set* as

$$\Theta \triangleq \bigcap_{\{\mathbf{x},d\} \in \mathcal{S}} \left\{ \mathbf{w} \in \mathbb{C}^{N+1} : |d - \mathbf{w}^{\mathrm{H}}\mathbf{x}| \leq \overline{\gamma} \right\}, \tag{1.30}$$

the SMF criterion can be summarized as finding a vector $\mathbf{w} \in \Theta$.

Considering online applications, the previous expression does not provide a practical way of determining Θ or a point in it since we do not have $\mathcal{S}$. The set-membership approach to iterative techniques, referred to as *set-membership*

[1] Misadjustment is defined as the ratio between the excess MSE and the minimum MSE, that is, $M = \frac{\xi(\infty) - \xi_{\min}}{\xi_{\min}}$ [1], where $\xi(\infty)$ corresponds to the steady-state MSE.

adaptive recursive techniques [25], is an alternative adaptive filtering formulation featuring a data-selective updating. This approach maintains the concept that, if the set of parameters leads to an error magnitude below a given threshold, no coefficient update is required at that particular iteration.

Consider a set of data pairs $\{\mathbf{x}(i), d(i)\}$, for $i \in \{0, 1, \ldots, k\}$, and define $\mathcal{H}(k)$ as the set containing all vectors $\mathbf{w}$ such that the associated output error at time instant k is upper bounded in magnitude by $\bar{\gamma}$, so that,

$$\mathcal{H}(k) = \{\mathbf{w} \in \mathbb{C}^{N+1} : |d(k) - \mathbf{w}^{\mathrm{H}}\mathbf{x}(k)| \leq \bar{\gamma}\}. \tag{1.31}$$

This set $\mathcal{H}(k)$ is called the *constraint set*. In the two-dimensional case ($N = 1$), the boundaries of $\mathcal{H}(k)$ are such that the error values are $\bar{\gamma}\mathrm{e}^{\mathrm{j}\phi}$, for $\phi \in [0, 2\pi)$, that is, $\mathcal{H}(k)$ comprises the region between the lines where $\left|d(k) - \mathbf{w}^{\mathrm{H}}\mathbf{x}(k)\right| = \bar{\gamma}$. For higher dimensions, hyperplanes delimit $\mathcal{H}(k)$; see [25–27] for details and [1, 28] for further developments.

Following the formulation of Section 1.5, the set-membership affine projection (SM-AP) algorithm, when updating, minimizes the following objective function:

$$\xi_{\mathrm{sm}}(k) = \|\mathbf{w}(k+1) - \mathbf{w}(k)\|^2_{\mathbf{P}(k)} + \|\mathbf{d}(k) - \mathbf{X}^{\mathrm{T}}(k)\mathbf{w}^*(k+1) - \bar{\boldsymbol{\gamma}}(k)\|^2_{\mathbf{A}(k)\mathbf{F}}, \tag{1.32}$$

where $\bar{\gamma}_i(k)$, the entries of $\bar{\boldsymbol{\gamma}}(k)$, are chosen such that $|\bar{\gamma}_i(k)| \leq \bar{\gamma}$ for $i \in \{1, \ldots, L{+}1\}$.

There are many ways to choose the entries of the constraint vector as long as they correspond to points represented by the adaptive-filter coefficients in $\mathcal{H}(k - i + 1)$, that is, $|\bar{\gamma}_i(k)| \leq \bar{\gamma}$. The choice of $\bar{\boldsymbol{\gamma}}(k)$ affects the overall computational complexity. By choosing $\bar{\gamma}_1(k) = e(k)/|e(k)|$ and all the remaining elements of $\bar{\boldsymbol{\gamma}}(k)$ as zeros, the solution is called a simple choice and results in a reduced computational complexity version of the SM-AP algorithm [16, 29].

Using $\mathbf{F} = \mathbf{I}$ and the simple constraint vector, the SM-AP algorithm has the following coefficient updating form:

$$\mathbf{w}(k+1) = \mathbf{w}(k) + \mu(k)\mathbf{X}(k)\left(\mathbf{X}^{\mathrm{H}}(k)\mathbf{X}(k) + \gamma\mathbf{I}\right)^{-1} e^*(k)\mathbf{u}_1, \tag{1.33}$$

where $\mathbf{u}_1^{\mathrm{T}} = [1\, 0 \cdots 0]$,

$$e(k) = d(k) - \mathbf{w}^{\mathrm{H}}(k)\,\mathbf{x}(k), \text{ and} \tag{1.34}$$

$$\mu(k) = \begin{cases} 1 - \frac{\bar{\gamma}}{|e(k)|} & \text{if } |e(k)| > \bar{\gamma}, \\ 0 & \text{otherwise;} \end{cases} \tag{1.35}$$

see [1] for details.

The algorithms presented in a simplified form in this chapter will appear in modified forms in the forthcoming chapters, each one adapted to the situation at hand.

Table 1.1 Expressions related to the misadjustment

Algorithm	M
LMS	$\frac{\mu \mathrm{tr}[\mathbf{R}]}{1-\mu \mathrm{tr}[\mathbf{R}]}$
NLMS	$\frac{\mu(N+2)}{(2-\mu)(N-1)}$
AP	$\frac{(L+1)\mu}{2-\mu}\frac{1-(1-\mu)^2}{1-(1-\mu)^{2(L+1)}}$
simple SM-AP	$\frac{(L+1)p_{\mathrm{up}}}{2-p_{\mathrm{up}}}\left(\frac{\bar{\gamma}^2}{\sigma_n^2}+1\right)$
RLS	$(N+1)\frac{1-\lambda}{2-(1-\lambda)}$

Table 1.2 Computational complexity in adaptive filtering algorithms

Algorithm	Multiplication	Addition	Division
LMS	$2N+3$	$2N+2$	0
NLMS	$2N+4$	$2N+5$	1
AP	$(2.5N+L+10)(L+1)$	$(1.5N+2.5)(L+1)+2N$	$L+2$
simple SM-AP[a]	$(1.5N+3L+7.5)(L+1)$	$(1.5N+7.5)(L+1)+0.5L$	$L+2$
RLS	$3N^2+11N+8$	$3N^2+7N+4$	1

[a] This estimate is an upper bound assuming that the updates occur all the time.

1.8 Performance and Computational Complexity

Table 1.1 lists the misadjustment expressions for the main adaptive filtering algorithms for a stationary environment. This information reveals the accuracy of the algorithms after convergence in stationary environments. The expression of the misadjustment related to the simple SM-AP algorithm includes the ratio between the threshold parameter and the background noise variance σ_n^2, as well as the probability of achieving an update, denoted as p_{up}. Reference [1] details how to derive these expressions.

As for the computational complexity, Table 1.2 lists the expressions (number of operations per update) for the classical adaptive filtering algorithms. This information is relevant to choose correctly the right family of algorithms, particularly in computationally sensitive applications.

1.9 Conclusion

This chapter introduced the classical adaptive filtering algorithms that are widely used in many applications. This presentation sets the conditions to extend and modify these algorithms to solve problems related to kernel adaptive filtering, sparsity aware learning, distributed learning, array processing, and learning on graphs. The algorithms were described in a concise format starting from their actual cost functions. We believe this is a shortcut to simplify the description of the more general algorithms addressed in the following chapters.

We summarize in Algorithm 1.1 all classical adaptive algorithms reviewed in this introduction. In the forthcoming chapters, real and complex arithmetic versions of these algorithms are utilized where appropriate.

Algorithm 1.1 The classical algorithms

Initialization

$\mathbf{w}_{\text{LMS}}(k) = \mathbf{w}_{\text{NLMS}}(k) = \mathbf{w}_{\text{AP}}(k) = \mathbf{w}_{\text{SMAP}}(k) = \mathbf{w}_{\text{RLS}}(k) = \mathbf{0}$

$\mathbf{u}_1^{\text{T}} = [1\,0\ldots 0]$

For $k > 0$ **do**

The LMS Algorithm

$e(k) = d(k) - \mathbf{w}_{\text{LMS}}^{\text{H}}(k)\mathbf{x}(k)$

$\mathbf{w}_{\text{LMS}}(k+1) = \mathbf{w}_{\text{LMS}}(k) + \mu e^*(k)\mathbf{x}(k)$

The NLMS Algorithm

$e(k) = d(k) - \mathbf{w}_{\text{NLMS}}^{\text{H}}(k)\mathbf{x}(k)$

$\mathbf{w}_{\text{NLMS}}(k+1) = \mathbf{w}_{\text{NLMS}}(k) + \mu \frac{e^*(k)}{\mathbf{X}^{\text{H}}(k)\mathbf{X}(k)+\gamma}\mathbf{x}(k)$

The Affine Projections Algorithm

$\mathbf{d}(k) = \left[d(k)\; d(k-1)\; \cdots\; d(k-L)\right]^{\text{T}}$

$\mathbf{X}(k) = \left[\mathbf{x}(k)\; \mathbf{x}(k-1)\; \cdots\; \mathbf{x}(k-L)\right]$

$\mathbf{e}^*(k) = \mathbf{d}^*(k) - \mathbf{X}^{\text{H}}(k)\mathbf{w}_{\text{AP}}(k)$

$\mathbf{w}_{\text{AP}}(k+1) = \mathbf{w}_{\text{AP}}(k) + \mu\mathbf{X}(k)\left[\mathbf{X}^{\text{H}}(k)\mathbf{X}(k) + \gamma\mathbf{I}\right]^{-1}\mathbf{e}^*(k)$

The Set-Membership Affine Projections Algorithm (simple choice)

$e(k) = d(k) - \mathbf{w}_{\text{SMAP}}^{\text{H}}\mathbf{x}(k)$

$$\mu(k) = \begin{cases} 1 - \frac{\bar{\gamma}}{|e(k)|} & \text{if } |e(k)| > \bar{\gamma} \\ 0 & \text{otherwise} \end{cases}$$

$\mathbf{w}_{\text{SMAP}}(k+1) = \mathbf{w}_{\text{SMAP}}(k) + \mu(k)\mathbf{X}(k)\left(\mathbf{X}^{\text{H}}(k)\mathbf{X}(k) + \gamma\mathbf{I}\right)^{-1} e^*(k)\mathbf{u}_1$

The RLS Algorithm

$e(k) = d(k) - \mathbf{w}_{\text{RLS}}^{\text{H}}(k)\mathbf{x}(k)$

$\boldsymbol{\psi}(k) = \mathbf{S}_{\text{D}}(k-1)\mathbf{x}(k)$

$\mathbf{S}_{\text{D}}(k) = \frac{1}{\lambda}\left[\mathbf{S}_{\text{D}}(k-1) - \frac{\boldsymbol{\psi}(k)\boldsymbol{\psi}^{\text{H}}(k)}{\lambda + \boldsymbol{\psi}^{\text{H}}(k)\mathbf{x}(k)}\right]$

$\mathbf{w}_{\text{RLS}}(k+1) = \mathbf{w}_{\text{RLS}}(k) + e^*(k)\mathbf{S}_{\text{D}}(k)\mathbf{x}(k)$

end

Problems

1.1 Warming up your mathematical skills: from Equation (1.6), compute the gradient of the LMS cost function with respect to $\mathbf{w}^*(k+1)$ and make it equal to the null vector, $\nabla_{\mathbf{w}^*(k+1)}\xi_{\text{LMS}}(k) = \mathbf{0}$, to obtain the LMS updating expression given in Equation (1.7).

1.2 In the system identification problem depicted in Figure 1.3, the linear and time-invariant unknown system has a pole in 0.8182 and a zero in −0.8182, that is, it can be represented by the difference equation

$$y(k) = x(k) + 0.8182x(k-1) + 0.8182y(k-1).$$

The input signal $x(k)$ corresponds to zero-mean white Gaussian noise (WGN) with variance $\sigma_x^2 = 1$, while the observation noise $n(k)$ is also zero-mean WGN

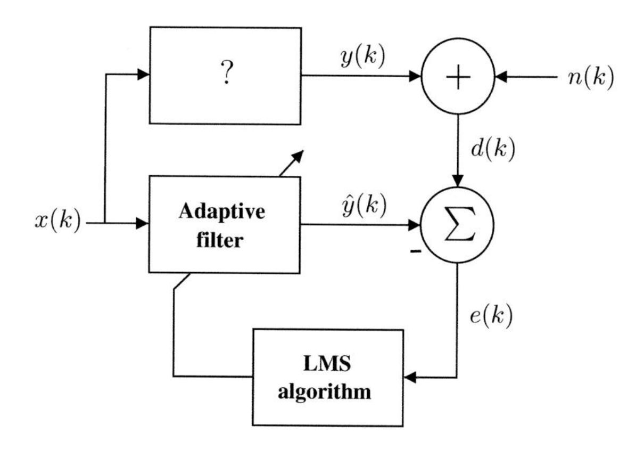

Figure 1.3 The LMS algorithm identifying an unknown system (Problem 1.2).

with variance $\sigma_n^2 = 10^{-6}$. Set the step size of the LMS algorithm to $\frac{1}{5(N+1)\sigma_x^2}$ and plot an estimate of the MSE (learning curve) with an ensemble of 100 independent runs (20 000 samples each run). Use order $N = 29$ or, equivalently, 30 coefficients. Also plot, in a dashed red line, the minimum theoretical MSE, $\xi_{\min}$, which corresponds $\mathbb{E}[e^2(k)]$ if we used $\mathbf{w}_{\mathrm{o}}$, the Wiener solution. Hint: The unknown system is an infinite impulse response filter and we are using a finite impulse response adaptive filter to identify it. Therefore, we do not have perfect modeling, and $\xi_{\min}$ shall be larger than σ_n^2. See Chapter 2 in [1] for more details on how to obtain $\xi_{\min} = \sigma_n^2 + \sigma_x^2 \sum_{k=N+1}^{\infty} h^2(k)$, with $h(k)$ being the impulse response of the unknown filter, in this case.

1.3 In the same setup of the previous problem, assume that $x(k)$ corresponds to a zero-mean unitary variance WGN $r(k)$ after passed through an autoregressive (AR) system with a pole in -0.9, that is, $x(k) = r(k) - 0.9x(k-1)$. Re-run the simulation with the LMS algorithm, the NLMS algorithm, and the AP algorithm with $L = 1$ (two hyperplanes, which corresponds to the Binormalized LMS (BNLMS) algorithm). Adjust the step sizes of the NLMS and the BNLMS algorithms such that they present the same misadjustment provided by the LMS algorithm with $\mu = \frac{1}{5(N+1)\sigma_x^2}$. Note that, now that the input signal is not white, the value of $\xi_{\min}$, or its estimate, is slightly more complicated to obtain.

1.4 In forward linear prediction, we try to predict a sample of a signal $s(k)$ under analysis from its past values, that is, $d(k) = s(k)$ and $x(k) = s(k-1)$. We can use an adaptive filter to estimate the prediction coefficients $\mathbf{w} = [w_0 \ \cdots \ w_N]^{\mathrm{T}}$ as indicated in Figure 1.4(a). For generating signal $s(k)$, we assume that the optimum coefficients are given by $\mathbf{w}_{\mathrm{o}} = [1.44 \ -0.68 \ -0.42 \ 0.24 \ 0.37 \ -0.35]^{\mathrm{T}}$. Signal $s(k)$ is synthesized as shown in Figure 1.4(b) where the excitation input $e(k)$ is a train of impulses with period 51 samples (meaning 6.375 ms or,

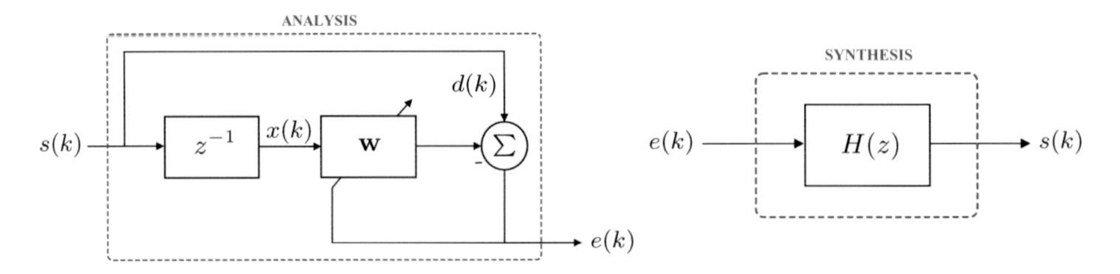

Figure 1.4 The adaptive filter in the signal prediction application.
(a) The overall filter (analysis) corresponds to $A(z) = 1 - w_0 z^{-1} \cdots - w_N z^{-(N+1)}$.
(b) Note that the synthesis filter is given as $H(z) = \frac{1}{A(z)}$.

equivalently, a "pitch" approximately equal to 156.86 Hz). The signal $s(k)$, obtained as previously described, corresponds to a segment of a synthetic vowel "ah."

Use the following commands to generate $\Delta t = 20$ ms of this voiced speech. We use Matlab® to describe the signal generation accurately but any other computer language can be easily adapted.

```
fs=8e3;            T=1/fs;
Deltat=20e-3;      t=0:T:Deltat;
exc=zeros(length(t),1);
period=51;         exc(1:period:end)=1;
wo=[1.44 -0.68 -0.42 0.24 0.37 -0.35];
A=[1 -wo];       s=filter(1,A,exc);
```

Run an adaptive filter using the LMS algorithm with step size $\mu = 0.1$, order $N = 5$, and data pair $\{d(k), x(k)\}$ as depicted in Figure 1.4. In the end, plot the optimum filter and the LMS result in the same figure. You will observe that they are quite different (the LMS did not have enough time to converge). Increase Δt and re-run the experiment; the LMS coefficient will get closer to the optimum coefficients. Further, increase Δt until the norm of the error in the coefficient vector is smaller than 0.2, that is, $\|\mathbf{w}_o - \mathbf{w}_{\text{LMS}}\| < 0.2$. How long did it take to converge? After convergence, check if the *a priori* prediction error $e(k)$ resembles the train of impulses with period 51 samples.

1.5 In Figure 1.5, we observe an adaptive signal enhancement where the interfering signal $n(k)$, after passing through an unknown room impulse response (RIR), corrupts the signal of interest (SOI) $s(k)$. In this application, the corrupted signal corresponds to the reference signal $d(k) = s(k) + n(k) * h_1(k)$. The input for an adaptive filter running the RLS algorithm is a signal also coming from the noise source but through another RIR, $x(k) = n(k) * h_2(k)$, such that it is not correlated with the SOI.

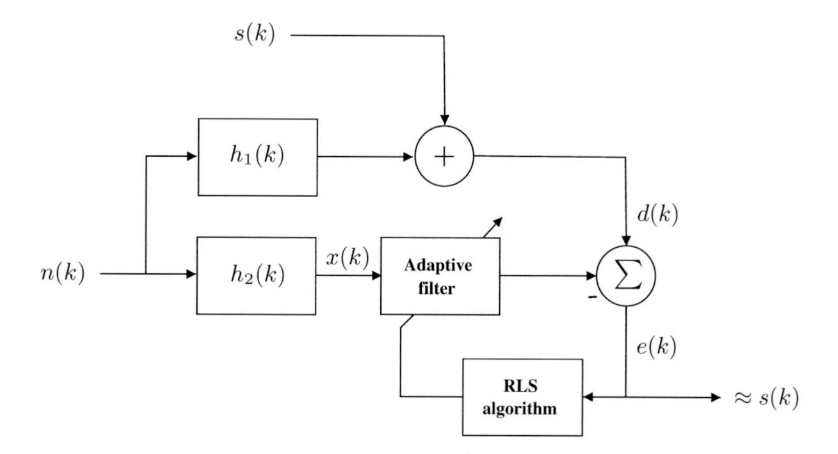

Figure 1.5 The RLS algorithm in a signal enhancement setup (Problem 1.5).

For this problem, assume that:

(a) The sampling frequency is $f_s = 8$ kHz.

(b) The SOI $s(k)$ corresponds to 4 s of speech signal (32 000 samples). Make it zero mean and unitary variance. You can record speech, use any recorded utterance, or synthesize the signal with a text-to-speech tool. If you are using Matlab® on MS® Windows operating system, you can use:
(see www.mathworks.com/matlabcentral/fileexchange/18091-text-to-speech)

```
txt='I like spring in California and winter in Rio de Janeiro.';
SV = actxserver('SAPI.SpVoice');
TK = invoke(SV,'GetVoices');
SV.Voice = TK.Item(1);
MS = actxserver('SAPI.SpMemoryStream');
MS.Format.Type = sprintf('SAFT8kHz16BitMono');
SV.AudioOutputStream = MS;
invoke(SV,'Speak',txt);
s = reshape(double(invoke(MS,'GetData')),2,[])';
s = (s(:,2)*256+s(:,1))/32768; s(s >= 1) = s(s >= 1)-2;
s=s(1:4*32000); s=s-mean(s); s=s/std(s);
```

(c) The interfering signal $n(k)$ corresponds to 4 s of the music signal. If you are using Matlab®, you can use `load handel`; make sure that the sampling frequency is the same (resample otherwise) and trim it to the same number of samples (32 000).

(d) The RIRs are $h_1(k) = \delta(k-1) + 0.5\delta(k-2)$ and $h_2(k) = \delta(k) - 0.25\delta(k-2)$.

(e) There is no observation noise.

(f) The filter order is $N = 9$ or, equivalently, the adaptive filter has 10 coefficients.

Run the RLS algorithm with $\lambda = 0.999$ and observe the $e(k)$ that should be approximately $s(k)$ after convergence. Compare the filter coefficients with $h_1(k)*h_2(k)$ which, in this case, corresponds to the optimal coefficients. Measure the SNR (in dB) of $d(k)$ (before enhancement) and of $e(k)$ (after enhancement). What is the SNR gain (in dB)?

References

[1] P. S. R. Diniz, *Adaptive Filtering: Algorithms and Practical Implementation*, 5th ed. (Cham, Switzerland, 2020).

[2] M. L. Honig and D. G. Messerschmitt, *Adaptive Filters: Structures, Algorithms, and Applications* (Kluwer Academic, Boston, 1984).

[3] S. T. Alexander, *Adaptive Signal Processing* (Springer, New York, 1986).

[4] M. Bellanger, *Adaptive Digital Filters*, 2nd ed. (Marcel Dekker, Inc., New York, 2001).

[5] B. Widrow and S. D. Stearns, *Adaptive Signal Processing* (Prentice Hall, Englewood Cliffs, 1985).

[6] J. R. Treichler, C. R. Johnson Jr., and M. G. Larimore, *Theory and Design of Adaptive Filters* (Wiley, New York, 1987).

[7] B. Farhang-Boroujeny, *Adaptive Filters: Theory and Applications* (Wiley, New York, 1998).

[8] S. Haykin, *Adaptive Filter Theory*, 4th ed. (Prentice Hall, Englewood Cliffs, 2002).

[9] A. H. Sayed, *Fundamentals of Adaptive Filtering* (Wiley, Hoboken, 2003).

[10] P. S. R. Diniz and B. Widrow, History of Adaptive Filters, in *A Short History of Circuits and Systems*, F. Maloberti and A. C. Davies (eds.) (River Publishers, Delft, 2016).

[11] B. Widrow and D. Park, History of Adaptive Signal Processing: Widrow's Group, in *A Short History of Circuits and Systems*, F. Maloberti and A. C. Davies (eds.) (River Publishers, Delft, 2016).

[12] C. F. Gauss, Theoria Combinationis Observationum Erroribus Minimis Obnoxiae: Pars Prior, Pars Posterior, Supplementum (Theory of the Combination of Observations Least Subject to Errors: Part One, Part Two, Supplement), in *Classics in Applied Mathematics*, G. W. Stewart (ed.) (SIAM, Philadelphia, 1995).

[13] J. A. Apolinário Jr. (ed.), *QRD-RLS Adaptive Filtering* (Springer, New York, 2009).

[14] F. T. Castoldi and M. L. R. de Campos, Application of a Minimum-Disturbance Description to Constrained Adaptive Filters, IEEE Signal Processing Letters **20**, pp. 1215–1218 (2013).

[15] K. Ozeki, *Theory of Affine Projection Algorithms for Adaptive Filtering* (Springer, New York, 2016).

[16] S. Werner, J. A. Apolinário Jr., and P. S. R. Diniz, Set Membership Proportionate Affine Projection Algorithms, EURASIP Journal on Audio, Speech, and Music Processing **2007**, pp. 1–10 (2007).

[17] S. Nagaraj, S. Gollamudi, S. Kapoor, and Y.-F. Huang, BEACON: An Adaptive Set-Membership Filtering Technique with Sparse Updates, IEEE Transactions on Signal Processing **47**, pp. 2928–2941 (1999).

[18] R. C. de Lamare and P. S. R. Diniz, Set-Membership Filtering Adaptive Algorithms Based on Time-Varying Error Bounds for CDMA Interference Suppression, IEEE Transactions on Vehicular Technology **58**, pp. 644–654 (2009).

[19] P. S. R. Diniz, On Data-Selective Adaptive Filtering, IEEE Transactions on Signal Processing **66**, pp. 4239–4252 (2018).

[20] F. Bouteille, P. Scalart, and M. Corazza, Pseudo Affine Projection Algorithm New Solution for Adaptive Identification. Proceedings of European Conference on Speech Communication and Technology, 1999, Vol. 1, pp. 427–430.

[21] F. Albu, M. Bouchard, and Y. Zakharov, Pseudo-Affine Projection Algorithms for Multichannel Active Noise Control, IEEE Transactions on Audio, Speech, and Language Processing **15**, pp. 1044–1052 (2007).

[22] J. Nagumo and A. Noda, A Learning Method for System Identification, IEEE Transactions on Automatic Control **12**, pp. 282–287 (1967).

[23] A. E. Albert and L. S. Gardner Jr., *Stochastic Approximation and Nonlinear Regression* (MIT Press, Cambridge, MA, 1967).

[24] B. Widrow and M. E. Hoff, Adaptive Switching Circuits, IRE WESCON Convention Record, **4**, pp. 96–104 (1960).

[25] S. Gollamudi, S. Nagaraj, S. Kapoor, and Y.-F. Huang, Set-Membership Filtering and a Set-Membership Normalized LMS Algorithm with an Adaptive Step Size, IEEE Signal Processing Letters **5**, pp. 111–114 (1998).

[26] P. S. R. Diniz and S. Werner, Set-Membership Binormalized Data Reusing LMS Algorithms, IEEE Transactions on Signal Processing **51**, pp. 124–134 (2003).

[27] S. Werner and P. S. R. Diniz, Set-Membership Affine Projection Algorithm, IEEE Signal Processing Letters **8**, pp. 231–235 (2001).

[28] M. V. S. Lima, T. N. Ferreira, W. A. Martins, and P. S. R. Diniz, Sparsity-Aware Data-Selective Adaptive Filters, IEEE Transactions on Signal Processing **62**, pp. 4557–4572 (2014).

[29] T. N. Ferreira, W. A. Martins, M. V. S. Lima, and P. S. R. Diniz, Convex Combination of Constraint Vectors for Set-Membership Affine Projection Algorithms. Proceedings of the IEEE International Conference on Acoustics, Speech and Signal Processing (ICASSP 2019), Brighton, UK, 2019, pp. 4858–4862.

2 Adaptive Filtering for Sparse Models

2.1 Introduction

In adaptive filtering, as well as in other areas requiring learning from data, there is a current trend of employing complex models relying on a large number of coefficients/parameters, meaning that the parameters form a vector, called *coefficient vector* or *parameter vector*, that lives in a high-dimensional space. In general, as the parameter space enlarges (has more dimensions), the training process becomes more difficult, and it is expected that the coefficient vector takes more time to converge to an acceptable estimate, that is, a longer training period is required. However, slow convergence is a critical issue in adaptive filtering, as its most interesting and practical applications very often deal with the tracking of nonstationary processes, thus requiring fast adaptation of the filter coefficients.

One way of overcoming this slow convergence issue is by exploiting some structure that the coefficients should have. The idea is adding prior knowledge about the problem so that the algorithm can explore the parameter space (or search space) wisely, like if the algorithm had a clue of where to search for the solution, instead of performing a simple uninformed search throughout the parameter space. Clearly, the impact on convergence speed provided by exploiting the parameters' structure increases with the number of parameters, that is, with the dimension of the search space.

There are many different parameter structures that can be exploited; some examples can be found in [1–6]. But one of the most important structures, which is the subject of this chapter, is the *sparsity*. Intuitively, a sparse parameter vector means that only a subset of the inputs is required to explain/model the data (later we will define the concepts of sparse and compressible vectors precisely) and, therefore, is very likely to happen when using complex models with a considerable amount of coefficients.

Exploiting sparsity in adaptive filtering has been a very active research topic since the early twenty-first century, as the classical algorithms have become inappropriate to deal with such a large set of coefficients. The goal of this chapter is to equip the reader with the fundamentals of this field. That is, instead of covering a huge number of adaptive filtering algorithms that exploit sparsity (there are hundreds of them), we opted for a unified presentation in which we divide these algorithms into two classes. We present each class's fundamental ideas, discuss the pros and cons, and introduce the most important or pioneer algorithms of each class.

This chapter is organized as follows. In Section 2.2, we provide a thorough explanation of the sparsity-promoting functions frequently used as regularizations, the so-called *sparsity-promoting regularizations*. This section is of paramount importance to fully understand the *algorithms employing sparsity-promoting regularizations*, which is an important class of algorithms that exploit sparsity, covered in Section 2.3. In Section 2.4, we present the *proportionate-type* algorithms, another class of algorithms widely used to exploit sparsity. The conclusions are drawn in Section 2.5. We hope the reader can appreciate the art of developing algorithms during the exposition of these two classes of algorithms. That is, while the development of the algorithms employing sparsity-promoting regularizations relies on optimization theory and geometric aspects, the development of the proportionate-type algorithms relies more on numerical analysis and intuition/heuristics.

2.2 Exploiting Sparsity through Regularization

Regularization has been widely used in areas like statistics, machine learning, and adaptive filtering. *Regularization can be regarded as a way of adding prior information about the model/parameters to be estimated or of encouraging them to have some structure.* Of course, adding a regularization changes the original (nonregularized) optimization problem and, therefore, the solutions of these two distinct problems are not necessarily equal. However, the discrepancy between these two solutions can be controlled by the regularization parameter, a positive real number that determines the weight given to the regularization term.

There are many benefits of using regularization, among which we highlight: *preventing the overfitting problem* of the parameters and *enhancing the problem conditioning*. Preventing the overfitting problem is crucial for the learning technique to perform well using new data, that is, data not belonging to the training set; this characteristic is known in machine learning as *generalization* [7]. The conditioning, or condition number, is related to the problem (and not the algorithm employed to solve it) and determines how the errors/uncertainties in the observations are propagated to the unknowns/estimates [8]. If the problem is ill-conditioned (i.e., has large condition number), then these errors can be severely amplified to the outputs no matter the algorithm that is used to solve it. In this case, it is better to make a slight modification to the original optimization problem, in order to enhance its conditioning, before solving it. Besides, the condition number also impacts the learning rate of gradient-based algorithms [9].

In addition to the aforementioned benefits, by employing regularization it might be possible to obtain estimators with reduced mean squared error (MSE) in comparison with unbiased solutions. As explained in [10], these estimators can "trade a little bias for a large reduction in variance."[1]

From the optimization point of view, the regularization term is simply a penalty function of the parameters. In adaptive filtering, in most cases, this penalty function is a vector norm, for example, the ℓ_2 norm of the coefficients $\mathbf{w}$. In this section, we are interested in the regularizations used to exploit sparsity, frequently called *sparsity-promoting regularizations* [11].

Recalling that a finite-dimensional vector, herein also called *signal*, is said to be sparse if it can be represented as a linear combination of a small number of basis vectors for some basis of the related vector space [12]. For instance, the vector $[0 \; \ldots \; 0 \; 1 \; 0 \ldots \; 0]^{\mathrm{T}}$ is sparse as it requires a single basis vector for its representation; the nonzero entry being equal to 1 or 1,000 does not change this fact. Therefore, it is natural to think of the ℓ_0 norm, which essentially counts the number of nonzero entries in a vector, when modeling or measuring the sparsity of a signal. However, minimizing this norm is challenging since it leads to an NP-hard problem, which turns its use prohibitive in many cases, especially in online applications [12]. Due to the practical limitations of using the ℓ_0 norm, several alternatives have been proposed, like the ℓ_1 norm, the reweighted ℓ_1 norm, and the approximation of the ℓ_0 norm [11]. In this section, all of these approaches to model sparsity are described, their gradient expressions are shown (as gradient-based optimization is widely used), and their pros and cons are discussed.

2.2.1 The ℓ_0 Norm

As previously explained, the ℓ_0 norm is the natural function to model/measure the sparsity of a vector. So, why searching for alternatives to the ℓ_0 norm instead of working directly with it? Here, we answer this question by pointing out the practical limitations of this norm.

The ℓ_0 norm of a given vector $\mathbf{w} = [w_0 \; w_1 \; \ldots \; w_N]^{\mathrm{T}} \in \mathbb{R}^{N+1}$ is defined as

$$\|\mathbf{w}\|_0 \triangleq \#\{i \in \mathcal{I} : w_i \neq 0\} \in \mathbb{N}, \tag{2.1}$$

where $\mathcal{I} \triangleq \{0, 1, \ldots, N\} \subset \mathbb{N}$ is the set of indexes for the entries of $\mathbf{w}$ and $\#$ denotes the cardinality (i.e., the number of elements) of a finite set. Thus, the ℓ_0 norm of a vector is the number of its nonzero entries. Observe that the ℓ_0 norm is not really a norm as the homogeneity property fails, that is, $\|\alpha\mathbf{w}\|_0 \neq |\alpha|\|\mathbf{w}\|_0$ in general.[2]

A major issue related to the ℓ_0 norm concerns the *computational complexity* involving its minimization. Indeed, the problem of minimizing $\|\mathbf{w}\|_0$ subject to some constraints on $\mathbf{w}$ is said to be NP-hard and, in general, requires a combinatorial search over all possible coordinate combinations; there are $(N+1)$ choose

[1] Recall that (MSE) = $(\text{bias})^2$ + (variance).

[2] Also, the ℓ_0 norm is not a pseudo-norm nor a quasi-norm, as sometimes stated erroneously, since both of these concepts relax other properties of a norm, but not homogeneity [13, 14].

k possible combinations, for each $k \in \{1, 2, \dots, N+1\}$. Clearly, a tough problem with a prohibitive complexity for vectors belonging to high-dimensional spaces or when dealing with online and real-time processing requirements.

Still considering the optimization problem, but from another perspective, the use of the ℓ_0 norm leads to an *ill-conditioned problem* since the ℓ_0 norm is very sensitive to nonzero coordinates due to its discontinuity. For example, consider the following two vectors belonging to $\mathbb{R}^{N+1}$

$$\mathbf{0} = [0 \ \dots \ 0]^{\mathrm{T}} \qquad \text{(the vector of zeros)},$$
$$\mathbf{1} = [1 \ \dots \ 1]^{\mathrm{T}} \qquad \text{(the vector of ones)}.$$

Clearly, $\|\mathbf{0}\|_0 = 0$. If you add a tiny perturbation to this vector, the result changes by a significant amount, that is,

$$\|\mathbf{0} + \delta\mathbf{1}\|_0 = N + 1,$$

for any $\delta \neq 0$. This characteristic hinders the use of the ℓ_0 norm in most practical applications since typically there are several sources of errors/uncertainties due to noise, quantization, numerical approximations, limited accuracy of the measurement devices, etc.

From the discussion in the previous paragraph, one may conclude that sparse signals (in the mathematical sense) are not frequently found in practice.[3] Actually, many times in the literature, the term sparse is used with a slightly different meaning; it is used for vectors having their energy concentrated mostly on a few entries, whereas their remaining elements have small or negligible energy. This kind of vector/signal is herein called *compressible* [12]. Roughly, the main difference between sparse and *compressible signals* is that the negligible coefficients of the latter are not required to be 0.

Even though there is a strong relationship between sparsity and the ℓ_0 norm, its practical use is very limited due to the severe drawbacks presented here. Next, we present the classical alternatives to the ℓ_0 norm.

2.2.2 The ℓ_1 Norm

Undoubtedly, the ℓ_1 norm is the most widely used substitute for the ℓ_0 norm. It has been used in areas like geophysics since the 1970s [15, 16], but it was in the 1990s that it became widely used due to the advent of techniques like the basis pursuit [17] and the least absolute shrinkage and selection operator (LASSO) [10, 18]. Since 2006, the ℓ_1 norm has also been used to recover sparse signals, the so-called compressed sensing (CS) [19]. The CS theory explained that under certain conditions, the ℓ_1 minimization is capable of finding the sparsest solution (i.e., the solution to the ℓ_0 minimization).

[3] Sparse signals are usually obtained by replacing the small coefficients of a real signal (like an image, audio, or the impulse response of a system) with zeros. That is, the real and compressible signal goes through a lossy compression process to generate a sparse signal.

The ℓ_1 norm of a given signal $\mathbf{w} \in \mathbb{R}^{N+1}$ is defined as

$$\|\mathbf{w}\|_1 \triangleq \sum_{i=0}^{N} |w_i|. \tag{2.2}$$

The ℓ_1 norm is, therefore, a continuous and convex function, meaning that there exist several efficient methods to solve the ℓ_1 minimization problem, and it is almost everywhere (a.e.) differentiable, thus turning gradient-based methods very suitable to its minimization. This simplicity and mathematical tractability have played a central role in the widespread use of the ℓ_1 norm.

The derivative of $\|\mathbf{w}\|_1$ with respect to a given entry w_i is[4]

$$\frac{\partial \|\mathbf{w}\|_1}{\partial w_i} = \text{sign}(w_i) \triangleq \begin{cases} 1 & \text{if } w_i > 0, \\ 0 & \text{if } w_i = 0, \\ -1 & \text{if } w_i < 0, \end{cases} \tag{2.3}$$

that is, the standard sign function applied to a real scalar. Thus, the gradient of $\|\mathbf{w}\|_1$ with respect to $\mathbf{w}$, denoted by $\nabla\|\mathbf{w}\|_1$ or $\mathbf{f}_{\ell_1}(\mathbf{w})$, is

$$\nabla\|\mathbf{w}\|_1 \triangleq \mathbf{f}_{\ell_1}(\mathbf{w}) = \text{sign}(\mathbf{w}) \triangleq [\text{sign}(w_0)\ \text{sign}(w_1)\ \ldots\ \text{sign}(w_N)]^{\text{T}}, \tag{2.4}$$

that is, when the operator sign is applied to a vector, it results in a vector with the standard sign function applied to each of its entries.

Despite the aforementioned advantages, the ℓ_1 norm has some critical drawbacks that motivated the search for other functions to exploit sparsity. These main drawbacks are as follows [11, 20–22]:

1 The ℓ_1 norm penalizes the larger coefficients more heavily than the smaller ones.

2 For gradient-based methods, the update term given by $-\nabla\|\mathbf{w}\|_1$ pushes all entries of $\mathbf{w}$ toward 0 with the same strength, instead of prioritizing those which are closer to 0.

3 The effectiveness of the ℓ_1 norm depends on the existence of high sparsity degree.

Observe that a good sparsity-promoting function should act in the small magnitude coefficients, pushing them toward 0 to encourage sparse solutions, without interfering in the large magnitude coefficients, which are relevant to explain/model the data [11]. The need to discriminate between low and high magnitude entries in order to force the former ones to 0 without shrinking the latter ones paved the way for further developments.

[4] Formally, $\|\mathbf{w}\|_1$ is not differentiable at the points $w_i = 0$; therefore, one must use the concept of subderivative at these points. For instance, the subdifferential of the function $|w|$ at $w = 0$ is the interval $[-1, 1]$, and any point $w_0 \in [-1, 1]$ is called a subderivative of $|w|$ at $w = 0$.

2.2.3 The Reweighted ℓ_1 Norm

The reweighted ℓ_1 ($r\ell_1$) norm proposed in 2008 aims to mitigate the problem of shrinking the large magnitude coefficients [20]. To do so, instead of penalizing each coefficient based on its magnitude $|w_i|$, as the ℓ_1 norm does, the $r\ell_1$ norm uses a function that grows less quickly for the large magnitude coefficients.

The $r\ell_1$ norm of a given signal $\mathbf{w} \in \mathbb{R}^{N+1}$, denoted by $F_{r\ell_1} : \mathbb{R}^{N+1} \longrightarrow \mathbb{R}_+$, is defined as

$$F_{r\ell_1}(\mathbf{w}) \triangleq \frac{1}{\log(1+\epsilon)} \sum_{i=0}^{N} \log(1+\epsilon|w_i|), \tag{2.5}$$

where $\epsilon \in \mathbb{R}_+$. Like the ℓ_0 norm, the $r\ell_1$ norm also is not really a norm. To better understand the role of ϵ, observe that $F_{r\ell_1}$ tends to the ℓ_1 norm as $\epsilon \to 0$, whereas $F_{r\ell_1}$ tends to the ℓ_0 norm as $\epsilon \to \infty$. The $r\ell_1$ norm is a continuous and a.e. differentiable function, but it is not convex anymore due to the logarithm. Nonetheless, gradient-based methods can still be used in its minimization [20].

The derivative of $F_{r\ell_1}(\mathbf{w})$ with respect to a given entry w_i is

$$\frac{\partial F_{r\ell_1}(\mathbf{w})}{\partial w_i} = \frac{\epsilon}{\log(1+\epsilon)} \times \frac{\text{sign}(w_i)}{1+\epsilon|w_i|}. \tag{2.6}$$

Notice that the first fraction on the right-hand side of Equation (2.6) does not depend on i and, therefore, it only modifies the length of the gradient vector, not its direction. Moreover, there already exists a parameter, *the regularization parameter*, which controls the weight given to $F_{r\ell_1}$, and thus the length of its gradient. Therefore, it is very common to discard this fraction and write the gradient vector as [19, 23]:

$$\nabla F_{r\ell_1}(\mathbf{w}) \triangleq \mathbf{f}_{r\ell_1}(\mathbf{w}) \triangleq [f_{r\ell_1}(w_0)\ f_{r\ell_1}(w_1)\ \ldots\ f_{r\ell_1}(w_N)]^{\mathrm{T}}, \tag{2.7}$$

where $f_{r\ell_1}(w_i)$ is a simplified version of Equation (2.6) given by

$$f_{r\ell_1}(w_i) = \frac{\text{sign}(w_i)}{1+\epsilon|w_i|}. \tag{2.8}$$

The $r\ell_1$ norm given in Equation (2.5), therefore, mitigates the problem of heavy penalization of the large magnitude coefficients. Besides, observe in Equation (2.8) that $f_{r\ell_1}(w_i)$ decreases as $|w_i|$ increases, meaning that gradient-based algorithms employing the $r\ell_1$ norm attract the smaller coefficients to 0 more strongly than the large magnitude ones. However, it still affects/shrinks the large magnitude coefficients, a problem that is solved by the function presented in Subsection 2.2.4.

2.2.4 The ℓ_0-Norm Approximation

The *ℓ_0-norm approximation* is also known as *smoothed ℓ_0 norm* because it is a continuous/smoothed version of the ℓ_0 norm. Although the functions commonly used as ℓ_0-norm approximations are known for a long time, only recently they have been used for the purpose of exploiting sparsity. For instance, the ℓ_0-norm approximation has been used in image processing [22], in medical image reconstruction [24], and in system identification [11, 21, 25]. Most of these works present results pointing out the superior performance of the ℓ_0-norm approximation over the ℓ_1 norm and the $r\ell_1$ norm.

The ℓ_0-norm approximation is a continuous and a.e. differentiable function $F_\beta : \mathbb{R}^{N+1} \longrightarrow [0, N+1] \subset \mathbb{R}_+$, in which $\beta \in \mathbb{R}_+$ is a parameter that controls the smoothness of the approximation. A common practice is to define analytically a continuous function F_β such that

$$\lim_{\beta \longrightarrow \infty} F_\beta(\mathbf{w}) = \|\mathbf{w}\|_0, \tag{2.9}$$

or equivalently,

$$\lim_{\beta \longrightarrow \infty} F_\beta(w_i \mathbf{e}_i) = \begin{cases} 1 & \text{if } w_i \neq 0, \\ 0 & \text{if } w_i = 0, \end{cases} \tag{2.10}$$

for all $\mathbf{w} \in \mathbb{R}^{N+1}$, where $\mathbf{e}_i$ is the i-th vector of the canonical basis of $\mathbb{R}^{N+1}$ (i.e., $\mathbf{e}_i$ has only zero elements, except for a 1 in its i-th coordinate, $i \in \mathcal{I}$). For finite values of β, we add the following property:

$$F_\beta(w_i \mathbf{e}_i) = \begin{cases} 1 & \text{if } |w_i| \gg 1/\beta, \\ 0 & \text{if } w_i = 0. \end{cases} \tag{2.11}$$

In words, a valid ℓ_0-norm approximation must be a continuous function whose value is equal to 0 at $w_i = 0$ and rapidly (depending on the value of β) saturates at 1 as w_i gets further from 0, for every coordinate w_i. Clearly, F_β is nonconvex. It is worth noticing that the reweighted ℓ_1 norm, denoted by $F_{r\ell_1}$, also converges to the ℓ_0 norm as $\epsilon \to \infty$. However, for finite ϵ, $F_{r\ell_1}$ increases with $|w_i|$, thus not satisfying the saturation property given in Equation (2.11). Therefore, $F_{r\ell_1}$ is not a particular case of F_β.

There are many functions F_β that can be used as ℓ_0-norm approximations, as illustrated in [11]. Here, we focus on the two most commonly used:

$$F_\beta(\mathbf{w}) = \sum_{i \in \mathcal{I}} \left(1 - \mathrm{e}^{-\beta|w_i|}\right), \tag{2.12a}$$

$$F_\beta(\mathbf{w}) = \sum_{i \in \mathcal{I}} \left(1 - \frac{1}{1 + \beta|w_i|}\right). \tag{2.12b}$$

The F_β given in Equation (2.12a) is the so-called Laplace function (LF) [24, 26], and is probably the most widely used ℓ_0-norm approximation. In addition, Equation (2.12b) describes the Geman–McClure function (GMF) [24, 27]. Notice that both approximations satisfy the properties in Equations (2.10) and (2.11).

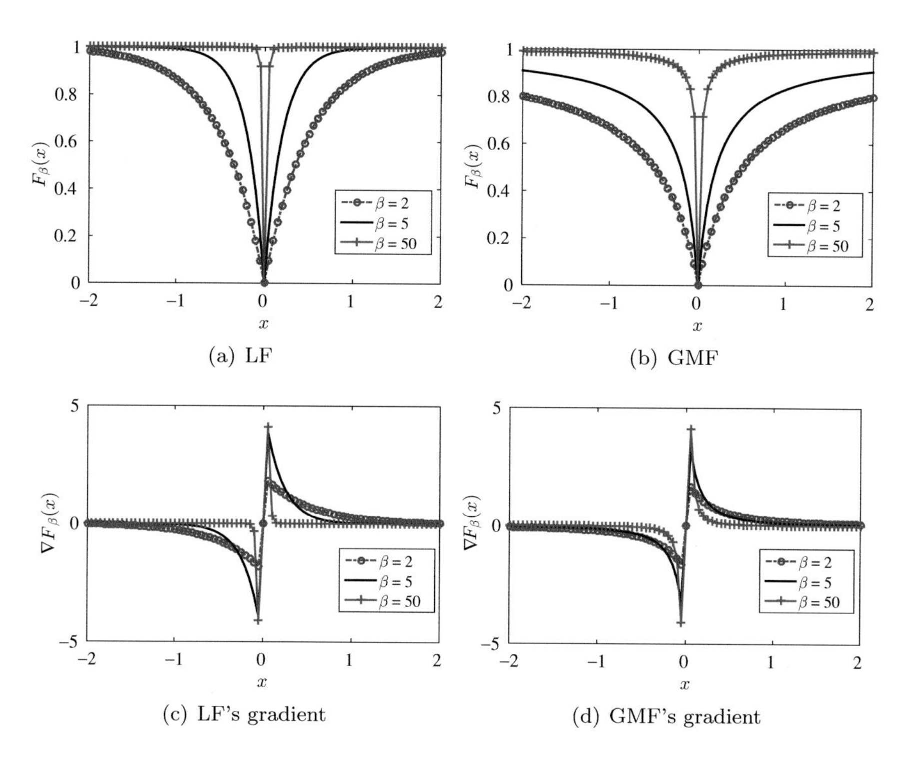

Figure 2.1 Functions $F_\beta(x)$ and their gradients $\nabla F_\beta(x)$, with $x \in \mathbb{R}$, for different values of β: (a) Laplace function (LF); (b) Geman–McClure function (GMF); (c) LF's gradient; and (d) GMF's gradient.

Defining $f_\beta(w_i) \triangleq \frac{\partial F_\beta(\mathbf{w})}{\partial w_i}$, the derivatives corresponding to Equation (2.12a) and Equation (2.12b) are, respectively,

$$f_\beta(w_i) = \beta \text{sign}(w_i) \mathrm{e}^{-\beta|w_i|}, \tag{2.13a}$$

$$f_\beta(w_i) = \frac{\beta \text{sign}(w_i)}{(1 + \beta|w_i|)^2}, \tag{2.13b}$$

where the function $\text{sign} : \mathbb{R} \longrightarrow \{-1, 0, 1\}$ is defined in Equation (2.3). Thus, we can define the gradient of $F_\beta(\mathbf{w})$ with respect to $\mathbf{w}$ as

$$\nabla F_\beta(\mathbf{w}) \triangleq \mathbf{f}_\beta(\mathbf{w}) \triangleq [f_\beta(w_0)\ f_\beta(w_1)\ \ldots\ f_\beta(w_N)]^{\mathrm{T}}. \tag{2.14}$$

Figure 2.1 depicts the LF, the GMF, and their respective gradients $\nabla F_\beta(x) = \mathrm{d}F_\beta(x)/\mathrm{d}x$, for $x \in \mathbb{R}$, using different values of β. Observe that the functions F_β are nonconvex and also that low values of β lead to smoother functions, whereas high values of β turn these functions more similar to the ℓ_0 norm. Besides, due to the exponential function, the LF goes from 0 to 1 faster than the GMF, at least for the same value of β, thus being a better approximation of the ℓ_0 norm. On the other hand, the GMF is smoother and simpler to compute. Also, observe that the gradients of both the LF and GMF are very strong (high

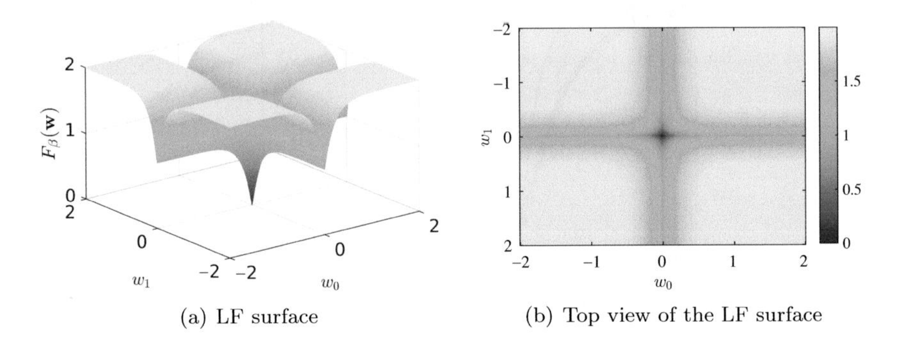

(a) LF surface (b) Top view of the LF surface

Figure 2.2 Laplace function F_β for $\beta = 5$ and $\mathbf{w} \in \mathbb{R}^2$: (a) LF surface and (b) top view of the LF surface.

magnitude) at values of x close to 0 and they vanish (i.e., $\nabla F_\beta(x) = 0$) as $|x|$ grows. Clearly, the gradient of the LF vanishes faster than that of the GMF, for a fixed β.

Figure 2.2 illustrates the surface of the LF $F_\beta(\mathbf{w})$, with $\mathbf{w} \in \mathbb{R}^2$ and $\beta = 5$. Observe that such function has a single global minimum at $\mathbf{w} = \mathbf{0}$ and there are infinitely many local minima at the axes, more precisely, when one of the coordinates is equal to 0, and the other one is sufficiently large. It is worth mentioning that most of the parameter space is covered by vast plateaus, whereas the concave part of the function occurs when any of the coefficients w_i approaches 0, as can be clearly observed in Figure 2.2(b).

From the previous discussion, it must be clear that the ℓ_0-norm approximation F_β does not have the drawbacks of the previous functions. Indeed, due to its saturation property, the high magnitude (relevant) coefficients are not heavily penalized, and a gradient-based minimization pushes only the small magnitude coefficients to 0, keeping the relevant coefficients unaltered. Also, the ℓ_0-norm approximation works well even when the sparsity degree is very low, as corroborated by the results and theoretical analysis in [28].

Gradient-Based Training

As illustrated in Figure 2.1, the parameter β represents a tradeoff between smoothness and quality of approximation. Therefore, this parameter can be used to avoid gradient-descent techniques from getting trapped in a local minimum of F_β, simply by reducing the value of β in order to obtain a smoother approximation. For instance, by decreasing the value of β in Figure 2.2, one can obtain a LF surface that does not exhibit local minima within the $[-2, 2] \times [-2, 2]$ region and, therefore, gradient-based techniques will work well within this region.[5] The price to be paid is to have a slower convergence due to the

[5] F_β will always have local minima, but as you set β closer to 0, these local minima occur much further from the origin. That is, one can always set β so that the local minima appear in a part of the parameter space that is of no interest.

smaller gradients since, for a given x close to 0, the magnitudes of the gradients decrease as β decreases, as can be observed in Figure 2.1.

To overcome the aforementioned issue, a fast algorithm for overcomplete sparse decomposition, called SL0, was proposed in [29]. Essentially, the SL0 algorithm uses an increasing sequence $\beta_1 < \beta_2 < \cdots < \beta_J$, applies the steepest-descent technique (or steepest-ascent, if the problem is written in the form of maximization, as in [29]) to solve approximately the minimization of F_{β_i}, and then uses the approximate solution of the i-th problem as the initialization of the next problem $F_{\beta_{i+1}}$. The process is repeated until F_{β_J} is solved. It is assumed that β_J is sufficiently large so that F_{β_J} is a good approximation of the ℓ_0 norm. The work in [29] also provides some theoretical guarantees that the SL0 solution is indeed the sparsest solution, that is, the one that minimizes the ℓ_0 norm. However, the choice of the sequence of β's is addressed only heuristically.

The same idea of the SL0 algorithm could be adapted for adaptive filtering by employing an iteration-dependent parameter $\beta(k)$, where $k \in \mathbb{Z}$ is the iteration index and defining some adaptation rule for it. Indeed, unlike in sparse decomposition applications, in which the SL0 is satisfactorily used, working with a predefined and always increasing sequence of β's is usually not possible in adaptive filtering since very often the problem involves nonstationary signals. In this way, a recursion capable of increasing and decreasing the value of $\beta(k)$ would be necessary, a topic that, to the best of our knowledge, has never been addressed. In what follows, we explain how β is usually chosen.

Choice of β

For the proper choice of β, we must introduce the concept of *zero-attraction region*, which is a region of the parameter space where the gradient is non-null, thereby any point inside this region is pushed/forced toward 0 by a gradient-descent technique. In Figure 2.2, for example, there are four plateaus in which $F_\beta(\mathbf{w}) = 2$, for any $\mathbf{w}$ belonging to the interior of these plateaus and, consequently, $\nabla F_\beta(\mathbf{w}) = 0$ for these $\mathbf{w}$'s. The remaining points of the parameter space, excluding the global minimum and the local minima, are such that $\nabla F_\beta(\mathbf{w}) \neq 0$ and, therefore, they belong to the zero-attraction region. In ultimate analysis, the choice of β concerns the size of the zero-attraction region, that is, it defines a threshold for the magnitude of the coefficients; coefficients below this threshold are pushed to 0, whereas the ones that are above this threshold remain unaltered.

Since $F_\beta(\mathbf{w})$ treats all entries w_i uniformly, it is easier to set β based on the univariate case depicted in Figure 2.1. Observe in Figures 2.1(c) and 2.1(d) that, for large values of β, the gradient $\nabla F_\beta(x)$ converges to 0 very fast as $|x|$ grows, meaning that the zero-attraction interval is short. On the other hand, the zero-attraction interval expands as β decreases. Thus, to set β properly, one must have some prior knowledge about the magnitude of the optimal coefficients. More precisely, one must have a clue about the magnitudes of the relevant and irrelevant coefficients. Ideally, β should be chosen such that the irrelevant coefficients fall inside the zero-attraction interval, whereas the relevant coefficients fall outside such range, thus not being affected by F_β.

When dealing with sparse vectors, it is easy to set β because we know that the irrelevant coefficients are precisely equal to 0 and, therefore, a small zero-attraction interval suffices. Although high values of β could be used in this case, it is preferable to use moderate values of β, like $\beta \in [5, 10]$ (typically $\beta = 5$ [11]), because smoothness is important to guarantee the effectiveness of gradient-based optimization methods, as previously explained.

For compressible signals, on the other hand, the task of choosing β is more complicated. In this case, one should set β such that the low magnitude coefficients are pushed to 0, but the relevant (high magnitude) coefficients should not be severely affected. As a simple example, suppose that the optimal coefficients have some negligible entries with magnitudes ranging from 0 to 0.5, whereas its relevant entries are greater than 1 in magnitude. In this example, considering the LF function depicted in Figure 2.1(a), one should not use $\beta = 50$. Actually, it would be better to choose β equal to 2 or 5 to guarantee that coefficients with magnitudes up to 0.5 are pushed to 0 while producing little or no effect in the relevant coefficients.

2.2.5 Remarks

Let us summarize the main points about sparsity-promoting functions covered in this section. First, we learned that although the ℓ_0 norm is the natural function to model sparsity, it has critical issues, due to its discontinuity, that prevent its use in practical applications. Then, we studied three functions that are frequently used to replace the ℓ_0 norm in optimization problems, namely, the ℓ_1 norm, the reweighted ℓ_1 norm $F_{r\ell_1}$, and the ℓ_0-norm approximation F_β, which are illustrated in Figure 2.3 for the case of a scalar independent variable $x \in \mathbb{R}$. These three functions are continuous and almost everywhere differentiable, thus allowing the use of gradient-based optimization techniques, meaning that the training technique is simple, an important feature when dealing with high-dimensional problems. However, only the ℓ_1 norm is convex, being, therefore, easier to deal with than the other two. On the other hand, the ℓ_1 norm has three significant drawbacks: (i) heavy penalization of the large coefficients, (ii) its gradient pushes all coefficients to 0 with equal strength which results in the shrinkage of relevant coefficients, and (iii) high sparsity degree is necessary to obtain its benefits. Next, we explained that $F_{r\ell_1}$ mitigates these issues, while F_β eliminates them. On the other hand, both of these functions are nonconvex, but only F_β satisfies the saturation property, as can be verified in Figure 2.3, which is key to solving the drawbacks (i) and (ii) mentioned above. These functions have been compared experimentally in some works, and the results are usually the same: The ℓ_0-norm approximation F_β provides the best performance, followed by $F_{r\ell_1}$, and the ℓ_1 norm comes in third place [11, 21, 22, 24, 25].

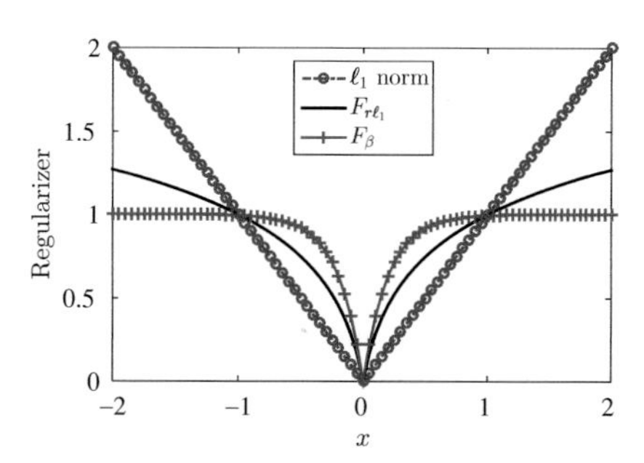

Figure 2.3 Sparsity-promoting regularizers applied on $x \in \mathbb{R}$: ℓ_1 norm; $F_{r\ell_1}$ with $\epsilon = 10$ [20]; F_β as the Laplace Function with $\beta = 5$ [11].

However, when working with nonconvex regularizers (that is, regularization functions), the possibility of introducing some undesired local minima or, even worse, changing the position of the global minimum exists. In this subsection, we elaborate more on this.

In Subsection 2.2.4, we showed an example of F_β surface containing some local minima (see Figure 2.2) and explained that by reducing β we could make these local minima as further from the origin as we want. We also mentioned the SL0 algorithm [29] as an efficient way to minimize $F_\beta(\mathbf{w})$, subject to some constraints on the parameters $\mathbf{w}$, using the steepest-descent method.

In adaptive filtering, however, F_β is usually employed as a regularizer, that is, a penalty function applied to the original objective function that we want to minimize, leading to the following regularized objective function $\mathcal{R}$:

$$\mathcal{R} = \left(\begin{array}{c} \text{Original objective} \\ \text{function} \end{array} \right) + \alpha \left(\begin{array}{c} \text{Sparsity-promoting} \\ \text{regularizer} \end{array} \right), \tag{2.15}$$

where $\alpha \in \mathbb{R}_+$ is the *regularization parameter*, responsible for determining the weight given to the regularizer. While the original objective function aims to find the coefficients $\mathbf{w}$ that best explain/model the data, the sparsity-promoting regularizer aims to sparsify $\mathbf{w}$, by maximizing the number of entries of $\mathbf{w}$ equal to 0. Since the regularization is used to help in the minimization of the original objective function (and not the other way around), the regularization must be much weaker in terms of magnitude; thus, α must be small. If this rule is not satisfied, then the regularization will have a high impact on $\mathcal{R}$ and will possibly generate a local or global minimum at $\mathbf{w} = \mathbf{0}$. It is worth noticing that, once again, the saturation property of F_β helps us since we know its maximum value, that is, for $\mathbf{w} \in \mathbb{R}^{N+1}$, $\max F_\beta(\mathbf{w}) = N + 1$. This implies that, for any point $\mathbf{w}$ in the parameter space, the value of $\mathcal{R}$ is equal to the value of the original objective function plus $\alpha \times (N + 1)$, in the worst case (i.e., this corresponds to the most significant change on the original objective function due to F_β). Therefore, a sufficient condition to guarantee that no undesired global minimum appears at $\mathbf{w} = \mathbf{0}$ is

$$\begin{pmatrix}\text{Original objective function}\\ \text{evaluated at } \mathbf{w}=\mathbf{0}\end{pmatrix} > \begin{pmatrix}\text{Original objective function}\\ \text{evaluated at } \mathbf{w}^\star\end{pmatrix} + \alpha(N+1),$$

where $\mathbf{w}^\star$ is the minimizer of the original objective function. Observe that $F_\beta(\mathbf{0}) = 0$, which means that $\mathcal{R}$ and the original objective function are equal at $\mathbf{w} = \mathbf{0}$, whereas $\alpha(N+1)$ corresponds to the maximum value of $\alpha F_\beta(\mathbf{w}^\star)$, which occurs only when $\mathbf{w}^\star$ does not have coefficients close to 0 (i.e., within the zero-attraction region).

In addition, adaptive filtering algorithms are very often derived based on an original objective function that is convex, for example, the MSE or the minimum disturbance criterion, both introduced in Chapter 1. The next example illustrates that by choosing α correctly, the regularized objective function $\mathcal{R}$ may remain convex, albeit F_β is nonconvex.

Example 2.1 (The Role of α)

Let us consider an example in which the original objective function is the MSE. It is widely known that the MSE surface is a paraboloid in the parameter space [30], and we assume that it has circular contours as depicted in Figure 2.4(a).[6] In addition, let us assume that the sparsity-promoting regularizer is the ℓ_0-norm approximation given in Equation (2.12a), that is, the LF with $\beta = 5$, and the minimum MSE solution is $\mathbf{w}_\text{o} = [1\ 0]^\text{T}$, marked with a circle in Figure 2.4, which depicts the surface of $\mathcal{R}$ for different values of the regularization parameter α.

In Figure 2.4(a), we have $\alpha = 0$ and, therefore, we observe only the MSE contours. As α increases from 0 to 0.1, we observe that the pattern of the contours change gradually, becoming similar to ellipses with minor axis aligned with the vertical coordinate w_1. This pattern is the effect of the regularizer F_β over the coordinate w_1, corresponding to the zero-entry of $\mathbf{w}_\text{o}$, while the coordinate w_0, which corresponds to the relevant/nonzero entry of $\mathbf{w}_\text{o}$, is almost not affected. In Figure 2.4(c), we set $\alpha = 1$ so that the regularization starts to compete against the original objective function, thus generating a local minimum at $\mathbf{w} = \mathbf{0}$, represented with a triangle in the figure. Finally, in Figure 2.4(d), we use $\alpha = 10$ to illustrate a case in which the regularization becomes dominant in $\mathcal{R}$, leading to a global minimum at $\mathbf{w} = \mathbf{0}$.

Observe that the colorbar ranges in Figure 2.4 can be used to indicate a bad choice of α. Indeed, when α is properly chosen, like in Figure 2.4(b), this range is very similar to the range obtained when $\alpha = 0$, see Figure 2.4(a), meaning that the regularization is not competing against the original objective function. On the other hand, if α is such that the regularization is capable of changing

[6] The contours of a surface correspond to the geometric places in which the surface has constant values. The MSE surface exhibits circular contours for uncorrelated inputs whose covariance matrix is $\sigma_x^2\mathbf{I}$, where σ_x^2 is the variance of the input signal and $\mathbf{I}$ is the identity matrix.

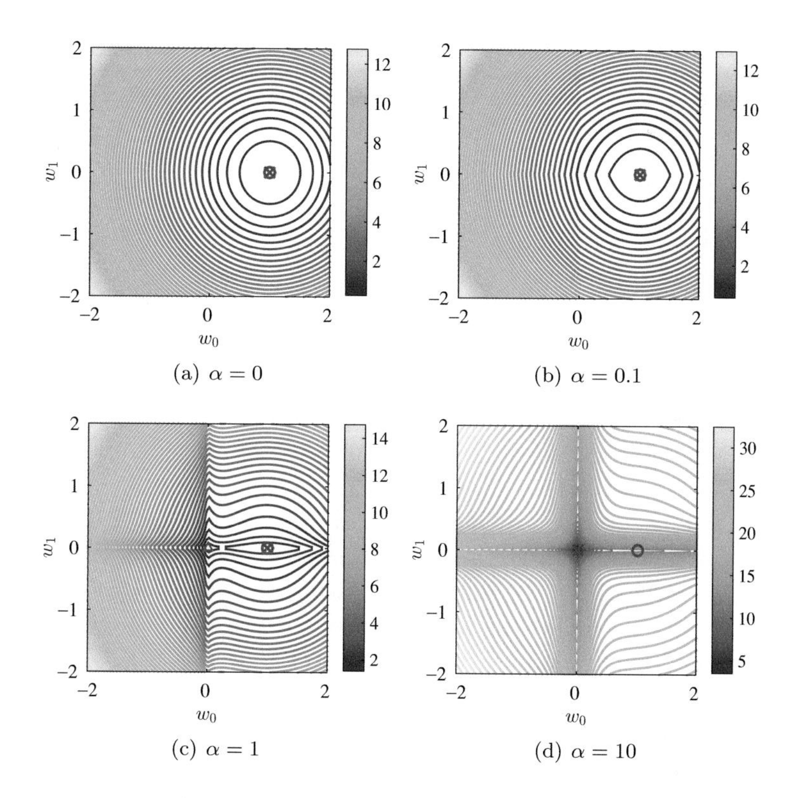

Figure 2.4 Contours of $\mathcal{R}$ for different values of the regularization parameter α: (a) $\alpha = 0$ (just the MSE contours); (b) $\alpha = 0.1$; (c) $\alpha = 1$; and (d) $\alpha = 10$. The circle denotes the minimum MSE solution $\mathbf{w}_o = [1\ 0]^T$, the cross is placed at the global minimum of $\mathcal{R}$, and the local minimum is represented by a triangle.

the colorbar range, like in Figures 2.4(c) and 2.4(d), then α might be too high and it would be wise reducing its value.

Note: The values of α used in Figure 2.4 are just for illustration purposes and should not be taken as guidelines. Usually, the choice of α changes from problem to problem. For instance, in the case of $\alpha = 0.1$, a local minimum might appear (just like in Figure 2.4(c)) if the value of β is increased or if the variance of the input signal σ_x^2 is decreased; in this case, one would have to reduce α to avoid the local minimum issue. Mathematically, to guarantee that $\mathcal{R}$ preserves the convexity of the original optimization problem, one must calculate its Hessian matrix and force it to be positive definite by selecting α and β properly.

2.3 Algorithms Employing Sparsity-Promoting Regularizations

In this section, we apply the sparsity-promoting regularization functions studied in Section 2.2 to some classical algorithms in order to enable them to exploit

Table 2.1 Summary of the sparsity-promoting regularizers $\mathcal{F}$ and their gradients

	$\mathcal{F}(\mathbf{w})$	i-th entry of $\mathbf{f}(\mathbf{w}) \triangleq \nabla\mathcal{F}(\mathbf{w})$
ℓ_1 norm	$\|\mathbf{w}\|_1$, see Equation (2.2)	$f_{\ell_1}(w_i) = \text{sign}(w_i)$
Reweighted ℓ_1 ($r\ell_1$) norm	$F_{r\ell_1}(\mathbf{w})$, see Equation (2.5)	$f_{r\ell_1}(w_i) = \dfrac{\text{sign}(w_i)}{1+\epsilon\|w_i\|}$
ℓ_0-norm approx.	$F_\beta(\mathbf{w})$, see Equation (2.12)	$f_\beta(w_i) = \begin{cases} \beta\text{sign}(w_i)\text{e}^{-\beta\|w_i\|} & \text{for the LF} \\ \dfrac{\beta\text{sign}(w_i)}{(1+\beta\|w_i\|)^2} & \text{for the GMF} \end{cases}$

the sparsity of a vector/signal. To provide a concise and unified exposition of regularization-based algorithms, we use the symbol $\mathcal{F}$ to denote any sparsity-promoting regularizer, and we postpone its choice as much as possible aiming at emphasizing the similarities among the several algorithms. For the reader's convenience, Table 2.1 summarizes the possible choices of $\mathcal{F}$ as well as their gradient expressions, covered in detail in Section 2.2.

The classical algorithms considered in this section are the LMS, the NLMS, and the AP algorithms, each of which was briefly discussed in Chapter 1. To generate the regularized versions of these algorithms, we follow the standard procedure: We take the original objective function used to derive each of these algorithms and we add $\mathcal{F}(\mathbf{w})$ as a penalty function. We also define the gradient of $\mathcal{F}(\mathbf{w})$ as

$$\nabla\mathcal{F}(\mathbf{w}) \triangleq \mathbf{f}(\mathbf{w}) = [f(w_0)\ f(w_1)\ \dots\ f(w_N)]^{\text{T}} \in \mathbb{R}^{N+1}. \tag{2.16}$$

After selecting the function $\mathcal{F}$, we include a subscript on $\mathbf{f}$ to inform of this choice. For example, if we choose $\mathcal{F} = F_\beta(\mathbf{w})$, then its gradient vector will be denoted by $\mathbf{f}_\beta(\mathbf{w})$ whose i-th entry is $f_\beta(w_i)$, as given in Table 2.1.

In the adaptive filtering literature, the prefixes *zero-attractor* (ZA), *reweighted zero-attractor* (RZA), and ℓ_0 are very often used to indicate that the ℓ_1 norm, the $r\ell_1$ norm, and the ℓ_0-norm approximation, respectively, were chosen as the sparsity-promoting regularization. For the LMS algorithm, for example, this leads to the following three sparsity-aware LMS-based algorithms: the *zero-attractor* LMS (ZA-LMS), the *reweighted zero-attractor* LMS (RZA-LMS), and the *ℓ_0-norm approximation* LMS (ℓ_0-LMS) algorithms.

2.3.1 Regularized LMS Algorithms

As explained in Chapter 1, the LMS algorithm uses a gradient-descent method to minimize the squared error $|e(k)|^2$, which can be regarded as an instantaneous and stochastic approximation of the MSE [30]. Thus, the regularized objective function corresponding to the LMS algorithm is given by

$$\mathcal{R}_{\text{LMS}}(\mathbf{w}(k)) \triangleq |e(k)|^2 + \alpha\mathcal{F}(\mathbf{w}(k)), \tag{2.17}$$

where $\mathcal{F}$ is any of the sparsity-promoting regularizers given in Table 2.1 and $\alpha \in \mathbb{R}_+$ is the regularization parameter, which is usually chosen as a small number in order for the regularization to be less important than the original objective function in the regularized problem $\mathcal{R}_{\text{LMS}}$, as explained in the Subsection 2.2.5.

Since the LMS algorithm employs a gradient-descent method, its recursion depends on the gradient of $\mathcal{R}_{\text{LMS}}$ with respect to $\mathbf{w}(k)$, denoted by $\nabla\mathcal{R}_{\text{LMS}}(\mathbf{w}(k))$, and is given by

$$\begin{aligned}
\mathbf{w}(k+1) &= \mathbf{w}(k) - \mu\ \nabla\mathcal{R}_{\text{LMS}}(\mathbf{w}(k)) \\
&= \mathbf{w}(k) - \mu\ \nabla|e(k)|^2 - \mu\alpha\ \nabla\mathcal{F}(\mathbf{w}(k)) \\
&= \mathbf{w}(k) + \underbrace{2\mu e(k)\mathbf{x}(k)}_{\text{LMS correction}} - \underbrace{\mu\alpha\mathbf{f}(\mathbf{w}(k)),}_{\text{Regularization correction}}
\end{aligned} \tag{2.18}$$

where we recognize a correction term corresponding to the standard LMS algorithm, whereas the regularization introduces a new correction term that encourages the filter coefficients $\mathbf{w}(k+1)$ to be sparse. The expressions for the gradient vector $\mathbf{f}(\mathbf{w}(k)) \triangleq \nabla\mathcal{F}(\mathbf{w}(k))$ can be found in Table 2.1 for the ℓ_1 norm, the $r\ell_1$ norm, and the ℓ_0-norm approximation. The LMS-based algorithms employing the ℓ_1 norm, the $r\ell_1$ norm, and the ℓ_0-norm approximation are known as ZA-LMS [23], RZA-LMS [23], and ℓ_0-LMS [25], respectively.[7] These algorithms are summarized in Algorithm 2.2.

Algorithm 2.2 The sparsity-aware LMS algorithms

Initialization:

$\mathbf{x}(0) = \mathbf{w}(0) = [0\ 0\ \dots\ 0]^{\mathrm{T}}$

choose μ and α as small positive numbers

choose the regularizer $\mathcal{F}$ (and related parameters):

$$\text{If } \mathcal{F} = \begin{cases} \|\mathbf{w}(k)\|_1, & \text{then we have the ZA-LMS algorithm [23]} \\ F_{r\ell_1}(\mathbf{w}(k)), & \text{then we have the RZA-LMS algorithm [23]} \\ F_\beta(\mathbf{w}(k)) & \text{then we have the } \ell_0\text{-LMS algorithm [25]} \end{cases}$$

For $k \geq 0$ (i.e., for every iteration) **do**

$e(k) = d(k) - \mathbf{w}^{\mathrm{T}}(k)\mathbf{x}(k)$

Compute $\mathbf{f}(\mathbf{w}(k))$ corresponding to the selected $\mathcal{F}$ (see Table 2.1)

$\mathbf{w}(k+1) = \mathbf{w}(k) + 2\mu e(k)\mathbf{x}(k) - \mu\alpha\mathbf{f}(\mathbf{w}(k))$

---

[7] The ℓ_0-LMS algorithm proposed in [25] uses the LF as F_β. However, instead of using the gradient $\mathbf{f}_\beta(\mathbf{w}(k))$, whose entries are specified in Table 2.1, it uses the Taylor series to approximate the exponential function by a first-order polynomial, for small values of $|w_i(k)|$. In this case, the algorithm exchanges quality of approximation (of the ℓ_0 norm), thus performance, for reduced computational burden.

Note: When we provide the step-by-step of an algorithm, like in Algorithm 2.2, we are focusing on didactic rather than computational efficiency. For example, we could avoid two multiplications at every iteration by defining and storing the auxiliary variables $\mu' = 2\mu$ and $\alpha' = \mu\alpha$ during the initialization stage.

2.3.2 Regularized AP Algorithms

Let us start by revisiting the notation of the main variables related to the AP algorithm (for more details, check Chapter 1):

$$\begin{array}{rclr}
\mathbf{x}(k) & = & [x(k)\; x(k-1)\; \ldots\; x(k-N)]^{\mathrm{T}} & \in \mathbb{R}^{N+1}, \\
\mathbf{X}(k) & = & [\mathbf{x}(k)\; \mathbf{x}(k-1)\; \ldots\; \mathbf{x}(k-L)] & \in \mathbb{R}^{(N+1)\times(L+1)}, \\
\mathbf{w}(k) & = & [w_0(k)\; w_1(k)\; \ldots\; w_N(k)]^{\mathrm{T}} & \in \mathbb{R}^{N+1}, \\
\mathbf{d}(k) & = & [d(k)\; d(k-1)\; \ldots\; d(k-L)]^{\mathrm{T}} & \in \mathbb{R}^{L+1}, \\
\mathbf{e}(k) & = & [e_0(k)\; e_1(k)\; \ldots\; e_L(k)]^{\mathrm{T}} & \in \mathbb{R}^{L+1},
\end{array} \tag{2.19}$$

where $\mathbf{x}(k)$ is the input vector (in tapped-delay line format), $\mathbf{X}(k)$ is the input matrix that includes the current input vector $\mathbf{x}(k)$ and also L previous input vectors, $\mathbf{w}(k)$ is the adaptive filter coefficient vector, $\mathbf{d}(k)$ is the desired vector, and $\mathbf{e}(k) = \mathbf{d}(k) - \mathbf{X}^{\mathrm{T}}(k)\mathbf{w}(k)$ is the error vector. In addition, N is the order of the adaptive filter, and L is the data reuse factor, that is, the amount of data from previous iterations that is to be used in the current iteration. Observe that $e_0(k)$, the 0-th entry of $\mathbf{e}(k)$, is computed using the data from the current iteration $(d(k), \mathbf{x}(k))$ and, therefore, it is equivalent to $e(k)$ in the LMS algorithm.

As explained in Chapter 1, the AP algorithm is derived by minimizing the minimum disturbance criterion (or principle) subject to some constraints related to the *a posteriori* errors being equal to 0. Thus, the regularized AP algorithms solve the following optimization problem

$$\begin{aligned}
&\text{minimize } \mathcal{R}_{\mathrm{AP}}(\mathbf{w}(k+1)) \triangleq \|\mathbf{w}(k+1) - \mathbf{w}(k)\|_2^2 + \alpha\mathcal{F}(\mathbf{w}(k+1)), \\
&\text{subject to } \mathbf{d}(k) - \mathbf{X}^{\mathrm{T}}(k)\mathbf{w}(k+1) = \mathbf{0},
\end{aligned} \tag{2.20}$$

where $\mathcal{R}_{\mathrm{AP}}(\mathbf{w}(k+1))$ represents the regularized objective function related to the AP algorithm. As explained in Subsection 2.3.1, $\mathcal{F}$ is any of the sparsity-promoting regularizers given in Table 2.1 and $\alpha \in \mathbb{R}_+$ is the regularization parameter.

In order to solve this optimization problem, we form the Lagrangian $\mathcal{L}$ as

$$\begin{aligned}
\mathcal{L}(\mathbf{w}(k+1), \boldsymbol{\lambda}) = & \|\mathbf{w}(k+1) - \mathbf{w}(k)\|_2^2 + \alpha\mathcal{F}(\mathbf{w}(k+1)) \\
& + \boldsymbol{\lambda}^{\mathrm{T}}\left[\mathbf{d}(k) - \mathbf{X}^{\mathrm{T}}(k)\mathbf{w}(k+1)\right],
\end{aligned} \tag{2.21}$$

which is a function of both the parameters $\mathbf{w}(k+1)$ and the Lagrange multipliers $\boldsymbol{\lambda} \in \mathbb{R}^{L+1}$. Then, we differentiate $\mathcal{L}$ with respect to $\mathbf{w}(k+1)$ and $\boldsymbol{\lambda}$ and equate the results to 0 (i.e., $\nabla \mathcal{L} = \mathbf{0}$) to obtain

$$\mathbf{w}(k+1) = \mathbf{w}(k) + \mathbf{X}(k)\frac{\boldsymbol{\lambda}}{2} - \frac{\alpha}{2}\,\nabla \mathcal{F}\left(\mathbf{w}(k+1)\right), \tag{2.22}$$

$$\mathbf{X}^{\mathrm{T}}(k)\mathbf{w}(k+1) = \mathbf{d}(k), \tag{2.23}$$

respectively. Then, the left multiplication of Equation (2.22) by $\mathbf{X}^{\mathrm{T}}(k)$ followed by the substitution of Equation (2.23) into the resulting equation leads to

$$\begin{aligned}\frac{\boldsymbol{\lambda}}{2} &= \left(\mathbf{X}^{\mathrm{T}}(k)\mathbf{X}(k)\right)^{-1}\mathbf{e}(k)\\ &\quad + \frac{\alpha}{2}\left(\mathbf{X}^{\mathrm{T}}(k)\mathbf{X}(k)\right)^{-1}\mathbf{X}^{\mathrm{T}}(k)\,\nabla \mathcal{F}\left(\mathbf{w}(k+1)\right),\end{aligned} \tag{2.24}$$

where it is assumed that $\mathbf{X}^{\mathrm{T}}(k)\mathbf{X}(k)$ is invertible. By substituting Equation (2.24) into Equation (2.22) we obtain

$$\begin{aligned}\mathbf{w}(k+1) &= \mathbf{w}(k) + \mathbf{X}(k)\left(\mathbf{X}^{\mathrm{T}}(k)\mathbf{X}(k)\right)^{-1}\mathbf{e}(k)\\ &\quad + \frac{\alpha}{2}\left[\mathbf{X}(k)\left(\mathbf{X}^{\mathrm{T}}(k)\mathbf{X}(k)\right)^{-1}\mathbf{X}^{\mathrm{T}}(k) - \mathbf{I}\right]\mathbf{f}(\mathbf{w}(k+1)),\end{aligned} \tag{2.25}$$

where $\mathbf{f}(\mathbf{w}(k+1)) \triangleq \nabla \mathcal{F}\left(\mathbf{w}(k+1)\right)$, as already defined. To transform the previous equation in a recursion, we replace $\mathbf{f}(\mathbf{w}(k+1))$ with $\mathbf{f}(\mathbf{w}(k))$ to obtain the recursion of the regularized AP algorithms:

$$\mathbf{w}(k+1) = \mathbf{w}(k) + \underbrace{\mu\mathbf{X}(k)\mathbf{S}(k)\mathbf{e}(k)}_{\text{AP correction}} + \underbrace{\mu\frac{\alpha}{2}\left[\mathbf{X}(k)\mathbf{S}(k)\mathbf{X}^{\mathrm{T}}(k) - \mathbf{I}\right]\mathbf{f}(\mathbf{w}(k)),}_{\text{Regularization correction}} \tag{2.26}$$

where, just like in the classical AP algorithm, we included a relaxation parameter μ that acts like a step size, and we defined $\mathbf{S}(k) \triangleq \left(\mathbf{X}^{\mathrm{T}}(k)\mathbf{X}(k) + \gamma\mathbf{I}\right)^{-1}$, where $\gamma \in \mathbb{R}_{+}$ is a small positive number used to prevent matrix $\mathbf{S}(k)$ from becoming singular or ill-conditioned. In Equation (2.26), we recognize a correction term corresponding to the standard AP algorithm, whereas the regularization introduces a new correction term that encourages the filter coefficients $\mathbf{w}(k+1)$ to be sparse.

Once again, the expressions for the gradient vector $\mathbf{f}(\mathbf{w}(k))$ can be found in Table 2.1 for the ℓ_1 norm, the $r\ell_1$ norm, and the ℓ_0-norm approximation. The AP-based algorithms employing the ℓ_1 norm, the $r\ell_1$ norm, and the ℓ_0-norm approximation are known as *zero-attractor* AP (ZA-AP) [31], *reweighted zero-attractor* AP (RZA-AP) [31], and *ℓ_0-norm approximation* AP (ℓ_0-AP) [21], respectively.[8] These algorithms are summarized in Algorithm 2.3.

[8] Actually, the authors in [21, 31] use ZA-APA and RZA-APA, instead of ZA-AP and RZA-AP, where APA stands for *affine projection algorithm*. Moreover, the herein called ℓ_0-AP algorithm was proposed in [21] by the name of *affine projection algorithm for sparse system identification*, abbreviated both as AP-SSI or APA-SSI in [11, 28]. In this chapter, we opted for using ℓ_0-AP instead of APA-SSI since the latter name is not informative of which sparsity-promoting regularizer is being used.

Algorithm 2.3 The sparsity-aware AP algorithms

Initialization:

$\mathbf{x}(0) = \mathbf{w}(0) = [0\ 0\ \ldots\ 0]^{\mathrm{T}}$

choose $\mu \in (0, 1]$

choose α and γ as small positive numbers

choose the regularizer $\mathcal{F}$ (and related parameters):

$$\text{If } \mathcal{F} = \begin{cases} \|\mathbf{w}(k)\|_1, & \text{then we have the ZA-AP algorithm [31]} \\ F_{r\ell_1}(\mathbf{w}(k)), & \text{then we have the RZA-AP algorithm [31]} \\ F_{\beta}(\mathbf{w}(k)) & \text{then we have the } \ell_0\text{-AP algorithm [21]} \end{cases}$$

For $k \geq 0$ (i.e., for every iteration) **do**

$\mathbf{e}(k) = \mathbf{d}(k) - \mathbf{X}^{\mathrm{T}}(k)\mathbf{w}(k)$

$\mathbf{S}(k) = \left(\mathbf{X}^{\mathrm{T}}(k)\mathbf{X}(k) + \gamma\mathbf{I}\right)^{-1}$

Compute $\mathbf{f}(\mathbf{w}(k))$ corresponding to the selected $\mathcal{F}$ (see Table 2.1)

$\mathbf{w}(k+1) = \mathbf{w}(k) + \mu\mathbf{X}(k)\mathbf{S}(k)\mathbf{e}(k) + \mu\frac{\alpha}{2}\left[\mathbf{X}(k)\mathbf{S}(k)\mathbf{X}^{\mathrm{T}}(k) - \mathbf{I}\right]\mathbf{f}(\mathbf{w}(k))$

It is interesting to interpret the optimization problem given in Equation (2.20) from the geometric point of view. Its solution $\mathbf{w}(k+1)$ must satisfy the constraint $\mathbf{X}^{\mathrm{T}}(k)\mathbf{w}(k+1) = \mathbf{d}(k)$, meaning that $\mathbf{w}(k+1)$ must lie on the intersection of $(L+1)$ hyperplanes, denoted by $\Pi_{(L+1)}$. If $\alpha = 0$, then $\mathbf{w}(k+1)$ is generated as an orthogonal projection of $\mathbf{w}(k)$ onto $\Pi_{(L+1)}$, which is the minimum Euclidean norm solution; this corresponds to the standard AP update (with $\mu = 1$). For $\alpha \neq 0$, the update process can be explained as a two-step procedure: (i) the standard AP update transports $\mathbf{w}(k)$ to a point $\mathbf{w}' \in \Pi_{(L+1)}$ and (ii) the regularization correction maps $\mathbf{w}'$ to $\mathbf{w}(k+1) \in \Pi_{(L+1)}$ by encouraging $\mathbf{w}(k+1)$ to be sparser than $\mathbf{w}'$, but still restricted to belong to $\Pi_{(L+1)}$. The fact that the sparse solution $\mathbf{w}(k+1)$ is confined to $\Pi_{(L+1)}$ can be seen as an advantage since it facilitates the choice of α (we postpone this explanation to Subsection 2.3.3), but it can also be seen as a disadvantage since it reduces the length of $\mathbf{f}(\mathbf{w}(k))$, as the length of a vector is always greater than or equal to the length of its projection, and requires more computations, as explained in [11, 21]. This issue motivated the development of a new algorithm, which eliminates the matrix $\mathbf{X}(k)\mathbf{S}(k)\mathbf{X}^{\mathrm{T}}(k)$, leading to the following recursion:

$$\mathbf{w}(k+1) = \mathbf{w}(k) + \underbrace{\mu\mathbf{X}(k)\mathbf{S}(k)\mathbf{e}(k)}_{\text{AP correction}} - \underbrace{\mu\frac{\alpha}{2}\mathbf{f}(\mathbf{w}(k))}_{\text{Regularization correction}}. \tag{2.27}$$

Observe that, for $\mu = 1$ and $\gamma = 0$, that is, eliminating the constants that were artificially introduced to increase the algorithm robustness, the $\mathbf{w}(k+1)$ in Equation (2.26) satisfies the constraint in Equation (2.20), but the $\mathbf{w}(k+1)$ in Equation (2.27) does not. Therefore, algorithms employing the recursion given in Equation (2.27) should not be called AP. Here, we call them *quasi-AP* (qAP) following the work that first noticed this fact [21]. The qAP algorithms employing sparsity-promoting regularizations are summarized in Algorithm 2.4.

Algorithm 2.4 The sparsity-aware qAP algorithms [11, 21]

Initialization:

$\mathbf{x}(0) = \mathbf{w}(0) = [0\ 0\ \ldots\ 0]^{\mathrm{T}}$

choose $\mu \in (0, 1]$

choose α and γ as small positive numbers

choose the regularizer $\mathcal{F}$ (and related parameters):

$$\text{If } \mathcal{F} = \begin{cases} \|\mathbf{w}(k)\|_1, & \text{then we have the ZA-qAP algorithm} \\ F_{r\ell_1}(\mathbf{w}(k)), & \text{then we have the RZA-qAP algorithm} \\ F_{\beta}(\mathbf{w}(k)) & \text{then we have the } \ell_0\text{-qAP algorithm [21]} \end{cases}$$

For $k \geq 0$ (i.e., for every iteration) **do**

$\mathbf{e}(k) = \mathbf{d}(k) - \mathbf{X}^{\mathrm{T}}(k)\mathbf{w}(k)$

$\mathbf{S}(k) = \left(\mathbf{X}^{\mathrm{T}}(k)\mathbf{X}(k) + \gamma\mathbf{I}\right)^{-1}$

Compute $\mathbf{f}(\mathbf{w}(k))$ corresponding to the selected $\mathcal{F}$ (see Table 2.1)

$\mathbf{w}(k+1) = \mathbf{w}(k) + \mu\mathbf{X}(k)\mathbf{S}(k)\mathbf{e}(k) - \mu\frac{\alpha}{2}\mathbf{f}(\mathbf{w}(k))$

Hint: If α is properly chosen, both AP and qAP algorithms provide similar performance in terms of convergence speed and MSE, but the qAP algorithm is more efficient. However, when a qAP algorithm is used with nonstationary input $\mathbf{x}(k)$, the problem of choosing α is considerably more difficult and maybe even impossible. This issue will be explained in Subsection 2.3.3.

Note: To the best of our knowledge, there is no article proposing the ZA-qAP and RZA-qAP algorithms. We opted for introducing them here only to maintain the pattern we have been following.

2.3.3 Regularized NLMS Algorithms

As explained in Chapter 1, the NLMS algorithm is a particular case of the AP algorithm, where $L = 0$, that is, only the data from the current iteration is used. Therefore, in this subsection, we present only its related optimization problem and the algorithms recursions. The derivation of these recursions is omitted here because it follows precisely the same steps used in Subsection 2.3.2.

The regularized NLMS algorithms solve the following optimization problem:

$$\begin{aligned} &\text{minimize } \mathcal{R}_{\mathrm{NLMS}}(\mathbf{w}(k+1)) \triangleq \|\mathbf{w}(k+1) - \mathbf{w}(k)\|_2^2 + \alpha\mathcal{F}(\mathbf{w}(k+1)), \\ &\text{subject to } d(k) - \mathbf{x}^{\mathrm{T}}(k)\mathbf{w}(k+1) = 0, \end{aligned} \tag{2.28}$$

where $\mathcal{R}_{\mathrm{NLMS}}(\mathbf{w}(k+1))$ represents the regularized objective function related to the NLMS algorithm, $\mathcal{F}$ is any of the sparsity-promoting regularizers given in Table 2.1, and $\alpha \in \mathbb{R}_+$ is the regularization parameter.

The solution to the optimization problem in Equation (2.28) leads to the following recursion:

Algorithm 2.5 The sparsity-aware NLMS algorithms

Initialization:

$\mathbf{x}(0) = \mathbf{w}(0) = [0\ 0\ \ldots\ 0]^{\mathrm{T}}$

choose $\mu \in (0, 1]$

choose α and γ as small positive numbers

choose the regularizer $\mathcal{F}$ (and related parameters):

$$\text{If } \mathcal{F} = \begin{cases} \|\mathbf{w}(k)\|_1, & \text{then we have the ZA-NLMS algorithm [31]} \\ F_{r\ell_1}(\mathbf{w}(k)), & \text{then we have the RZA-NLMS algorithm [31]} \\ F_\beta(\mathbf{w}(k)) & \text{then we have the } \ell_0\text{-NLMS algorithm [21]} \end{cases}$$

For $k \geq 0$ (i.e., for every iteration) **do**

$e(k) = \mathbf{d}(k) - \mathbf{x}^{\mathrm{T}}(k)\mathbf{w}(k)$

Compute $\mathbf{f}(\mathbf{w}(k))$ corresponding to the selected $\mathcal{F}$ (see Table 2.1)

$$\mathbf{w}(k+1) = \mathbf{w}(k) + \frac{\mu e(k)}{\|\mathbf{x}(k)\|_2^2 + \gamma}\mathbf{x}(k) + \mu\frac{\alpha}{2}\left[\frac{\mathbf{x}(k)\mathbf{x}^{\mathrm{T}}(k)}{\|\mathbf{x}(k)\|_2^2 + \gamma} - \mathbf{I}\right]\mathbf{f}(\mathbf{w}(k))$$

$$\mathbf{w}(k+1) = \mathbf{w}(k) + \underbrace{\frac{\mu e(k)}{\|\mathbf{x}(k)\|_2^2 + \gamma}\mathbf{x}(k)}_{\text{NLMS correction}} + \underbrace{\mu\frac{\alpha}{2}\left[\frac{\mathbf{x}(k)\mathbf{x}^{\mathrm{T}}(k)}{\|\mathbf{x}(k)\|_2^2 + \gamma} - \mathbf{I}\right]\mathbf{f}(\mathbf{w}(k))}_{\text{Regularization correction}}, \tag{2.29}$$

where, as explained in Subsection 2.3.2, μ can be understood both as a relaxation factor, since it actually relaxes the constraint in Equation (2.28), or as a step size and should be chosen as $\mu \in (0, 1]$, and $\gamma \in \mathbb{R}_+$ is a small positive number used to avoid numerical problems that occur when $\|\mathbf{x}(k)\|_2^2 \to 0$. Algorithm 2.5 summarizes the sparsity-aware NLMS algorithms employing different regularizations.

Following the same reasoning that motivated the qAP algorithms in Subsection 2.3.2, for the NLMS algorithm we can eliminate the matrix $\frac{\mathbf{x}(k)\mathbf{x}^{\mathrm{T}}(k)}{\|\mathbf{x}(k)\|_2^2+\gamma}$ in order to reduce the number of arithmetic operations required by the recursion and also to let the sparsity-promoting term $\mathbf{f}(\mathbf{w}(k))$ to be unconstrained, leading to the recursion of the so-called *quasi-NLMS* (qNLMS) algorithms given by

$$\mathbf{w}(k+1) = \mathbf{w}(k) + \underbrace{\frac{\mu e(k)}{\|\mathbf{x}(k)\|_2^2 + \gamma}\mathbf{x}(k)}_{\text{NLMS correction}} - \underbrace{\mu\frac{\alpha}{2}\mathbf{f}(\mathbf{w}(k))}_{\text{Regularization correction}}. \tag{2.30}$$

The qNLMS algorithms employing sparsity-promoting regularizations are summarized in Algorithm 2.6.

Note: The ℓ_0-qNLMS algorithm was proposed in [25] under the name of ℓ_0-NLMS algorithm, but this is not an accurate name since its update equation for $\mu = 1$ and $\gamma = 0$, see Equation (2.30), does not satisfy the constraint in Equation (2.28). Moreover, from the numerical point of view, just the NLMS correction term is normalized by the energy of $\mathbf{x}(k)$ (assuming γ is negligible compared to $\|\mathbf{x}(k)\|_2^2$), whereas the regularization correction is not normalized

Algorithm 2.6 The sparsity-aware qNLMS algorithms [11, 21]

Initialization:

$\mathbf{x}(0) = \mathbf{w}(0) = [0\ 0\ \ldots\ 0]^{\mathrm{T}}$

choose $\mu \in (0, 1]$

choose α and γ as small positive numbers

choose the regularizer $\mathcal{F}$ (and related parameters):

$$\text{If } \mathcal{F} = \begin{cases} \|\mathbf{w}(k)\|_1, & \text{then we have the ZA-qNLMS algorithm} \\ F_{r\ell_1}(\mathbf{w}(k)), & \text{then we have the RZA-qNLMS algorithm} \\ F_\beta(\mathbf{w}(k)) & \text{then we have the } \ell_0\text{-qNLMS algorithm [21, 25]} \end{cases}$$

For $k \geq 0$ (i.e., for every iteration) **do**

$e(k) = \mathbf{d}(k) - \mathbf{x}^{\mathrm{T}}(k)\mathbf{w}(k)$

Compute $\mathbf{f}(\mathbf{w}(k))$ corresponding to the selected $\mathcal{F}$ (see Table 2.1)

$$\mathbf{w}(k+1) = \mathbf{w}(k) + \frac{\mu e(k)}{\|\mathbf{x}(k)\|_2^2 + \gamma}\mathbf{x}(k) - \mu\frac{\alpha}{2}\mathbf{f}(\mathbf{w}(k))$$

and, as a result, it might be difficult to set α, especially in situations where the energy of $\mathbf{x}(k)$ may vary. Think of this as follows. There are two correction terms in the recursion: (i) the NLMS correction, which aims to model the data, thus reducing the squared error and (ii) the regularization correction, which encourages sparse solutions. For the NLMS versions described in Algorithm 2.5, since both correction terms are normalized by the energy of $\mathbf{x}(k)$, one needs to set just a single parameter α to determine the balance between modeling the data and finding sparse solutions. As already explained, modeling the data should be "more important" than sparsifying the solution, which justifies the use of low values of α. The point is, for the qNLMS versions given in Algorithm 2.6, as only one of these correction terms is normalized by $\|\mathbf{x}(k)\|_2^2$, the "degree of importance" related to this correction term is time-varying, whereas the "degree of importance" of the other term is constant and equal to α. For example, if at a given iteration k, the energy $\|\mathbf{x}(k)\|_2^2$ becomes very large, then the NLMS correction may become very small (tending to 0), whereas the regularization correction is unaltered. This situation would configure a case in which we put more emphasis on finding sparse solutions than on modeling the data and, as a consequence, it might generate local or global minimum points as discussed in Subsection 2.2.5.[9]

2.3.4 Numerical Experiments

Here, we compare the different sparsity-promoting regularizers considering that optimal coefficients have varying degrees of sparsity. The experiments consist

[9] Observe that this normalization issue in the qNLMS algorithm also happens in the qAP algorithm. We opted for explaining it here because the NLMS algorithm is easier to understand since it is closely related to the LMS algorithm and also because of the issue related to the name of the algorithm.

of identifying an unknown system $\mathbf{w}_o$ comprised of 16 coefficients given by the following [21, 31]:

- In Experiment 1, we set its fourth tap equal to 1, whereas the others are equal to 0, thus the optimal coefficients have a very high sparsity degree.
- In Experiment 2, the odd taps are set to 1, while the even taps are equal to 0, thus $\mathbf{w}_o$ has a sparsity degree of 50%.
- In Experiment 3, all the 16 taps are equal to 1, that is, the sparsity degree in $\mathbf{w}_o$ is 0% (this kind of signal is also called *dispersive*).

As for the adaptive filter, the number of coefficients is also 16 and the following algorithms are tested: the AP, the ℓ_0-AP, the ℓ_0-qAP, the ZA-AP, and the RZA-AP. The parameters of these algorithms were set so that the algorithms differ only by the regularization function. Thus, we set the step size $\mu = 0.9$, the regularization factor $\gamma = 10^{-12}$, the data-reuse factor $L = 4$, F_β as the GMF with $\beta = 5$, and we use $\alpha = 5 \times 10^{-3}$ and $\epsilon = 100$, in accordance with the values used in [31]. Moreover, the reference signal $d(k)$ is assumed to be corrupted by an additive white Gaussian measurement noise with variance $\sigma_n^2 = 0.01$.

Figure 2.5 depicts the MSE results for Experiments 1–3. One can observe that all the algorithms exhibited similar convergence speeds, as they were adjusted with the same values of μ, γ, L, and α. In Experiment 1, depicted in Figure 2.5(a), all the sparsity-aware algorithms outperformed the AP algorithm due to the high sparsity degree involved in this experiment. Moreover, the ones

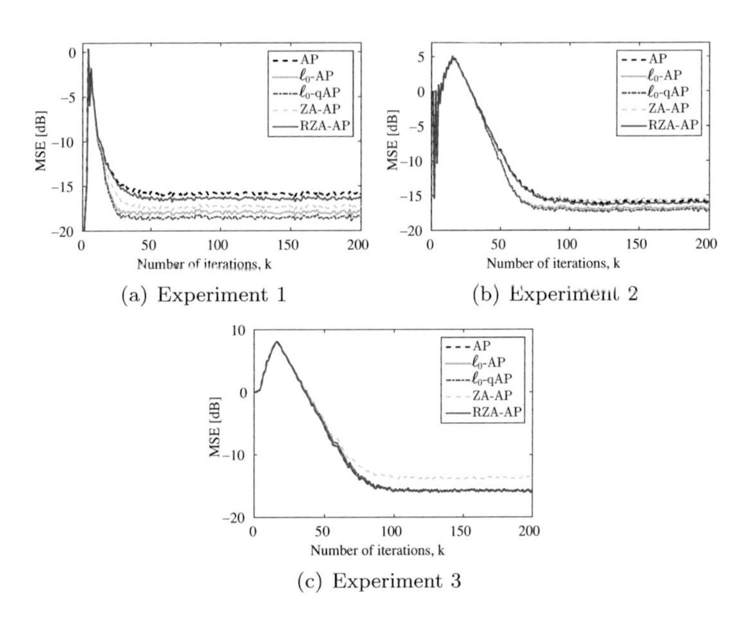

Figure 2.5 MSE learning curves for some sparsity-aware AP (and qAP) algorithms considering the optimal coefficients have different sparsity degrees: (a) Experiment 1 (very sparse); (b) Experiment 2 (moderately sparse); and (c) Experiment 3 (dispersive).

that use the ℓ_0-norm approximation (i.e., the ℓ_0-AP and the ℓ_0-qAP algorithms) achieved the best results. In Experiment 2, depicted in Figure 2.5(b), one can notice that the ℓ_0-AP and the ℓ_0-qAP algorithms were still better than the others, but the improvement over the AP algorithm was reduced since the unknown system is not very sparse. One should observe that, in this case, the algorithms employing the ℓ_1 norm and the $r\ell_1$ did not perform better than the AP algorithm. This indicates that these two regularizers require higher values of sparsity degree in order to bring some advantage. Finally, in Experiment 3, we tested these regularizers in a case where the unknown system does not have any coefficient equal to 0. One can observe that the algorithms employing the $r\ell_1$ norm and the ℓ_0-norm approximation achieved the same MSE results as the AP algorithm, that is, these regularizers were robust to the absence of sparsity (in the sense that they do not introduce any harm if the vector to be identified happens to be nonsparse/dispersive). This behavior is justified by the fact that the entries of the gradient vector corresponding to these two regularizers converge to 0 as $|w_i|$ grows, as can be verified in Table 2.1. The ZA-AP algorithm, on the other hand, employs the ℓ_1 norm, which shrinks the large coefficients leading to the worst MSE result, as depicted in Figure 2.5(c).

2.3.5 Further Readings

In this section, we applied the sparsity-promoting regularizers studied in Section 2.2 to the following three classical algorithms: the LMS, the NLMS, and the AP algorithms. We also provided some numerical experiments to illustrate the effect of each regularizer as the sparsity degree of the optimal coefficients vary. Clearly, our presentation focused on the regularizers, thus explaining how they can be combined to allow the exploitation of sparsity by the algorithms and also illustrating the benefits introduced by each of the regularizers. However, the reader must be aware that sparsity-promoting regularizations can be applied to many other algorithms, including those not belonging to the adaptive filtering field. Here, we provide some notes to complement the content of this section.

Still in the adaptive filtering context, sparsity-promoting regularizers have been combined with the recursive least squares (RLS) algorithm [32] and also with the set-membership (SM) framework. In [11], the SM versions of the ℓ_0-AP and ℓ_0-qAP algorithms were proposed (under the name of SSM-AP and QSSM-AP, respectively). These algorithms combine the advantages of the ℓ_0-norm approximation with the data selection scheme used in the SM approach, which confers robustness against uncertainties and noise [33–35]. Indeed, in the comprehensive set of simulations provided in [11], which encompasses many algorithms as well as sparse and compressible signals, these algorithms achieved remarkable MSE and misalignment results. Also, notice that in the same way we used the fact that the NLMS algorithm can be obtained from the AP algorithm by setting the data-reuse factor as $L = 0$, sparsity-promoting versions of the binormalized LMS (BNLMS) algorithm can be derived by making $L = 1$.

Recently, the sparsity-promoting regularizers described in Section 2.2 have been applied to the so-called *feature adaptive filtering* [1, 2] to exploit the hidden sparsity in the optimal parameters. Although most works in this area employ the ℓ_1 norm due to its simplicity, the ℓ_0-norm approximation has proven to be more advantageous [36].

In the context of nonlinear adaptive filtering, sparsity-promoting regularizers have been used in Volterra filters [37]. In fact, the Volterra series provides an excellent basis to apply these regularizers, as the number of coefficients increase very fast with the filter and memory orders of the Volterra series [38, 39]. Indeed, since in practical applications the nonlinear components are often much fewer than the number of coefficients in a truncated Volterra series, the optimal coefficients tend to be very sparse [37]. In [37], the results using the ℓ_0-norm approximation were superior to the results employing the ℓ_1 norm.

In the context of regression as well as in neural networks, it is very common to employ some of these regularizers [7, 10, 18, 40, 41]. Recently, they have also been applied in the context of distributed learning; see [42] and references therein.

In addition to the sparsity-promoting regularizers presented in this chapter, there also exist the so-called *mixed norms* that have been used for block-sparse systems, that is, systems in which the nonzero coefficients appear in a small number of clusters [43, 44]. In this approach, the coefficients are separated in several clusters, but determining an adequate size for these clusters can be challenging. Besides, to the best of our knowledge, this approach has not been tested using real data and compressible signals yet.

2.4 Proportionate-Type Algorithms

In the proportionate-type algorithms, the update term applied to each adaptive filter coefficient $w_i(k)$ is *proportional* to its own magnitude $|w_i(k)|$, meaning that large magnitude coefficients update rapidly (through large steps), whereas small magnitude coefficients update slowly (with small steps). These algorithms are quite interesting in applications involving the identification of systems whose coefficients have a wide range of absolute values, and one has prior information about the value of their small magnitude coefficients, as is the case of a sparse system identification in which we know beforehand that the small coefficients are equal to 0. Such prior information is of paramount importance in these algorithms as they update the small magnitude coefficients very slowly and, therefore, the adaptive filter coefficients must be initialized properly (at least, the coefficients w_i whose optimal values are small in magnitude), as explained in [9].

The proportionate-type algorithms are extensions of the classical algorithms, like the NLMS and AP algorithms, in which this proportionate-update principle is introduced through the so-called *proportionate matrix* $\mathbf{G}(k)$. Matrix $\mathbf{G}(k)$ is diagonal with nonzero entries given by

$$g_i(k) = \mathcal{G}(|w_i(k)|) \ , \text{ for } i \in \{0, 1, \ldots, N\} \tag{2.31}$$

that is, $g_i(k)$ is a function of $|w_i(k)|$. Essentially, the different proportionate-type algorithms are obtained by making different choices of $g_i(k)$, or equivalently, of this function $\mathcal{G}$.

In this section, we cover some of the most important proportionate-type algorithms. We begin with the simpler ones, which are based on the classical NLMS algorithm, and then we move to more general algorithms based on the SM and affine-projection (AP) ideas. We start with the first algorithm of its kind, the proportionate NLMS (PNLMS) algorithm [45]. Then, we cover the improved PNLMS (IPNLMS) algorithm, the μ-law PNLMS (MPNLMS) algorithm, and the improved MPNLMS (IMPNLMS) algorithm. One should notice that these proportionate-type algorithms have become "less proportional" along the years in order to increase the algorithm robustness, as the results reported for the PNLMS algorithm pointed out several cases in which its performance suffers a significant degradation [9, 46, 47]. Indeed, while in the PNLMS algorithm, we have $\mathcal{G}$ in Equation (2.31) equal to the identity function (in general), in the other algorithms mentioned above, a function $\mathcal{G}$ that grows less quickly with $|w_i(k)|$ is used. We also present the set-membership PNLMS (SM-PNLMS) algorithm.

2.4.1 Proportionate-Type Algorithms Based on the NLMS Recursion

In this subsection, we provide a unified treatment of the proportionate-type algorithms based on the NLMS recursion, herein collectively called proportionate-type NLMS (Pt-NLMS) algorithms, which include the PNLMS, the IPNLMS, the MPNLNS, and the IMPNLMS, among many other algorithms. Specifically, we present their general update equation, the related optimization problem, geometric interpretations, the role of the proportionate matrix $\mathbf{G}(k)$, and we address the choice of the common parameters. Besides, the material presented in this subsection can be adapted to the SM-PNLMS and SM-PAP algorithms easily.

The update equation (recursion) of the proportionate-type algorithms based on the NLMS recursion is given by

$$\mathbf{w}(k+1) = \mathbf{w}(k) + \mu \frac{e(k)\mathbf{G}(k)\mathbf{x}(k)}{\mathbf{x}^{\mathrm{T}}(k)\mathbf{G}(k)\mathbf{x}(k) + \delta}, \tag{2.32}$$

where $\mathbf{w}(k), \mathbf{x}(k) \in \mathbb{R}^{N+1}$ are the adaptive filter coefficient (weight) vector and the input vector applied to the adaptive filter at iteration k, respectively. Denoting the desired or reference signal (also known as target) by $d(k) \in \mathbb{R}$, the error signal at the k-th iteration is defined as

$$e(k) = d(k) - \mathbf{w}^{\mathrm{T}}(k)\mathbf{x}(k). \tag{2.33}$$

The regularization parameter $\delta \in \mathbb{R}_+$ is a small nonnegative number used to avoid numerical issues when $\mathbf{x}^{\mathrm{T}}(k)\mathbf{G}(k)\mathbf{x}(k)$ tends to 0. The step size or learning rate parameter is denoted by $\mu \in \mathbb{R}_+$. Just like in the NLMS algorithm, the step size should be chosen in the range $0 < \mu \leq 1$, and it represents a trade-off between convergence speed and steady-state error, that is, higher (lower) values of μ lead to faster (slower) convergence, but higher (lower) levels of steady-state MSE. Finally, $\mathbf{G}(k)$ is the proportionate matrix, which is a diagonal matrix whose elements on the main diagonal are denoted by $g_i(k), i \in \{0, 1, \ldots, N\}$. As previously explained, the only difference among the proportionate-type algorithms based on the NLMS recursion lies on the definition of this matrix, that is, on the relation between $g_i(k)$ and $|w_i(k)|$ in Equation (2.31), a topic addressed in the subsections to come.

First, observe that the recursion given in Equation (2.32) resembles that of the NLMS algorithm. Indeed, if we choose $\mathbf{G}(k) = \mathbf{I}$ (the identity matrix), then the two recursions coincide. However, unlike the NLMS algorithm, whose update direction is given by $\mathbf{x}(k)$, the update direction in Equation (2.32) is determined by vector $\mathbf{x}'(k) = \mathbf{G}(k)\mathbf{x}(k)$. If $\mathbf{G}(k)$ is chosen properly (i.e., according to the proportional-update principle), then the entries of $\mathbf{x}'(k)$ are amplified or attenuated, in relation to the corresponding entries of $\mathbf{x}(k)$, by a function of the magnitude of its corresponding entry $|w_i(k)|$. The practical effect is that the adaptive filter coefficients $w_i(k)$ with larger (smaller) magnitudes are updated with larger (shorter) steps. The matrix $\mathbf{G}(k)$ appearing in the denominator is responsible for keeping the algorithm normalized.

Second, the recursion given in Equation (2.32) is related to the solution of the following constrained optimization process:

$$\begin{aligned} &\min \|\mathbf{w}(k+1) - \mathbf{w}(k)\|^2_{\mathbf{G}^{-1}(k)} \\ &\text{subject to } d(k) - \mathbf{x}^{\mathrm{T}}(k)\mathbf{w}(k+1) = 0 \,, \end{aligned} \tag{2.34}$$

where $\|\mathbf{w}(k+1) - \mathbf{w}(k)\|^2_{\mathbf{G}^{-1}(k)} = [\mathbf{w}(k+1) - \mathbf{w}(k)]^{\mathrm{T}}\mathbf{G}^{-1}(k)[\mathbf{w}(k+1) - \mathbf{w}(k)]$ is a norm induced by the matrix $\mathbf{G}^{-1}(k)$. This is a minimum disturbance problem with linear constraint. The constraint means $\mathbf{w}(k+1)$ must yield an *a posteriori* error equal to 0, that is, it must fit the data $(\mathbf{x}(k), d(k))$ exactly. The cost function forces $\mathbf{w}(k+1)$ to be close to $\mathbf{w}(k)$ in some sense, but not in the usual Euclidean sense, in order to maintain a good fit of the prior data $(\mathbf{x}(i), d(i))$, with $i < k$.

To solve the optimization problem given in Equation (2.34), we apply the method of Lagrange multipliers and form the Lagrangian function

$$\mathcal{L}(\mathbf{w}(k+1), \lambda) = \|\mathbf{w}(k+1) - \mathbf{w}(k)\|^2_{\mathbf{G}^{-1}(k)} + \lambda \left[d(k) - \mathbf{x}^{\mathrm{T}}(k)\mathbf{w}(k+1)\right], \tag{2.35}$$

where $\lambda \in \mathbb{R}$ is the Lagrange multiplier. Then, we differentiate $\mathcal{L}$ with respect to $\mathbf{w}(k+1)$ and λ and equate the results to 0 to obtain

$$\frac{\partial \mathcal{L}}{\partial \mathbf{w}(k+1)} = \mathbf{0} \quad \therefore \quad \mathbf{w}(k+1) = \mathbf{w}(k) + \frac{\lambda}{2}\mathbf{G}(k)\mathbf{x}(k), \tag{2.36}$$

$$\frac{\partial \mathcal{L}}{\partial \lambda} = 0 \quad \therefore \quad \mathbf{x}^{\mathrm{T}}(k)\mathbf{w}(k+1) = d(k), \tag{2.37}$$

in which we have assumed that $\mathbf{G}(k)$ exists (i.e., it is nonsingular). If we premultiply Equation (2.36) by $\mathbf{x}^{\mathrm{T}}(k)$, use Equation (2.37), and the definition of the error signal $e(k)$, then we get

$$\frac{\lambda}{2} = \frac{e(k)}{\mathbf{x}^{\mathrm{T}}(k)\mathbf{G}(k)\mathbf{x}(k)}. \tag{2.38}$$

Substituting this relation in Equation (2.36), we obtain the following recursion:

$$\mathbf{w}(k+1) = \mathbf{w}(k) + \frac{e(k)\mathbf{G}(k)\mathbf{x}(k)}{\mathbf{x}^{\mathrm{T}}(k)\mathbf{G}(k)\mathbf{x}(k)}. \tag{2.39}$$

Finally, to obtain Equation (2.32), we just need to include two parameters in Equation (2.39): the regularization factor δ in the denominator and the step size parameter μ, also known as *relaxation factor* since, from the optimization point of view, it relaxes the problem constraint by not forcing the *a posteriori* error to be equal to 0 (which happens only when $\mu = 1$).

In this subsection, we addressed some topics common to every proportionate-type algorithm based on the NLMS recursion. In the following subsections, we focus on the main ingredient of the proportionate-type algorithms: the proportionate matrix $\mathbf{G}(k)$.

2.4.2 The PNLMS Algorithm

The PNLMS algorithm was proposed by Duttweiler [45] in 2000, and it inspired the development of many other algorithms following the same proportionate-update principle, but using slightly different proportionate matrices. In [45], the PNLMS algorithm is shown to converge faster than the NLMS algorithm when estimating/identifying impulse responses corresponding to echo paths. Such impulse responses are usually very long (the higher the sampling rate, the longer they are), but most of their energy is concentrated in a few samples, meaning that echo paths are examples of sparse systems found in practice [48].

In the PNLMS recursion given in Equation (2.32), each entry on the main diagonal of the proportionate matrix $\mathbf{G}(k) = \mathsf{Diag}\{[g_0(k)\ g_1(k)\ \cdots\ g_N(k)]\}$ is given by

$$g_i(k) = \frac{\gamma_i(k)}{\sum_{j=0}^{N} |\gamma_j(k)|}, \tag{2.40}$$

$$\gamma_i(k) = \max\left\{ \underbrace{|w_i(k)|}_{1^{\text{st}}\text{ term}},\ \underbrace{\rho \max\{\delta_\rho,\ |w_0(k)|,\ \cdots,\ |w_N(k)|\}}_{2^{\text{nd}}\text{ term}} \right\}, \tag{2.41}$$

Algorithm 2.7 The PNLMS algorithm

Initialization:

$\mathbf{x}(0) = \mathbf{w}(0) = [0\ 0\ \ldots\ 0]^{\mathrm{T}}$
choose μ in the range $0 < \mu \leq 1$
choose ρ in the range $0 < \rho \ll 1$
choose δ and δ_p as small positive constants

For $k \geq 0$ (i.e., for every iteration) **do**

$e(k) = d(k) - \mathbf{w}^{\mathrm{T}}(k)\mathbf{x}(k)$
$\gamma_i(k) = \max\{|w_i(k)|\ ,\ \rho \max\{\delta_p, |w_0(k)|, ..., |w_{\mathrm{N}}(k)|\}\}$, for all i
$g_i(k) = \dfrac{\gamma_i(k)}{\sum_{j=0}^{\mathrm{N}} \gamma_j(k)}$, for all i
$\mathbf{G}(k) = \mathsf{Diag}\{[g_0(k)\ g_1(k)\ \ldots\ g_N(k)]\}$
$\mathbf{w}(k+1) = \mathbf{w}(k) + \mu \dfrac{e(k)\mathbf{G}(k)\mathbf{x}(k)}{\mathbf{x}^{\mathrm{T}}(k)\mathbf{G}(k)\mathbf{x}(k) + \delta}$

where $\max\{\cdot\}$ returns the maximum among the elements of a given set, $\delta_\rho \in \mathbb{R}_+$ is used to prevent $\mathbf{w}(k)$ from stalling during the initialization stage, in case the coefficients are initialized as $\mathbf{w}(0) = \mathbf{0}$ and its usual value is $\delta_\rho = 0.01$, and $\rho \in \mathbb{R}_+$ is used to prioritize the first term in Equation (2.41) and its typical value is $\rho = 0.01$. However, we should emphasize that the proper choices of δ_ρ and ρ depend on prior knowledge about the application (more precisely, about the magnitudes of the optimal coefficients). The complete description of the PNLMS algorithm is given in Algorithm 2.7.

Remarks:

1 The first term in Equation (2.41) conveys the proportional-update idea, but the second term is necessary to address the cases in which $w_i(k) = 0$. That is, if we were to use only the first term, then the update process of a given coefficient $w_i(k)$ would stall whenever $w_i(k) = 0$, thus hindering the initialization and also the tracking of time-varying systems.

2 To better understand the proportionate-update principle, let us consider just the first term in Equation (2.41), which leads to

$$g_i(k) = \frac{|w_i(k)|}{\sum_{j=0}^{N} |w_j(k)|} = \frac{|w_i(k)|}{\|\mathbf{w}(k)\|_1}, \forall i, \tag{2.42}$$

where $\|\cdot\|_1$ denotes the ℓ_1 norm. In addition, observe that the denominator of Equation (2.42) does not change with i and, therefore, it vanishes when Equation (2.42) is substituted in Equation (2.39) due to the presence of matrix $\mathbf{G}(k)$ on both the numerator and denominator of this equation. Consequently, we can think of Equation (2.42) as $g_i(k) \propto |w_i(k)|$

(proportional relation), which means that the update step applied to each coefficient $w_i(k)$ is proportional to its own magnitude. Moreover, since

$$\sum_{i=0}^{N} g_i(k) = 1, \tag{2.43}$$

the product $\mu g_i(k)$ can be interpreted as an "equivalent step size" for the i-th coefficient, that is, the fraction of μ corresponding to the update of $w_i(k)$. This interpretation allows us to conclude that in the proportionate-update framework, the step size μ is unevenly distributed across the coefficients, privileging those with higher magnitudes, but impairing the lower magnitude ones.

3 If the filter coefficients are initialized as $\mathbf{w}(0) = \mathbf{0}$, then $\gamma_i(0) = \rho\delta_\rho$ for all i, meaning that

$$g_i(k) = \frac{1}{N+1}, \quad \forall i, \tag{2.44}$$

which means that the step size is evenly distributed across the coefficients, just like in the NLMS algorithm.

4 The PNLMS algorithm works well when the impulse response of the system to be identified is very sparse, resembling the Kronecker's delta function. On the other hand, as the sparsity degree decreases, the performance of the PNLMS algorithm deteriorates, in comparison with the standard algorithms.

2.4.3 The IPNLMS Algorithm

The IPNLMS algorithm proposed in 2002 [46] is more robust to lower sparsity degrees than the PNLMS algorithm, providing good results even when applied to dispersive systems. The IPNLMS algorithm benefits from the combination of the proportional update term with the standard update term of the NLMS algorithm, which works better than the PNLMS algorithm in the estimation of dispersive systems.

In the IPNLMS recursion given in Equation (2.32), each entry on the main diagonal of the proportionate matrix $\mathbf{G}(k) = \mathsf{Diag}\{[g_0(k)\ g_1(k)\ \cdots\ g_N(k)]\}$ is given by

$$g_i(k) = \underbrace{\frac{1-\alpha}{2(N+1)}}_{\text{NLMS term}} + \underbrace{\frac{(1+\alpha)|w_i(k)|}{2\|\mathbf{w}(k)\|_1 + \delta}}_{\text{PNLMS term}}, \tag{2.45}$$

where $\delta \in \mathbb{R}_+$ is also a regularization factor used to avoid numerical problems when $\|\mathbf{w}(k)\|_1$ tends to 0 and $\alpha \in \mathbb{R}$ is an adjustable parameter that represents a tradeoff between the NLMS and PNLMS terms in Equation (2.45). In fact, α should be chosen in the range $-1 \leq \alpha < 1$. If $\alpha = -1$, then the IPNLMS is equivalent to the NLMS algorithm, whereas for $\alpha \approx 1$, its update resembles that of the PNLMS algorithm. Typically, this parameter is chosen close to $\alpha = -0.5$

Algorithm 2.8 The IPNLMS algorithm

Initialization:

$\mathbf{x}(0) = \mathbf{w}(0) = [0\ 0\ \ldots\ 0]^{\mathrm{T}}$

choose μ in the range $0 < \mu \leq 1$

choose δ as a small positive constant

choose α in the range $-1 \leq \alpha < 1$

For $k \geq 0$ (i.e., for every iteration) **do**

$e(k) = d(k) - \mathbf{w}^{\mathrm{T}}(k)\mathbf{x}(k)$

$g_i(k) = \dfrac{1-\alpha}{2(N+1)} + \dfrac{(1+\alpha)|w_i(k)|}{2\|\mathbf{w}(k)\|_1 + \delta}$

$\mathbf{G}(k) = \mathsf{Diag}\{[g_0(k)\ g_1(k)\ \ldots\ g_N(k)]\}$

$\mathbf{w}(k+1) = \mathbf{w}(k) + \mu \dfrac{e(k)\mathbf{G}(k)\mathbf{x}(k)}{\mathbf{x}^{\mathrm{T}}(k)\mathbf{G}(k)\mathbf{x}(k) + \delta}$

so that the IPNLMS algorithm behaves more like an NLMS than a PNLMS algorithm. It is interesting to observe that the NLMS term in Equation (2.45) already prevents the IPNLMS coefficients from stalling, which is another reason for not allowing $\alpha = 1$. The IPNLMS algorithm is given in Algorithm 2.8.

2.4.4 The MPNLMS Algorithm

When applied to the identification of sparse impulse responses, the PNLMS algorithm usually provides a very fast convergence speed in the early iterations, but later it slows down. The MPNLMS algorithm proposed in 2005 [47] addresses this issue. The key idea of the MPNLMS algorithm is to employ step sizes proportional to a function of the magnitude of the coefficients; such function must grow rapidly for low magnitude coefficients, increasing the resolution in this range, but must grow slowly for high magnitude coefficients, thus preventing numerical issues [47].

In the MPNLMS recursion given in Equation (2.32), each entry on the main diagonal of the proportionate matrix $\mathbf{G}(k) = \mathsf{Diag}\{[g_0(k)\ g_1(k)\ \cdots\ g_N(k)]\}$ is given by

$$g_i(k) = \frac{\hat{\gamma}_i(k)}{\sum\limits_{j=0}^{N} |\hat{\gamma}_j(k)|}, \tag{2.46}$$

$$\hat{\gamma}_i(k) = \max\{F(|w_i(k)|),\ \rho \max\{\delta_\rho,\ F(|w_0(k))|,\ \cdots,\ F(|w_N(k)|)\}\}, \tag{2.47}$$

$$F(|w_i(k)|) = \ln(1 + \mu_F |w_i(k)|), \tag{2.48}$$

where $\delta_\rho \in \mathbb{R}_+$ is used to prevent $\mathbf{w}(k)$ from stalling during the initialization stage in case the coefficients are set as $\mathbf{w}(0) = \mathbf{0}$ and its typical value is $\delta_\rho = 0.01$, $\rho \in \mathbb{R}_+$ is used to prioritize the term $F(|w_i(k)|)$ in Equation (2.47) and its typical value is $\rho = 0.01$, and $\mu_F \in \mathbb{R}_+$ is a large positive number

Algorithm 2.9 The MPNLMS algorithm

Initialization:

$\mathbf{x}(0) = \mathbf{w}(0) = [0\ 0\ \ldots\ 0]^{\mathrm{T}}$

choose μ in the range $0 < \mu \leq 1$

choose ρ in the range $0 < \rho \ll 1$

choose δ and δ_p as small positive constants

choose μ_F as a large positive number

For $k \geq 0$ (i.e., for every iteration) **do**

$e(k) = d(k) - \mathbf{w}^{\mathrm{T}}(k)\mathbf{x}(k)$

$F(|w_i(k)|) = \ln(1 + \mu_F |w_i(k)|)$, for all i

$\hat{\gamma}_i(k) = \max\{F(|w_i(k)|),\ \rho \max\{\delta_\rho,\ F(|w_0(k))|,\ \cdots,\ F(|w_N(k)|)\}\}, \forall i$

$g_i(k) = \dfrac{\hat{\gamma}_i(k)}{\sum_{j=0}^{N} |\hat{\gamma}_j(k)|}$, for all i

$\mathbf{G}(k) = \mathsf{Diag}\{[g_0(k)\ g_1(k)\ \ldots\ g_N(k)]\}$

$\mathbf{w}(k+1) = \mathbf{w}(k) + \mu \dfrac{e(k)\mathbf{G}(k)\mathbf{x}(k)}{\mathbf{x}^{\mathrm{T}}(k)\mathbf{G}(k)\mathbf{x}(k) + \delta}$

related to the identification accuracy requirement, which is typically chosen as $\mu_F = 1000$. Observe that the main difference between the MPNLMS and PNLMS algorithms is the use of a natural logarithmic function by the former one to achieve the benefits mentioned in the previous paragraph. The complete description of the MPNLMS algorithm is given in Algorithm 2.9.

Remarks:

1 As in the PNLMS algorithm, proper selection of the parameters ρ, δ_ρ, and μ_F require some prior knowledge about the application/impulse response to be estimated.

2 If the filter coefficients are initialized as $\mathbf{w}(0) = \mathbf{0}$, then $\hat{\gamma}_i(0) = \rho\delta_\rho$ for all i, meaning that

$$g_i(k) = \frac{1}{N+1}, \forall i,$$

and the MPNLMS algorithm will act just like an NLMS algorithm.

3 In normal operation, $\hat{\gamma}_i(k)$ should be equal to the first term in Equation (2.47) leading to

$$g_i(k) = \frac{F(|w_i(k)|)}{\sum_{j=0}^{N} F(|w_j(k)|)},$$

that is, in the MPNLMS algorithm, the step sizes applied to each coefficient are proportional to $F(|w_i(k)|)$, and not $|w_i(k)|$ as in the PNLMS algorithm.

2.4.5 The IMPNLMS Algorithm

Just like the PNLMS, the MPNLMS algorithm also has poor performance when identifying systems that are not very sparse. A natural idea to improve this algorithm is to follow the same approach used in the IPNLMS algorithm, that is, to use "equivalent step sizes" formed by the combination of two terms: one corresponding to the NLMS and the other corresponding to the MPNLMS algorithm. In addition to adapting the IPNLMS idea, the IMPNLMS algorithm proposed in 2008 [49] also employs a measure of the sparsity degree in the coefficient vector $\mathbf{w}(k)$ in order to select the parameter $\alpha(k)$ automatically (the parameter that determines the weight given to each of the two terms above).

In the IMPNLMS recursion given in Equation (2.32), each entry on the main diagonal of the proportionate matrix $\mathbf{G}(k) = \mathsf{Diag}\{[g_0(k)\ g_1(k)\ \cdots\ g_N(k)]\}$ is given by

$$g_i(k) = \underbrace{\frac{1-\alpha(k)}{2(N+1)}}_{\text{NLMS term}} + \underbrace{\frac{(1+\alpha(k))F(|w_i(k)|)}{2\|F(|\mathbf{w}(k)|)\|_1+\delta}}_{\text{MPNLMS term}}, \tag{2.49}$$

where

$$F(|w_i(k)|) = \ln\left(1+\mu_F|w_i(k)|\right), \tag{2.50}$$

$$\alpha(k) = 2\xi(k) - 1, \tag{2.51}$$

$$\xi(k) = (1-\lambda)\xi(k-1) + \lambda\xi_{\mathbf{w}}(k), \tag{2.52}$$

$$\xi_{\mathbf{w}}(k) = \frac{N+1}{N+1-\sqrt{N+1}}\left(1 - \frac{\|\mathbf{w}(k)\|_1}{\sqrt{N+1}\|\mathbf{w}(k)\|_2}\right), \tag{2.53}$$

$F(|\mathbf{w}(k)|) = [F(|w_0(k)|)\ F(|w_1(k)|)\ \ldots\ F(|w_N(k)|)]^{\mathrm{T}} \in \mathbb{R}_+^{N+1}$, $\delta \in \mathbb{R}_+$ is a regularization factor, $\mu_F \in \mathbb{R}_+$ is a large positive number related to the identification accuracy requirement and is typically chosen as $\mu_F = 1000$, and $\lambda \in \mathbb{R}_+$ should be chosen as $0 < \lambda \ll 1$. Low values of λ privilege the past data, whereas high values of λ prioritize the instantaneous estimate of the sparsity degree given by $\xi_{\mathbf{w}}(k)$. Observe that if $\xi(k) \approx 1$, corresponding to a vector $\mathbf{w}(k)$ with high sparsity degree, then $\alpha(k) \approx 1$, meaning that the NLMS term in $g_i(k)$ vanishes and, consequently, the algorithm acts like the MPNLMS algorithm. On the contrary, if $\xi(k) \approx 0$, corresponding to a dispersive vector $\mathbf{w}(k)$, then $\alpha(k) \approx -1$ and the MPNLMS term vanishes.

The complete description of the IMPNLMS algorithm is given in Algorithm 2.10.

2.4.6 The SM-PNLMS Algorithm

The SM-PNLMS algorithm proposed in [50] combines the proportionate-update principle with the SM filtering concept, which allows the adaptive filter to update its coefficients only when the input data brings enough innovation [30, 60].

Algorithm 2.10 The IMPNLMS algorithm

Initialization:

$\mathbf{x}(0) = \mathbf{w}(0) = [0\ 0\ \ldots\ 0]^{\mathrm{T}}$

$\xi(0) = 0$

choose μ in the range $0 < \mu \leq 1$

choose δ as a small positive number

choose μ_F as a large positive number

choose λ in the range $0 < \lambda \ll 1$

For $k \geq 0$ (i.e., for every iteration) **do**

$e(k) = d(k) - \mathbf{w}^{\mathrm{T}}(k)\mathbf{x}(k)$

$\xi_{\mathbf{w}}(k) = \dfrac{N+1}{N+1-\sqrt{N+1}}\left(1 - \dfrac{\|\mathbf{w}(k)\|_1}{\sqrt{N+1}\|\mathbf{w}(k)\|_2}\right)$

$\xi(k) = (1-\lambda)\xi(k-1) + \lambda\xi_{\mathbf{w}}(k)$

$\alpha(k) = 2\xi(k) - 1$

$F(|w_i(k)|) = \ln(1 + \mu_F|w_i(k)|)$, for all i

$g_i(k) = \dfrac{1-\alpha(k)}{2(N+1)} + \dfrac{(1+\alpha(k))F(|w_i(k)|)}{2\|F(|\mathbf{w}(k)|)\|_1 + \delta}$, for all i

$\mathbf{G}(k) = \mathsf{Diag}\{[g_0(k)\ g_1(k)\ \ldots\ g_N(k)]\}$

$\mathbf{w}(k+1) = \mathbf{w}(k) + \mu\dfrac{e(k)\mathbf{G}(k)\mathbf{x}(k)}{\mathbf{x}^{\mathrm{T}}(k)\mathbf{G}(k)\mathbf{x}(k) + \delta}$

The update equation of the SM-PNLMS algorithm is slightly different from that in Equation (2.32), as the former uses a time-varying step-size parameter $\mu(k)$, and can be written as

$$\mathbf{w}(k+1) = \mathbf{w}(k) + \mu(k)\frac{e(k)\mathbf{G}(k)\mathbf{x}(k)}{\mathbf{x}^{\mathrm{T}}(k)\mathbf{G}(k)\mathbf{x}(k) + \delta}, \tag{2.54}$$

where $\mathbf{G}(k) = \mathsf{Diag}\{[g_0(k)\ \cdots\ g_N(k)]\}$,

$$g_i(k) = \frac{1-\kappa\mu(k)}{N+1} + \frac{\kappa\mu(k)|w_i(k)|}{\|\mathbf{w}(k)\|_1 + \delta}, \tag{2.55}$$

$$\mu(k) = \begin{cases} 1 - \frac{\overline{\gamma}}{|e(k)|} & \text{if } |e(k)| > \overline{\gamma} \\ 0 & \text{otherwise.} \end{cases} \tag{2.56}$$

Thus, the update equation can be rewritten in a more direct form as follows:

$$\mathbf{w}(k+1) = \begin{cases} \mathbf{w}(k) + \left(1 - \dfrac{\overline{\gamma}}{|e(k)|}\right)\dfrac{e(k)\mathbf{G}(k)\mathbf{x}(k)}{\mathbf{x}^{\mathrm{T}}(k)\mathbf{G}(k)\mathbf{x}(k) + \delta} & \text{if } |e(k)| > \overline{\gamma}, \\ \mathbf{w}(k) & \text{otherwise.} \end{cases} \tag{2.57}$$

where $\delta \in \mathbb{R}_+$ is a *regularization factor* used to prevent divisions by 0, $\overline{\gamma} \in \mathbb{R}_+$ is the prescribed threshold that defines how much residual error is acceptable, and $\kappa \in [0, 1)$ represents a compromise between the NLMS update ($\kappa \to 0$) and the proportionate update ($\kappa \to 1$) and is usually chosen as $\kappa = 0.5$. Observe

Algorithm 2.11 The SM-PNLMS algorithm

Initialization:

$\mathbf{x}(0) = \mathbf{w}(0) = [0\ 0\ \ldots\ 0]^{\mathrm{T}}$

choose δ as a small positive constant

choose κ in the range $0 \leq \kappa < 1$

For $k \geq 0$ (i.e., for every iteration) **do**

$e(k) = d(k) - \mathbf{w}^{\mathrm{T}}(k)\mathbf{x}(k)$

If $|e(k)| > \overline{\gamma}$

$\mu(k) = 1 - \frac{\overline{\gamma}}{|e(k)|}$

$g_i(k) = \dfrac{1-\kappa\mu(k)}{N+1} + \dfrac{\kappa\mu(k)|w_i(k)|}{\|\mathbf{w}(k)\|_1 + \delta}$, for all i

$\mathbf{G}(k) = \mathsf{Diag}\{[g_0(k)\ g_1(k)\ \ldots\ g_N(k)]\}$

$\mathbf{w}(k+1) = \mathbf{w}(k) + \mu(k)\dfrac{e(k)\mathbf{G}(k)\mathbf{x}(k)}{\mathbf{x}^{\mathrm{T}}(k)\mathbf{G}(k)\mathbf{x}(k) + \delta}$

Else

$\mathbf{w}(k+1) = \mathbf{w}(k)$

End if

that the choice of $\overline{\gamma}$ depends on some previous knowledge about the problem uncertainties. For applications in which the desired signal is corrupted by an additive noise $n(k)$ with variance denoted by σ_n^2, a common choice is to select $\overline{\gamma} = \sqrt{5\sigma_n^2}$ [51].

Notice that the elements $g_i(k)$ in the SM-PNLMS algorithm are very similar to the ones used in the IPNLMS algorithm. Consequently, the SM-PNLMS algorithm's performance is usually similar to that of the IPNLMS algorithm, but the former algorithm has two advantages: (i) it does not update at every iteration, thus saving computational resources and (ii) it uses a variable step-size $\mu(k)$ which is automatically chosen. The full description of the SM-PNLMS algorithm is given in Algorithm 2.11.

2.4.7 Numerical Experiments

Here, the PNLMS, IPNLMS, and MPNLMS algorithms are used to identify an unknown system whose coefficients are denoted by $\mathbf{w}_o$. We also consider two other algorithms: the NLMS, representing the baseline for our comparisons, and the ℓ_0-NLMS, representing an algorithm that uses a sparsity-promoting regularizer.

The specific parameters of these Pt-NLMS algorithms were set according to their typical values, described in the subsections where these algorithms were explained. As for the ℓ_0-NLMS algorithm, we used the Laplacian function with $\beta = 5$. For all algorithms, we set $\delta = 10^{-12}$, $\alpha = 2 \times 10^{-3}$, and the step size μ is informed later. The input signal is a zero-mean white Gaussian noise (WGN) with unitary variance. The measurement noise is also zero-mean WGN with variance $\sigma_n^2 = 10^{-2}$ and is uncorrelated with the input signal. We assume the

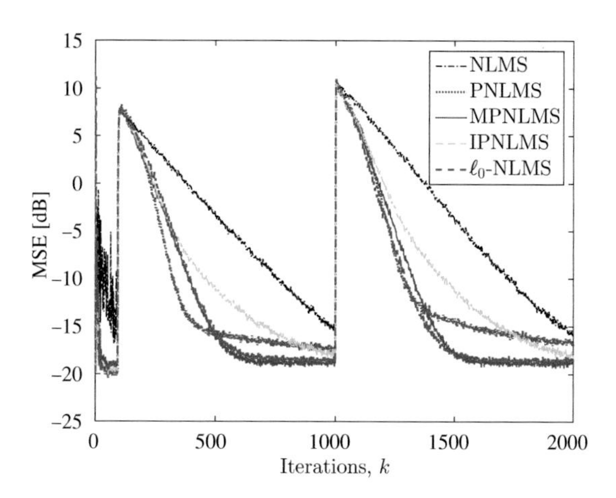

Figure 2.6 MSE learning curve in a nonstationary scenario.

adaptive filter and the unknown systems have the same number of coefficients. The MSE learning curves for each algorithm are generated by averaging the outcomes of 1000 independent trials.

In Figure 2.6, we consider a nonstationary scenario in which the optimal coefficients are given by $\mathbf{w}_{\text{o}}^{(1)}$ during the first 1000 iterations, and then they change to $\mathbf{w}_{\text{o}}^{(2)}$, defined as

$$\mathbf{w}_{\text{o},i}^{(1)} = \begin{cases} 0, & \text{for } 0 \leq i \leq 93, \\ 1, & \text{for } 94 \leq i \leq 99, \end{cases} \qquad \mathbf{w}_{\text{o},i}^{(2)} = \begin{cases} 1, & \text{for } 0 \leq i \leq 5, \\ 0, & \text{for } 6 \leq i \leq 99. \end{cases} \tag{2.58}$$

Both $\mathbf{w}_{\text{o}}^{(1)}$ and $\mathbf{w}_{\text{o}}^{(2)}$ have 6 nonzero coefficients out of 100. The algorithms were initialized with $\mathbf{w}(0) = \mathbf{0}$, and the step sizes of the NLMS, PNLMS, IPNLMS, MPNLNS and ℓ_0-NLMS algorithms were set as 0.4, 0.15, 0.4, 0.6, and 0.99, respectively, so that they could achieve the same steady-state MSE level as fast as possible. In this figure, both the MPNLMS and ℓ_0-NLMS algorithms were the fastest ones to reach a specific MSE level; actually, the ℓ_0-NLMS algorithm was a bit faster in the tracking of $\mathbf{w}_{\text{o}}^{(2)}$. The PNLMS algorithm started very fast but then its convergence speed slowed down at a given iteration; this behavior is typical since the PNLMS algorithm accelerates the convergence of the high-magnitude coefficients in detriment of the low-magnitude ones. Finally, the results of the IPNLMS algorithm correspond to a balance between the PNLMS and NLMS algorithms.

2.4.8 Further Readings

The literature about proportionate type algorithms is vast. For different versions of Pt-NLMS algorithms, one may refer to [52–55]. Analogously, for several versions of proportionate AP algorithms, one may refer to [56–58]. Recently, an approach called *coefficient vector reusing* has received some attention [59]. In the same way that the AP algorithm generalizes the NLMS algorithm by reusing previous input vectors, the SM-PAP algorithm proposed in [60] generalizes the SM-PNLMS algorithm.

2.5 Conclusion

This chapter presented many adaptive filtering algorithms incorporating tools to exploit the sparsity inherent to some models in real applications. The chapter started by covering in some detail the sparsity-promoting regularizations frequently used to exploit sparsity. Combined with the classical adaptive filtering algorithms, these regularizations form a set of solutions giving rise to several sparsity-aware online algorithms. Another class of algorithms that can exploit sparsity is the proportionate-type adaptive filtering algorithms, in which coefficient updates are proportional, to a certain extent (depending on the algorithm), to their own magnitudes. When comparing these two classes of algorithms, one can observe how different their principles are. While the algorithms employing sparsity-promoting regularizations push some coefficients to 0 in order to encourage sparse solutions (in the case of the ℓ_0-norm approximation, for example, only the low magnitude coefficients suffer from this zero-attraction effect, whereas the relevant coefficients remain unaltered), the proportionate-type algorithms act by realizing an uneven distribution of the step size μ among the coefficients, which accelerates the convergence of high magnitude coefficients, but slows down the convergence of low magnitude coefficients. That is, the algorithms employing sparsity-promoting regularizations focus on the low magnitude coefficients, whereas the proportionate-type algorithms focus on the high magnitude coefficients.

From our experience with the algorithms described in this chapter, we can state that:

1 Considering the algorithms employing sparsity-promoting regularizers, the ℓ_0-norm approximation has always yielded the best results in our tests, and it is only slightly more complex than the ℓ_1 norm.
2 As for the algorithms following the proportional-update principle, the ones whose recursions are "more proportional" to the magnitude of the coefficients usually work very well in highly sparse scenarios, but their performance degrade severely as the sparsity degree decreases. In our tests, the MPNLMS and SM-PNLMS algorithms are usually more robust to different simulation setups than the other Pt-NLMS algorithms.
3 Another advantage of algorithms employing the ℓ_0-norm approximation is that they always capitalize on sparse vectors, even when the sparsity degree

is low (confer the MSE analysis of the ℓ_0-AP algorithm in [28]). The same behavior is not observed in other regularizers nor in the proportionate-type algorithms. Besides, parameter β provides an easy mechanism to adjust the ℓ_0-norm approximation to deal with compressible vectors.

Finally, we should say that the online learning algorithms that exploit sparse models presented in this chapter are far from representing the entire set of solutions. Some examples of alternative approaches and applications can be found in [61–66]. Also, in [67], an algorithm combining both the sparsity-promoting regularization and the proportional-update principle has shown some interesting results.

Problems

2.1 For under-determined problems of the type $\mathbf{Ax} = \mathbf{b}$, it is a common practice to include some constraint on $\mathbf{x}$ in order to find a single solution. Considering the data:

$$\mathbf{A} = \begin{bmatrix} 1 & 2 & 0 & 2 \\ 1 & 2 & 2 & 0 \\ 1 & 3 & 0 & 0 \end{bmatrix} \text{ and } \mathbf{b} = \begin{bmatrix} 1 \\ 1 \\ 2 \end{bmatrix}$$

(a) Compute the minimum ℓ_2 (Euclidean) norm solution $\hat{\mathbf{x}}_2 = \mathbf{A}^\dagger \mathbf{b}$, where $\mathbf{A}^\dagger = \mathbf{A}^{\mathrm{T}} \left(\mathbf{A}\mathbf{A}^{\mathrm{T}}\right)^{-1}$ is known as pseudo-inverse (or right inverse of $\mathbf{A}$).
(b) Compute the minimum ℓ_1-norm solution $\hat{\mathbf{x}}_1$.
(c) Compute the minimum ℓ_0-norm (sparsest) solution $\hat{\mathbf{x}}_0$. Is there a closed-form solution? Is this solution unique?
Hint: Since $\mathbf{x} \in \mathbb{R}^4$ lives in a low-dimensional space, you can write a program to test all possible combinations of the columns of $\mathbf{A}$. This is called brute-force solution or exhaustive search. Observe that for $\mathbf{A} \in \mathbb{R}^{m \times n}$, this procedure leads to $\sum\limits_{p=1}^{n} \binom{n}{p}$ possibilities that need to be tested. Since large values of n are frequently used in practical applications, the exhaustive search solution is not feasible.

2.2 Prove that the $r\ell_1$ norm, defined in Equation (2.5), converges to the ℓ_0 norm as $\epsilon \to \infty$ and converges to the ℓ_1 norm as $\epsilon \to 0$.

2.3 For $\mu = 1$ and $\gamma = 0$, show that the recursion of the sparsity-aware AP algorithms, given in Equation (2.26), leads to a $\mathbf{w}(k+1)$ that satisfies the constraint in Equation (2.20). Also, show that the corresponding qAP algorithms, whose general recursion is given in Equation (2.27), do not satisfy the constraint in Equation (2.20).

2.4 The PAP algorithm is characterized by the recursion

$$\mathbf{w}(k+1) = \mathbf{w}(k) + \mu\mathbf{G}(k)\mathbf{X}(k)\mathbf{S}(k)\mathbf{e}(k),$$

where $\mathbf{S}(k) = \left(\mathbf{X}^{\mathrm{T}}(k)\mathbf{G}(k)\mathbf{X}(k) + \delta\mathbf{I}\right)^{-1}$. Show that this PAP recursion with $\mu = 1$ and $\delta = 0$ is obtained as the solution of the following optimization problem:

$$\begin{aligned} &\min \|\mathbf{w}(k+1) - \mathbf{w}(k)\|^2_{\mathbf{G}^{-1}(k)} \\ &\text{subject to } \mathbf{d}(k) - \mathbf{X}^{\mathrm{T}}(k)\mathbf{w}(k+1) = \mathbf{0}\ . \end{aligned}$$

2.5 Consider the problem of identifying a highly sparse unknown system/vector $\mathbf{w}_{\mathrm{o}}$ comprised of 100 coefficients. Assume that the input signal is a zero-mean WGN with unitary variance, the measurement noise is also a zero-mean WGN with variance 0.01, and these signals are uncorrelated with each other. Also, the adaptive filter is comprised of 100 coefficients initialized as $\mathbf{w}(0) = \mathbf{0}$. Plot and compare the MSE learning curves for the NLMS, PNLMS, IPNLMS, MPNLMS, IMPNLMS, and ℓ_0-NLMS algorithms considering that:
(a) Only the first coefficient of $\mathbf{w}_{\mathrm{o}}$ is non-null and equal to 1.
(b) Only the 50th coefficient of $\mathbf{w}_{\mathrm{o}}$ is non-null and equal to 1.
(c) Only the last coefficient of $\mathbf{w}_{\mathrm{o}}$ is non-null and equal to 1.
Observe that the sparsity degree is precisely the same in all these cases. So, why the MSE results change in each of them?
Hint: In order to compare the algorithms in a fair manner, adjust their parameters so that they have the same convergence/learning rate and compare the steady-state MSE level they achieve. Do not forget to report the parameters used in each algorithm!

2.6 Repeat the item (a) of Problem 2.5, but for different initialization of the adaptive filter coefficients: $\mathbf{w}(0) = c \times \mathbf{1}$, where $\mathbf{1}$ is the vector of ones and $c \in \{0, 0.1, 0.5, 2\}$. Explain why some of these algorithms suffer a significant performance degradation as c increases.

2.7 Repeat the Problem 2.5, but considering that $\mathbf{w}_{\mathrm{o}}$ is a compressible vector. To allow some control over the "low magnitude" coefficients, take the vectors $\mathbf{w}_{\mathrm{o}}$ given in Problem 2.5 and replace their null coefficients with $p \in \{0.001, 0.005, 0.01, 0.02\}$. Verify which algorithms are mostly impaired by the increase of p.

2.8 Repeat the Problem 2.5, but considering that $\mathbf{w}_\text{o}$ is a compressible vector. To do so, the entries of $\mathbf{w}_\text{o}$ that are equal to 0 must be replaced by a perturbation modeled by a uniform distribution over the interval $[-p, p]$, with $p \in \{0.001, 0.005, 0.01, 0.05\}$. Verify which algorithms are mostly impaired by the increase of p.

2.9 Repeat the Problem 2.5, but varying the degree of sparsity in $\mathbf{w}_\text{o}$ by making:

(a) Only the first coefficient of $\mathbf{w}_\text{o}$ is non-null and equal to 1.
(b) The first 10 coefficients of $\mathbf{w}_\text{o}$ are non-null and equal to 1.
(c) The first half of the coefficients of $\mathbf{w}_\text{o}$ are non-null and equal to 1.
(d) All the 100 coefficients of $\mathbf{w}_\text{o}$ are non-null and equal to 1.

Verify which algorithms are mostly impaired by the reduction of the sparsity degree.

References

[1] P. S. R. Diniz, H. Yazdanpanah, and M. V. S. Lima, Feature LMS algorithms. Proceedings of the IEEE International Conference on Acoustics, Speech and Signal Processing (ICASSP 2018), Calgary, AB, 2018, pp. 4144–4148.

[2] H. Yazdanpanah, P. S. R. Diniz, and M. V. S. Lima, Feature adaptive filtering: Exploiting hidden sparsity, IEEE Transactions on Circuits and Systems I: Regular Papers **67**, pp. 2358–2371 (2020).

[3] H. Yazdanpanah and J. A. Apolinário Jr., The extended feature LMS algorithm: Exploiting hidden sparsity for systems with unknown spectrum, Circuits, Systems, and Signal Processing **40**, pp. 174–192 (2021).

[4] D. B. Haddad, L. O. dos Santos, L. F. Almeida, G. A. S. Santos, and M. R. Petraglia, ℓ_2-norm feature least mean square algorithm, Electronics Letters **56**, pp. 516–519 (2020).

[5] P. S. R. Diniz, H. Yazdanpanah, and M. V. S. Lima, Feature LMS algorithm for bandpass system models. Proceedings of the 27th European Signal Processing Conference (EUSIPCO), A Coruna, Spain, 2019, pp. 1–5.

[6] H. Yazdanpanah, P. S. R. Diniz, and M. V. S. Lima, Low-complexity feature stochastic gradient algorithm for block-lowpass systems, IEEE Access **7**, 141587–141593 (2019).

[7] C. M. Bishop, *Pattern Recognition and Machine Learning* (Springer, New York, 2006).

[8] L. N. Trefethen and D. Bau, III, *Numerical Linear Algebra* (SIAM, Philadelphia, 1997).

[9] M. V. S. Lima, G. S. Chaves, T. N. Ferreira, and P. S. R. Diniz, Do proportionate algorithms exploit sparsity?, ArXiv: http://arxiv.org/abs/2108.00840.

[10] T. Hastie, R. Tibshirani, and J. Friedman, *The Elements of Statistical Learning: Data Mining, Inference, and Prediction*, 2nd ed. (Springer, New York, 2017).

[11] M. V. S. Lima, T. N. Ferreira, W. A. Martins, and P. S. R. Diniz, Sparsity-aware data-selective adaptive filters, IEEE Transactions on Signal Processing **62**, pp. 4557–4572 (2014).
[12] Y. Eldar and G. Kutyniok, *Compressed Sensing: Theory and Applications* (Cambridge University Press, Cambridge, 2012).
[13] N. Bourbaki, *Topological Vector Spaces* (Springer, New York, 1987).
[14] G. Kothe, *Topological Vector Spaces I* (Springer, New York, 1983).
[15] J. F. Claerbout and F. Muir, Robust modeling with erratic data, Geophysics **38**, pp. 826–844 (1973).
[16] F. Santosa and W. W. Symes, Linear inversion of band-limited reflection seismograms, SIAM Journal on Scientific Computing **7**, pp. 1307–1330 (1986).
[17] S. S. Chen, D. L. Donoho, and M. A. Saunders, Atomic decomposition by basis pursuit, SIAM Journal on Scientific Computing **20**, pp. 33–61 (1998).
[18] R. Tibshirani, Regression shrinkage and selection via the Lasso, Journal of the Royal Statistical Society, Series B (Methodological) **58**, pp. 267–288 (1996).
[19] E. J. Candès and M. B. Wakin, An introduction to compressive sampling, IEEE Signal Processing Magazine **25**, pp. 21–30 (2008).
[20] E. J. Candès, M. B. Wakin, and S. P. Boyd, Enhancing sparsity by reweighted ℓ_1 minimization, Journal of Fourier Analysis and Applications **14**, pp. 877–905 (2008).
[21] M. V. S. Lima, W. A. Martins, and P. S. R. Diniz, Affine projection algorithms for sparse system identification. Proceedings of the IEEE International Conference on Acoustics, Speech and Signal Processing (ICASSP 2013), Vancouver, Canada, May 2013, pp. 5666–5670.
[22] L. Mancera and J. Portilla, L0-norm-based sparse representation through alternate projections. Proceedings of the IEEE International Conference on Image Processing (ICIP 2006), Atlanta, GA, USA, October 2006, pp. 2089–2092.
[23] Y. Chen, Y. Gu, and A. O. Hero, Sparse LMS for system identification. Proceedings of the IEEE International Conference on Acoustics, Speech and Signal Processing (ICASSP 2009), Taipei, Taiwan, April 2009, pp. 3125–3128.
[24] J. Trzasko and A. Manduca, Highly undersampled magnetic resonance image reconstruction via homotopic ℓ_0-minimization, IEEE Transactions on Medical Imaging **28**, 106–121 (2009).
[25] Y. Gu, J. Jin, and S. Mei, ℓ_0 norm constraint LMS algorithm for sparse system identification, IEEE Signal Processing Letters **16**, pp. 774–777 (2009).
[26] P. Huber, *Robust Statistics* (Wiley, New York, 1981).
[27] D. Geman and G. Reynolds, Nonlinear image recovery with half-quadratic regularization, IEEE Transactions on Image Processing **4**, 932–946 (1995).
[28] M. V. S. Lima, I. Sobron, W. A. Martins, and P. S. R. Diniz, Stability and MSE analyses of affine projection algorithms for sparse system identification. Proceedings of the IEEE International Conference on Acoustics, Speech and Signal Processing (ICASSP 2014), Florence, Italy, May 2014, pp. 6399–6403.
[29] H. Mohimani, M. Babaie-Zadeh, and C. Jutten, A fast approach for overcomplete sparse decomposition based on smoothed ℓ^0 norm, IEEE Transactions on Signal Processing **57**, 289–301 (2009).
[30] P. S. R. Diniz, *Adaptive Filtering: Algorithms and Practical Implementation*, 5th ed. (Springer, New York, 2020).

[31] R. Meng, R. C. de Lamare, and V. H. Nascimento, Sparsity-aware affine projection adaptive algorithms for system identification. Proceedings of the Sensor Signal Processing for Defense (SSPD 2011), London, UK, September 2011, pp. 1–5.

[32] H. Yazdanpanah and P. S. R. Diniz, Recursive least-squares algorithms for sparse system modeling. Proceedings of the 2017 IEEE International Conference on Acoustics Speech and Signal Processing, New Orleans, LA, March 2017, pp. 3878–3883.

[33] H. Yazdanpanah, M. V. S. Lima, and P. S. R. Diniz, On the robustness of set-membership adaptive filtering algorithms, EURASIP Journal on Advances in Signal Processing **2017**, pp. 1–12 (2017).

[34] P. L. Combettes, The foundations of set theoretic estimation, Proceedings of the IEEE **81**, pp. 182–208 (1993).

[35] M. V. S. Lima and P. S. R. Diniz, Fast learning set theoretic estimation. Proceedings of the 21st European Signal Processing Conference (EUSIPCO), Marrakesh, Morocco, 2013, pp. 1–5.

[36] H. Yazdanpanah, J. A. Apolinário Jr., P. S. R. Diniz, and M. V. S. Lima, ℓ_0-norm feature LMS algorithms. Proceedings of the IEEE Global Conference on Signal and Information Processing (GlobalSIP), Anaheim, CA, USA, 2018, pp. 311–315.

[37] H. Yazdanpanah, A. Carini, and M. V. S. Lima, L_0-norm adaptive Volterra filters. Proceedings of the 27th European Signal Processing Conference (EUSIPCO), A Coruna, Spain, 2019, pp. 1–5.

[38] V. Mathews and G. Sicuranza, *Polynomial Signal Processing* (Wiley, New York, 2000).

[39] A. Fermo, A. Carini, and G. L. Sicuranza, Low complexity nonlinear adaptive filters for acoustic echo cancellation, European Transactions on Telecommunications **14**, pp. 161–169 (2003).

[40] M. Mohri, A. Rostamizadeh, and A. Tawalkar, *Foundations of Machine Learning*, 2nd ed. (MIT Press, Cambridge, USA, 2018).

[41] C. C. Aggarwal, *Neural Networks and Deep Learning* (Springer, Switzerland, 2018).

[42] R. A. do Prado, R. M. Guedes, F. R. Henriques, F. M. da Costa, L. D. T. J. Tarrataca, and D. B. Haddad, On the analysis of the incremental ℓ_0-LMS algorithm for distributed systems, Circuits, Systems, and Signal Processing **40**, pp. 845–871 (2021).

[43] S. Jiang and Y. Gu, Block-sparsity-induced adaptive filter for multi-clustering system identification, IEEE Transactions on Signal Processing **63**, pp. 5318–5330 (2015).

[44] Y. Li, Z. Jiang, Z. Jin, X. Han, and J. Yin, Cluster-sparse proportionate NLMS algorithm with the hybrid norm constraint, IEEE Access **6**, pp. 47794–47803 (2018).

[45] D. L. Duttweiler, Proportionate normalized least-mean-squares adaptation in echo cancelers, IEEE Transactions on Speech and Audio Processing **8**, pp. 508–518 (2000).

[46] J. Benesty and S. L. Gay, An improved PNLMS algorithm. Proceedings of the IEEE International Conference on Acoustics, Speech and Signal Processing (ICASSP 2002), Dallas, USA, May 2002, pp. 1881–1884.

[47] H. Deng and M. Doroslovacki, Improving convergence of the PNLMS algorithm for sparse impulse response identification, IEEE Signal Processing Letters **12**, pp. 181–184 (2005).

[48] E. Hänsler and G. Schmidt, *Acoustic Echo and Noise Control: A Practical Approach* (Wiley, Hoboken, 2004).

[49] L. Ligang, M. Fukumoto, and S. Saiki, An improved mu-law proportionate NLMS algorithm. Proceedings of the IEEE International Conference on Acoustics, Speech and Signal Processing (ICASSP 2008), 2008 pp. 3797–3800.

[50] S. Werner, J. A. Apolinário Jr., P. S. R. Diniz, and T. I. Laakso, A set-membership approach to normalized proportionate adaptation algorithms. Proceeding of the European Signal Processing Conference (EUSIPCO 2005), Antalya, Turkey, September 2005, pp. 1–4.

[51] M. V. S. Lima and P. S. R. Diniz, Steady-state MSE performance of the set-membership affine projection algorithm, Circuits, Systems, and Signal Processing **32**, pp. 1811–1837 (2013).

[52] P. A. Naylor, J. Cui, and M. Brookes, Adaptive algorithms for sparse echo cancellation, Signal Processing **86**, pp. 1182–1192 (2006).

[53] F. C. de Souza, O. J. Tobias, R. Seara, and D. R. Morgan, A PNLMS algorithm with individual activation factors, IEEE Transactions on Signal Processing **58**, pp. 2036–2047 (2010).

[54] F. C. de Souza, R. Seara, and D. R. Morgan, An enhanced IAF-PNLMS adaptive algorithm for sparse impulse response identification, IEEE Transactions on Signal Processing **60**, pp. 3301–3307 (2012).

[55] A. W. H. Khong and P. A. Naylor, Efficient use of sparse adaptive filters. Proceedings of the Fortieth Asilomar Conference on Signals, Systems and Computers (ACSSC 2006), 2006, pp. 1375–1379.

[56] C. Paleologu, S. Ciochina, and J. Benesty, An efficient proportionate affine projection algorithm for echo cancellation, IEEE Signal Processing Letters **17**, pp. 165–168 (2010).

[57] Z. Zheng, Z. Liu, and Y. Dong, Steady-State and Tracking Analyses of the Improved Proportionate Affine Projection Algorithm, IEEE Transactions on Circuits and Systems II: Express Briefs **65**, pp. 1793–1797 (2018).

[58] R. Arablouei, K. Dogançay and S. Perreau, Proportionate affine projection algorithm with selective projections for sparse system identification. Proceedings of the Asia-Pacific Signal Information Processing Association Annual Summit Conference, 2010, pp. 362–366.

[59] J. V. G. de Souza, D. B. Haddad, F. R. Henriques, and M. R. Petraglia, Novel proportionate adaptive filters with coefficient vector reusing, Circuits, Systems, and Signal Processing **39**, pp. 2473–2488 (2020).

[60] S. Werner, J. A. Apolinário Jr., and P. S. R. Diniz, Set-membership proportionate affine projection algorithms, EURASIP Journal on Audio, Speech, and Music Processing **2007**, pp. 1–10 (2007).

[61] H. Yazdanpanah, P. S. R. Diniz, and M. V. S. Lima, Improved simple set-membership affine projection algorithm for sparse system modelling: Analysis and implementation, IET Signal Processing **14**, pp. 81–88 (2020).

[62] K. Nose-Filho, A. K. Takahata, R. Lopes, and J. M. T. Romano, Improving sparse multichannel blind deconvolution with correlated seismic data: Foundations and further results, IEEE Signal Processing Magazine **35**, pp. 41–50 (2018).

[63] Y. Kopsinis, K. Slavakis, and S. Theodoridis, Online sparse system identification and signal reconstruction using projections onto weighted l_1 balls, IEEE Transactions on Signal Processing **59**, pp. 936–952 (2011).

[64] K. S. Olinto, D. B. Haddad, and M. R. Petraglia, Transient analysis of ℓ_0-LMS and ℓ_0-NLMS algorithms, Signal Processing **127**, pp. 217–226 (2016).

[65] M. Pereyra et al., A survey of stochastic simulation and optimization methods in signal processing, IEEE Journal of Selected Topics in Signal Processing **10**, pp. 224–241 (2016).

[66] S. Chouvardas, K. Slavakis, Y. Kopsinis, and S. Theodoridis, A sparsity promoting adaptive algorithm for distributed learning, IEEE Transactions on Signal Processing **60**, pp. 5412–5425 (2012).

[67] T. N. Ferreira, M. V. S. Lima, W. A. Martins, and P. S. R. Diniz, Low-complexity proportionate algorithms with sparsity-promoting penalties. Proceedings of the 2016 IEEE International Symposium on Circuits and Systems, Montreal, Canada, May 2016, pp. 253–256.

3 Kernel-Based Adaptive Filtering

3.1 Introduction

Kernel-based methods, since its inception [1–4], have witnessed a growing interest in machine learning algorithms. Typically it utilizes Mercer kernels (to be defined) to generate nonlinear estimation algorithms inspired by their linear counterparts. These kernel functions are applied to pairs of input data vectors to measure their resemblance, representing an inner product in high-dimensional feature space. In particular, we are interested in Hilbert spaces, denoted as $\mathcal{H}$, which are vector spaces with an inner product that induces a complete metric space [5]. Hilbert spaces can be thought as generalizations of Euclidean spaces.

Many learning algorithms based on kernel methods are not suitable for online implementation since the computational cost increases with the size of the acquired data set. In this chapter, we discuss how to apply the kernel idea to derive nonlinear adaptive filtering algorithms meant to operate in online environments. After introducing the concept of kernel-based learning algorithms utilizing Mercer kernels, we will present the derivations of several kernel adaptive filtering algorithms such as the kernel LMS, kernel affine projection (KAP), kernel RLS (KRLS), and set-membership kernel affine projection (SM-KAP) algorithms.

3.2 Kernel Functions

The following discussions address the kernel functions and their key role in learning algorithms in general, emphasizing their application to online nonlinear adaptive filtering algorithms. This section presents a particular and condensed description of the kernel method, meant to be employed in the derivation of online adaptive filtering algorithms in a straightforward manner.

In typical learning algorithms, it is usually required to employ some type of similarity measure to compare two distinct vectors from a data set where one of the vectors represents a pattern. This kind of concept is widely diffused in machine learning and pattern recognition [9–13]. In an online setup, a given set of input data or observations, $\mathbf{x}(k) \in \mathbb{R}^{N+1}$ for $k = 0, 1, \ldots, K$, is paired with a set of outputs $y(k) \in \mathbb{R}$ for $k = 0, 1, \ldots, K$, where two input vectors are mapped into a real value through a function called *kernel function* in the process. This function has as arguments two samples of the input vector $\kappa(\mathbf{x}(k), \mathbf{x}(l)) \in \mathbb{R}$, where at least in the main cases we deal with in this chapter, the kernel function is non-negative and symmetric, that is, $\kappa(\mathbf{x}(k), \mathbf{x}(l)) \geq 0$ and $\kappa(\mathbf{x}(k), \mathbf{x}(l)) = \kappa(\mathbf{x}(l), \mathbf{x}(k))$, for all $\mathbf{x}(k), \mathbf{x}(l) \in \mathbb{R}^{N+1}$, respectively.[1]

A possible task of the kernel function is to serve as a measure of how the vectors $\mathbf{x}(k)$ and $\mathbf{x}(l)$ are related or similar. The feature of the kernel approach is to infer the similarity between the two vectors without processing in their original vector form. The kernel operation might be represented mathematically as

$$\begin{aligned}\kappa :\ & \mathbb{R}^{N+1} \times \mathbb{R}^{N+1} \to \mathbb{R} \\ & (\mathbf{x}(k), \mathbf{x}(l)) \mapsto \kappa(\mathbf{x}(l), \mathbf{x}(k)) .\end{aligned} \tag{3.1}$$

A typical way to measure the similarity between two vectors is the inner product where

$$\langle \mathbf{x}(k), \mathbf{x}(l) \rangle = \mathbf{x}^{\mathrm{T}}(k)\mathbf{x}(l) = \sum_{i=0}^{N} x_i(k) x_i(l), \tag{3.2}$$

also known as dot product. The inner product is useful for measuring the length of a vector, the length of the difference (distance), or the angle between two vectors.

Although in some applications, the inner product applied to the input vector directly is sufficient to measure the resemblance between the vectors $\mathbf{x}(k)$ and $\mathbf{x}(l)$ in their original forms, it is possible to get more useful information by applying a possibly nonlinear transformation to the input signal vectors. The strategy corresponds to map the input vector into the so-called *feature space* denoted as

$$\boldsymbol{\upsilon} : \mathbb{R}^{N+1} \to \mathcal{H}, \tag{3.3}$$

where the input vector is mapped in the new Hilbert space $\mathcal{H}$, as represented by

$$\mathbf{x}(k) \mapsto \boldsymbol{\chi}(k) := \boldsymbol{\upsilon}(\mathbf{x}(k)) . \tag{3.4}$$

The key for the success of the kernel method is the freedom to choose the mapping function $\boldsymbol{\upsilon}(\cdot)$, which will enable more powerful similarity measures between two vectors after mapping them in the feature space. These measures still rely on the inner product but this time performed at the feature space as follows:

$$\langle \boldsymbol{\upsilon}(\mathbf{x}(k)), \boldsymbol{\upsilon}(\mathbf{x}(l)) \rangle = \langle \boldsymbol{\chi}(k), \boldsymbol{\chi}(l) \rangle = \kappa(\mathbf{x}(k), \mathbf{x}(l)), \tag{3.5}$$

where the feature space is not necessarily in the same dimension $\mathbb{R}^{N+1}$ as the input signal data (it can even have infinite dimension). In order for us to justify

[1] This is not the general kernel type.

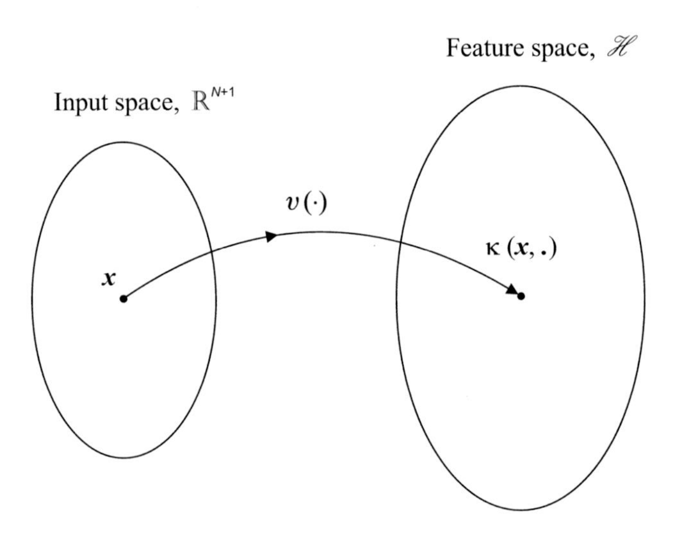

Figure 3.1 Kernel mapping representation.

this trick of computing inner products as kernel evaluations, we shall introduce the *reproducing kernel Hilbert space* (RKHS) concept.

A Hilbert space $\mathcal{H}$ whose vectors are real-valued functions is considered to be a RKHS if there exists a kernel function $\kappa(\cdot,\cdot)$ such that $\forall \mathbf{x} \in \mathbb{R}^{N+1}$, $\kappa(\mathbf{x},\cdot) \in \mathcal{H}$ and any other function $f(\cdot) \in \mathcal{H}$ can be evaluated at $\mathbf{x}$ as $f(\mathbf{x}) = \langle f(\cdot), \kappa(\mathbf{x},\cdot)\rangle$, which is called the *reproducing property* of the kernel. The reproducing property allows one to compute inner products via functional evaluations (*kernel trick*). Indeed, given $\mathbf{x}(i) \in \mathbb{R}^{N+1}$, one has that $\kappa(\mathbf{x}(i),\cdot)$ is a real-valued function which is also a vector in the RKHS $\mathcal{H}$, which means that it can be evaluated at $\mathbf{x}$ as $\kappa(\mathbf{x}(i),\mathbf{x}) = \langle \kappa(\mathbf{x}(i),\cdot), \kappa(\mathbf{x},\cdot)\rangle$. In particular, if we are given a function $f(\cdot) = \sum_{i=0}^{I} \alpha_i \kappa(\mathbf{x}(i),\cdot) \in \mathcal{H}$, then $f\mathbf{x} = \sum_{i=0}^{I} \alpha_i \kappa(\mathbf{x}(i),\mathbf{x}) \in \mathbb{R}$. Figure 3.1 depicts the mapping of data vectors in the feature space.

The question that remains is: Given a kernel function $\kappa(\cdot,\cdot)$, how one can determine whether this kernel can be used for constructing an RKHS $\mathcal{H}$ or not? Section 3.2.1 characterizes the kernels used to construct RKHS feature spaces.

3.2.1 Mercer Kernels

Aligned with the classic adaptive filtering theory where the input signal correlation function plays an active role, in the feature domain, we can also assemble the similarity among $I+1$ input signal vectors in a matrix $\boldsymbol{K}(k)$ called Gram matrix, defined by

$$\boldsymbol{K}(k) = \begin{bmatrix} \kappa\left(\mathbf{x}(k), \mathbf{x}(k)\right) & \kappa\left(\mathbf{x}(k), \mathbf{x}(k-1)\right) & \cdots & \kappa\left(\mathbf{x}(k), \mathbf{x}(k-I)\right) \\ \kappa\left(\mathbf{x}(k-1), \mathbf{x}(k)\right) & \kappa\left(\mathbf{x}(k-1), \mathbf{x}(k-1)\right) & \cdots & \kappa\left(\mathbf{x}(k-1), \mathbf{x}(k-I)\right) \\ \vdots & \vdots & \ddots & \vdots \\ \kappa\left(\mathbf{x}(k-I), \mathbf{x}(k)\right) & \kappa\left(\mathbf{x}(k-I), \mathbf{x}(k-1)\right) & \cdots & \kappa\left(\mathbf{x}(k-I), \mathbf{x}(k-I)\right) \end{bmatrix}, \tag{3.6}$$

where, in this chapter, $I+1$ represents the number of input signal vectors retained in the data dictionary, a feature inherent to many kernel-based algorithms. In our adaptive filtering environment, the Gram matrix must be positive definite, and those kernel matrices that satisfy this condition are called positive definite or *Mercer kernels*.

Assuming that the Gram matrix $\boldsymbol{K}(k)$ is symmetric, it has $I+1$ orthonormal eigenvectors $\mathbf{q}_i$ so that we can form a matrix $\mathbf{Q}$ whose columns are composed by the $\mathbf{q}_i$'s, as follows:

$$\mathbf{Q}^{\mathrm{T}}\boldsymbol{K}(k)\mathbf{Q} = \begin{bmatrix} \lambda_0 & 0 & \cdots & 0 \\ 0 & \lambda_1 & & \vdots \\ \vdots & 0 & \cdots & \vdots \\ \vdots & \vdots & & 0 \\ 0 & 0 & \cdots & \lambda_I \end{bmatrix} = \boldsymbol{\Lambda}, \tag{3.7}$$

where the eigenvalues $\lambda_i \geq 0$ given that $\boldsymbol{K}$ is also positive semi-definite. For simplicity, the iteration variable k is omitted. Each element of the Gram matrix can be calculated as follows:

$$\kappa\left(\mathbf{x}(k-i), \mathbf{x}(k-l)\right) = \left(\mathbf{Q}_{i,0:I}\boldsymbol{\Lambda}^{\frac{1}{2}}\right)\left(\mathbf{Q}_{l,0:I}\boldsymbol{\Lambda}^{\frac{1}{2}}\right)^{\mathrm{T}}. \tag{3.8}$$

If we define

$$\boldsymbol{\upsilon}\left(\mathbf{x}(k-i)\right) = \left(\mathbf{Q}_{i,0:I}\boldsymbol{\Lambda}^{\frac{1}{2}}\right)^{\mathrm{T}}, \tag{3.9}$$

it is possible to conclude that

$$\kappa\left(\mathbf{x}(k-i), \mathbf{x}(k-l)\right) = \boldsymbol{\upsilon}^{\mathrm{T}}\left(\mathbf{x}(k-i)\right)\boldsymbol{\upsilon}\left(\mathbf{x}(k-l)\right), \ \forall i, l \in \{0, 1, \ldots, I\}, \tag{3.10}$$

where it is observed that the entries of the Gram matrix can be calculated through the inner product of transformed vectors, which are related to the eigenvectors of the Gram matrix $\boldsymbol{K}(k)$ in an implicit form. This discussion allows us to infer that, given a kernel function such that any related Gram matrix is positive definite, it is possible to build a mapping in the feature space[2]

[2] This map is related to the eigendecomposition of $\boldsymbol{K}(k)$ using eigenvectors in the feature space $\mathcal{H}$.

such that Equation (3.5) holds. On the other hand, for any map $\boldsymbol{v}(\mathbf{x}) = \kappa(\mathbf{x}, \cdot)$ into an RKHS $\mathcal{H}$, the corresponding kernel matrices $\boldsymbol{K}$ are positive definite.

In summary, a key feature of the kernel methods is to devise algorithms that, *instead of* explicitly performing the inner product

$$\langle \mathbf{x}, \mathbf{x}(k-i) \rangle = \mathbf{x}^{\mathrm{T}} \mathbf{x}(k-i) \tag{3.11}$$

as a similarity measure, it employs a mapping to a feature space via the chosen kernel function $\kappa(\cdot, \cdot)$ that allows us to use the *kernel trick*

$$\langle \boldsymbol{\kappa}(\cdot, \mathbf{x}), \boldsymbol{\kappa}(\cdot, \mathbf{x}(k-i)) \rangle = \kappa(\mathbf{x}, \mathbf{x}(k-i)), \tag{3.12}$$

which is a natural reproducing property of a Mercer kernel [11].

3.2.2 Examples of Kernel Functions

The proper choice of the data transformation into $\mathcal{H}$ affects the quality of the estimation we are performing utilizing the kernel approach. Some typical and widely used kernel functions are following listed:

- Cosine similarity kernel:

$$\kappa(\mathbf{x}(k), \mathbf{x}(l)) = \frac{\mathbf{x}^{\mathrm{T}}(k)\mathbf{x}(l)}{\|\mathbf{x}(k)\|_2 \|\mathbf{x}(l)\|_2}, \tag{3.13}$$

  where the cosine kernel measures angles between the vectors $\mathbf{x}(k)$ and $\mathbf{x}(l)$.

- Sigmoid kernel:

$$\kappa(\mathbf{x}(k), \mathbf{x}(l)) = \tanh(a\mathbf{x}^{\mathrm{T}}(k)\mathbf{x}(l) + b), \tag{3.14}$$

  where a and b are real constants. This kernel function does not lead to a positive semidefinite Gram matrix, but it is quite often found in the literature in an attempt to relate it to the neural networks, particularly the multilayer perceptron.

- Polynomial kernel:

$$\kappa(\mathbf{x}(k), \mathbf{x}(l)) = (a\mathbf{x}^{\mathrm{T}}(k)\mathbf{x}(l) + b)^n, \tag{3.15}$$

  where $n \in \mathbb{N}$ and $b \geq 0$, and if $b \neq 0$, this is an inhomogeneous polynomial kernel. Also, the non-negative value of b is required to guarantee that the kernel matrix is positive definite, and in any case, the choice of parameters requires that the diagonal entries of the kernel matrix be positive. Let's take a simple example where the data vectors have two entries, that is, $N = 1$, and the polynomial power is $n = 2$. In this case,

$$\begin{aligned}\kappa\left(\mathbf{x}(k),\mathbf{x}(l)\right) =& (a\mathbf{x}^{\mathrm{T}}(k)\mathbf{x}(l)+b)^n \\ =& b^2 + 2ab(x_0(k)x_0(l) + x_1(k)x_1(l)) \\ &+ a^2(x_0^2(k)x_0^2(l) + x_1^2(k)x_1^2(l)) + 2a^2x_0(k)x_0(l)x_1(k)x_1(l) \\ =& \left[\, b \;\; \sqrt{2ab}x_0(k) \;\; \sqrt{2ab}x_1(k) \;\; ax_0^2(k) \;\; ax_1^2(k) \;\; \sqrt{2}ax_0(k)x_1(k)\right] \\ &\times \begin{bmatrix} b \\ \sqrt{2ab}x_0(l) \\ \sqrt{2ab}x_1(l) \\ ax_0^2(l) \\ ax_1^2(l) \\ \sqrt{2}ax_0(l)x_1(l) \end{bmatrix} \\ =& \langle \boldsymbol{\upsilon}\left(\mathbf{x}(k)\right), \boldsymbol{\upsilon}\left(\mathbf{x}(l)\right)\rangle \\ =& \boldsymbol{\upsilon}^{\mathrm{T}}\left(\mathbf{x}(k)\right)\boldsymbol{\upsilon}\left(\mathbf{x}(l)\right). \end{aligned} \tag{3.16}$$

As can be observed, the polynomial kernel is described as the inner product of the mapping function $\boldsymbol{\upsilon}(\cdot)$ evaluated at $\mathbf{x}(k)$ and $\mathbf{x}(l)$, respectively. This is usually the case of Mercer kernels, where in this particular case, the feature space has a dimension equal to six. Note that the feature mapping function has some connections with the second-order Volterra series representation. It is straightforward to verify that if this is homogeneous, that is $b = 0$, the polynomial kernel has only second-order terms.

The polynomial kernel might face numerical instability for large values of the exponent n, since the kernel value might tend to 0 or infinity in the cases where $|a\mathbf{x}^{\mathrm{T}}(k)\mathbf{x}(l) + b|$ is less or greater than 1, respectively.

- Gaussian kernel:

$$\begin{aligned}\kappa\left(\mathbf{x}(k),\mathbf{x}(l)\right) =& \mathrm{e}^{\frac{1}{2}(\mathbf{X}(k)-\mathbf{X}(l))^{\mathrm{T}}\boldsymbol{\Sigma}^{-1}(\mathbf{X}(k)-\mathbf{X}(l))} \\ =& \mathrm{e}^{-\frac{1}{2}\sum_{i=0}^{I}\frac{1}{\sigma_i^2}(x_i(k)-x_i(l))^2}, \end{aligned} \tag{3.17}$$

where $\boldsymbol{\Sigma}$ is a diagonal matrix whose entries are given by σ_i^2, for $i = 0, 1, \ldots, I$. In many situations we utilize a fixed $\sigma_i^2 = \sigma^2$, where the quantity σ^2 is also called kernel bandwidth. The Gaussian kernel can be shown to be a Mercer kernel, where for the Gaussian case, the feature space has dimension equal to infinity. It is important to highlight that this Gaussian denomination here does not match its definition applied to probability density function, as can be observed by the lack of a normalization term.

- Laplacian kernel:

$$\kappa\left(\mathbf{x}(k),\mathbf{x}(l)\right) = \mathrm{e}^{-\frac{\|(\mathbf{X}(k)-\mathbf{X}(l))\|}{\sigma}}. \tag{3.18}$$

There is an infinite amount of possible kernels that one can come up with, although the generation of Mercer type of kernels requires particular attention. Luckily some combination rules can be applied to Mercer kernels that result in straightforward derivation of more sophisticated kernels still satisfying the condition that the Gram matrix is positive definite. Simple rules are the addition of and the product of Mercer kernels leading to a Mercer kernel; see [9, 10] for more details. It is worth mentioning that the kernel matrix entries are not necessarily positive numbers, except for the diagonal elements. These elements $\kappa(\mathbf{x}(l), \mathbf{x}(l))$ are in many cases normalized to one.

3.3 The Kernel LMS algorithm

In this section, we will describe the online LMS kernel based (KLMS) algorithm, originally presented in [7, 8]. In the context of adaptive filtering, the idea of the kernel approach is to map the input signal vector into a high-dimensional feature space through the feature function $\boldsymbol{v}(\mathbf{x})$. Defining a reference center $\mathbf{x}$, we can interpret that the mapping of the incoming signal vector to the reference center is represented by the kernel denoted as $\kappa(\mathbf{x}, \cdot)$. Using this idea we can infer that adaptive coefficients will generate an estimate of the desired signal in the high-dimensional space, such that it is expected good estimate performance through an inner product as follows:

$$y(k) = \langle \mathbf{w}(k), \boldsymbol{v}(\mathbf{x}(k)) \rangle, \tag{3.19}$$

where $\mathbf{w}(k)$ represents the nonlinear filter model based on the kernel approach. As usual $y(k)$ represents the adaptive-filter output signal.

The standard approach to derive the LMS algorithm is to use an instantaneous estimate of the mean-square error (MSE), defined as $\xi(k) = \mathbb{E}[e^2(k)]$. This instantaneous square error estimate is given by

$$\begin{aligned} e^2(k) &= (d(k) - y(k))^2 \\ &= d^2(k) - 2d(k)y(k) + y^2(k) \\ &= d^2(k) - 2d(k)\langle \mathbf{w}(k), \boldsymbol{v}(\mathbf{x}(k)) \rangle + \langle \mathbf{w}(k), \boldsymbol{v}(\mathbf{x}(k)) \rangle^2. \end{aligned} \tag{3.20}$$

The kernel LMS algorithm is then given by

$$\begin{aligned} \mathbf{w}(k+1) &= \mathbf{w}(k) - \mu \nabla_{\mathbf{w}} e^2(k) \\ &= \mathbf{w}(k) - 2\mu e(k) \frac{\partial e(k)}{\partial \mathbf{w}(k)} \\ &= \mathbf{w}(k) + 2\mu e(k) \boldsymbol{v}(\mathbf{x}(k)). \end{aligned} \tag{3.21}$$

By assuming $\mathbf{w}(0) = \mathbf{0}$, Equation (3.21) above can be rewritten as

$$\begin{aligned} \mathbf{w}(k+1) &= \mathbf{w}(0) + 2\mu \sum_{i=0}^{k} e(i) \boldsymbol{v}(\mathbf{x}(i)) \\ &= 2\mu \sum_{i=0}^{k} e(i) \boldsymbol{v}(\mathbf{x}(i)). \end{aligned} \tag{3.22}$$

It should be reminded that since the dimension of the feature space is high, and possibly infinity as in the case of the Gaussian kernel, it is not feasible to work with the update Equation (3.21). The strategy to deal with this drawback is to apply $\mathbf{w}(k+1)$ to a feature vector mapping a reference center $\mathbf{x}(k)$, related to the new incoming vector, as follows:

$$\langle \mathbf{w}(k+1), \boldsymbol{\upsilon}(\mathbf{x}(k))\rangle = 2\mu \sum_{i=0}^{k} e(i)\langle \boldsymbol{\upsilon}(\mathbf{x}(i)), \boldsymbol{\upsilon}(\mathbf{x}(k))\rangle. \tag{3.23}$$

By defining $g(i) = \langle \mathbf{w}(i+1), \boldsymbol{\upsilon}(\mathbf{x}(i))\rangle$ and utilizing the kernel trick, we obtain

$$\begin{aligned} g(k) &= 2\mu \sum_{i=0}^{k} e(i)\kappa\left(\mathbf{x}(i), \mathbf{x}(k)\right) \\ &= 2\mu \sum_{i=0}^{k-1} e(i)\kappa\left(\mathbf{x}(i), \mathbf{x}(k)\right) + 2\mu e(k)\kappa\left(\mathbf{x}(k), \mathbf{x}(k)\right), \end{aligned} \tag{3.24}$$

where the error is given by

$$e(k) = d(k) - g(k). \tag{3.25}$$

As can be observed in Equation (3.24), the number of centers in the kernel dictionary tends to grow as the number of samples increases. Note that the computation of $g(k)$ entails the use of previously selected input-data vectors that should be stored for this purpose. In the following subsection, we discuss strategies on how to include or not a new incoming data in the kernel dictionary.

3.3.1 Model Reduction

A regular feature of the kernel learning is the increase in the model order with the amount of data. Whenever new data arises, it is expected that its information should be incorporated in the learning process. As a result, for online applications, it is crucial to come up with some strategies to reduce the model order according to some meaningful criterion. The model reduction can also be interpreted as a sparsification or a data selection strategy.

A possible strategy to select the data is to compare the new incoming data $\mathbf{x}(k)$ with previously selected data $\mathbf{x}(k-l)$ for $l \in \mathcal{L}_I(k-1)$, where $\mathcal{L}_I(k-1) = \{l_0(k-1), l_1(k-1), \ldots, l_I(k-1)\}$ represents indexes of the data set available at the instant k. The indexes $\{l_j(k-1)\}_{j=0}^{I}$ are chosen from the set $\{0, 1, \ldots, k-1\}$ representing the previous input data presented to the adaptive algorithm.

The sparsification method based on the novelty criterion evaluates the distance between the new incoming data with respect to those stored in the data indexed set, so that if

$$\underset{l \in \mathcal{L}_I(k-1)}{\text{minimum}} \|\mathbf{x}(k) - \mathbf{x}(k-l)\|^2 < \gamma_{\text{d}}, \tag{3.26}$$

the new data $\mathbf{x}(k)$ is discarded, otherwise another test is performed. The parameter $\gamma_{\rm d}$ represents a prescribed threshold. The second test verifies if the output error in absolute value is small enough, that is

$$|e(k)| < \gamma_{\rm e}, \tag{3.27}$$

and confirming that this is the case the data is definitely discarded. Otherwise, the data is either included in the set of stored data, as in the classical approach or could replace an entry of the set. The latter strategy would be particularly useful for online embedded systems since the amount of stored data would not grow with time and could stick to a maximum number of entries. For a learning algorithm operating in highly nonstationary environments, a simple strategy is to discard the oldest data indexed as $l = l_I(k-1)$.

Another criterion for sparsification is the so-called coherence approach where the new data is included in the dictionary if

$$\underset{l \in \mathcal{L}_I(k-1)}{\text{maximum}} |\kappa\left(\mathbf{x}(k), \mathbf{x}(k-l)\right)| \leq \gamma_{\rm c}, \tag{3.28}$$

where $0 \leq \gamma_{\rm c} < 1$ is a given threshold to measure the acceptable coherence between the new incoming data and the previously stored ones. According to [14], the coherence criterion will lead asymptotically to a finite dictionary, whenever $\mathbb{R}^{I+1}$ is a subspace of a Banach space $\mathbb{R}^N$ for $N > I+1$.[3] In the coherence definition above, the kernel functions, sometimes called kernel atoms, are considered normalized. Otherwise, normalization is required.

The stochastic analysis of the KLMS algorithm is quite cumbersome, and some results can be found in [15] for the Gaussian kernel case. Considering the coherence approach for sparsification, Algorithm 3.12 fully describes the kernel LMS algorithm, where we first decide if the new incoming data brings about enough innovation and proceed with its inclusion in the dictionary. Otherwise, by saving the position of the entry already in the dictionary that closely resembles the new input entry, that is saving l corresponding to the maximum as $l_{\rm max}$; we can perform a correction on the corresponding error $e(l_{\rm max})$ induced by the limited innovation inherent to $\mathbf{x}(k)$. The correction is performed by

$$e(l_{\rm max}) := e(l_{\rm max}) + \mu e(k). \tag{3.29}$$

It is worth noting that in Algorithm 3.12, we are employing a sparsification strategy to avoid that the dictionary grows at every iteration and, in addition, limits the dictionary size to $I_{\rm max}$. The size of the dictionary is usually much larger than the dimension of the input signal vector.

[3] In functional analysis, Cauchy sequences are those which converge to a point $\mathbf{x} \in \mathcal{H}$ if $\|\mathbf{x}(k) - \mathbf{x}\| \to 0$ as $k \to \infty$, whereas the Banach space is a normed space that is complete since all Cauchy sequences in this space converge. The Hilbert space, in addition, is a complete inner product space [5, 10]; therefore, it is a Banach space with an inner product.

Algorithm 3.12 Kernel LMS algorithm

Initialization

$\mathbf{x}(0) = [0\,0\ldots0]^{\mathrm{T}}$

$e(0) = 0$

$\kappa\left(\mathbf{x}(l_0(0)), \mathbf{x}(0)\right) = \kappa\left(\mathbf{x}(0), \mathbf{x}(0)\right)$ choose $\gamma_c < 1$

For do $k \geq 0$

$$g(k) = 2\mu \sum_{l \in \mathcal{L}_I(k-1)} e(l)\kappa\left(\mathbf{x}(k-l), \mathbf{x}(k)\right) + 2\mu\kappa\left(\mathbf{x}(k), \mathbf{x}(k)\right) \tag{3.24}$$

$$e(k) = d(k) - g(k) \tag{3.25}$$

If $\underset{l \in \mathcal{L}_I(k-1)}{\text{maximum}} |\kappa\left(\mathbf{x}(k), \mathbf{x}(k-l)\right)| \leq \gamma_c$, and save l corresponding to the maximum as $l_{\max}$ (3.28)

$I = I + 1$

If $I \leq I_{\max}$

$l_0(k+1) = k+1$, including $\mathbf{x}(k)$ in the dictionary

Else

$l_0(k+1) = k+1$, include $\mathbf{x}(k)$ in the dictionary and remove $\mathbf{x}(l_{\max})$

End if

Else

$$e(l_{\max}) = e(l_{\max}) + \mu e(k) \tag{3.29}$$

End if

3.3.2 Random Fourier Feature

The random Fourier feature is a technique to map the data in randomized feature space with low dimension, first proposed in [26]. This tool might apply to online learning as long as the inherent sampling procedure has fast enough implementation. Most kernel-based methods do not scale well for big-data sets, requiring large storage and high computational cost. A proposed solution is to replace the kernel evaluation by an approximation entailing the inner product of two vectors, each of them representing the outcome of a randomized feature map from $\mathbb{R}^{N+1}$ to $\mathbb{R}^{D+1}$, for $N > D$. The random Fourier feature method consists of a tool to reduce the computation in kernel-based applications that are widely used. It entails the approximation of a kernel function through an inner product as

$$\kappa\left(\mathbf{x}(l), \mathbf{x}(k)\right) \approx \mathbf{z}^{\mathrm{T}}\left(\mathbf{x}(k)\right)\mathbf{z}\left(\mathbf{x}(l)\right), \tag{3.30}$$

where $\mathbf{z}(\cdot) : \mathbb{R}^{N+1} \rightarrow \mathbb{R}^{D+1}$.

Assume that the Fourier transform of the kernel is $\mathcal{K}(\omega)$ and that we desire to build feature spaces that approximate shift-invariant kernels $\kappa\left(\mathbf{x}(l) - \mathbf{x}(k)\right)$ within reasonable accuracy [26]. The literature's most widely used shift-invariant kernels are the Gaussian, the Laplacian, and the Cauchy kernels. The choice of the shift-invariant kernel family benefits from the Bochner's theorem, see [26, 27], stating that the Fourier transform of any positive-definite function,

$\kappa(\mathbf{x}(l) - \mathbf{x}(k))$, is nonnegative. That is, for $\kappa(\mathbf{x}(l), \mathbf{x}(k)) = \kappa(\mathbf{x}(l) - \mathbf{x}(k))$, it follows that:

$$\kappa(\mathbf{x}(l) - \mathbf{x}(k)) = \int_{\mathbb{R}^{D+1}} \mathcal{K}(\boldsymbol{\omega})\, \mathrm{e}^{\mathrm{j}\boldsymbol{\omega}^{\mathrm{T}}(\mathbf{x}(l)-\mathbf{x}(k))} \mathrm{d}\boldsymbol{\omega}, \tag{3.31}$$

where $\mathcal{K}(\boldsymbol{\omega}) \geq 0, \forall \boldsymbol{\omega}$. The equation above represents a probability measure if the kernel is normalized, that is, $\kappa(\mathbf{0}) = 1$, see [29, 30] for proofs and further discussions.

In random Fourier case, the similarity between two vectors is given by

$$\kappa(\mathbf{x}(l), \mathbf{x}(k)) \approx \mathbf{z}^{\mathrm{T}}(\mathbf{x}(k))\mathbf{z}(\mathbf{x}(l)) = \sum_{i=0}^{D} z_i(\mathbf{x}(k))\, z_i(\mathbf{x}(l)), \tag{3.32}$$

which, in this equation, the inner product is measuring an approximation of a kernel in the feature space.

In some sense, the success of the random Fourier method lies on how well $\mathbf{z}(\mathbf{x})$ approximates $\boldsymbol{v}(\mathbf{x})$, defined in Equation (3.4), in the sense that

$$\begin{aligned} f(\mathbf{x}) &= \sum_{i=0}^{I} \alpha_i \kappa(\mathbf{x}(x), \mathbf{x}), \\ &= \sum_{i=0}^{I} \alpha_i \boldsymbol{v}^{\mathrm{T}}(\mathbf{x}(i))\boldsymbol{v}(\mathbf{x}), \\ &\approx \sum_{i=0}^{I} \alpha_i \mathbf{z}^{\mathrm{T}}(\mathbf{x}(i))\mathbf{z}(\mathbf{x}). \end{aligned} \tag{3.33}$$

A popular random Fourier embedding method has the form [28]:

$$\mathbf{z}(\mathbf{x}) = \sqrt{\frac{1}{D+1}} \begin{bmatrix} \sin\left(\boldsymbol{\omega}_1^{\mathrm{T}}\mathbf{x}\right) \\ \cos\left(\boldsymbol{\omega}_1^{\mathrm{T}}\mathbf{x}\right) \\ \sin\left(\boldsymbol{\omega}_2^{\mathrm{T}}\mathbf{x}\right) \\ \cos\left(\boldsymbol{\omega}_2^{\mathrm{T}}\mathbf{x}\right) \\ \vdots \\ \sin\left(\boldsymbol{\omega}_{\frac{D+1}{2}}^{\mathrm{T}}\mathbf{x}\right) \\ \cos\left(\boldsymbol{\omega}_{\frac{D+1}{2}}^{\mathrm{T}}\mathbf{x}\right) \end{bmatrix}, \tag{3.34}$$

where the random vectors $\boldsymbol{\omega}_i$, for $i \in \{1, 2, \ldots, \frac{D+1}{2}\}$, are drawn from a distribution represented by $\mathcal{K}(\boldsymbol{\omega})$.

Another common embedding method is [28]

$$\mathbf{z}(\mathbf{x}) = \sqrt{\frac{1}{D+1}} \begin{bmatrix} \cos\left(\boldsymbol{\omega}_1^{\mathrm{T}}\mathbf{x} + b_1\right) \\ \cos\left(\boldsymbol{\omega}_2^{\mathrm{T}}\mathbf{x} + b_2\right) \\ \vdots \\ \cos\left(\boldsymbol{\omega}_{D+1}^{\mathrm{T}}\mathbf{x} + b_{D+1}\right) \end{bmatrix}, \tag{3.35}$$

where $\boldsymbol{\omega}_i$ is drawn from a distribution represented by $\mathcal{K}(\boldsymbol{\omega})$, and b_i is drawn from a uniform distribution in the range $[0, 2\pi]$, for $i \in \{1, 2, \ldots, D+1\}$. By comparing the methods of Equations (3.34) and (3.35), it is possible to observe that the former requires fewer samples of vector $\boldsymbol{\omega}$. By contrast, the latter also involves the addition of shift-varying noise represented by b. The best choice depends on the approximation achieved in Equation (3.33).

For both embedding methods presented, it is possible to show that

$$\mathbf{z}^{\mathrm{T}}(\mathbf{x}(i))\,\mathbf{z}(\mathbf{x}(l)) \approx \mathbb{E}_{\boldsymbol{\omega}}\left[\cos \boldsymbol{\omega}^{\mathrm{T}}(\mathbf{x}(i) - \mathbf{x}(l))\right] = \kappa(\mathbf{x}(i), \mathbf{x}(l)) \tag{3.36}$$

for any non-negative $\mathcal{K}(\boldsymbol{\omega})$. The proof uses some trigonometric identities and the assumptions related to $\boldsymbol{\omega}$ and b, see Problem 3.16.

Define a mapping $\zeta : \mathbf{x} \mapsto \mathrm{e}^{\mathrm{j}\boldsymbol{\omega}^{\mathrm{T}}\mathbf{x}}$, if the entries of $\boldsymbol{\omega}$ are drawn from i.i.d. zero-mean Gaussian processes, it is possible to show that

$$\mathbb{E}_{\boldsymbol{\omega}}\left[\zeta(\mathbf{x}(i))\zeta^*(\mathbf{x}(l))\right] = \mathrm{e}^{-\frac{1}{2}(\mathbf{x}(i)-\mathbf{x}(l))^{\mathrm{T}}(\mathbf{x}(i)-\mathbf{x}(l))} = \kappa(\mathbf{x}(l) - \mathbf{x}(k)), \tag{3.37}$$

which is nothing but the Gaussian kernel, see Problem 3.17.

In terms of the kernel matrix, the random Fourier feature achieves the following approximation:

$$\begin{aligned}\boldsymbol{K}(k) &= \begin{bmatrix} \kappa(\mathbf{x}(k), \mathbf{x}(k)) & \kappa(\mathbf{x}(k), \mathbf{x}(k-1)) & \cdots & \kappa(\mathbf{x}(k), \mathbf{x}(k-I)) \\ \kappa(\mathbf{x}(k-1), \mathbf{x}(k)) & \kappa(\mathbf{x}(k-1), \mathbf{x}(k-1)) & \cdots & \kappa(\mathbf{x}(k-1), \mathbf{x}(k-I)) \\ \vdots & \vdots & \ddots & \vdots \\ \kappa(\mathbf{x}(k-I), \mathbf{x}(k)) & \kappa(\mathbf{x}(k-I), \mathbf{x}(k-1)) & \cdots & \kappa(\mathbf{x}(k-I), \mathbf{x}(k-I)) \end{bmatrix}, \\ &\approx \begin{bmatrix} \mathbf{z}^T(\mathbf{x}(k))\,\mathbf{z}(\mathbf{x}(k)) & \mathbf{z}^T(\mathbf{x}(k))\,\mathbf{z}(\mathbf{x}(k-1)) & \cdots & \mathbf{z}^T(\mathbf{x}(k))\,\mathbf{z}(\mathbf{x}(k-I)) \\ \mathbf{z}^T(\mathbf{x}(k-1))\,\mathbf{z}(\mathbf{x}(k)) & \mathbf{z}^T(\mathbf{x}(k-1))\,\mathbf{z}(\mathbf{x}(k-1)) & \cdots & \mathbf{z}^T(\mathbf{x}(k-1))\,\mathbf{z}(\mathbf{x}(k-I)) \\ \vdots & \vdots & \ddots & \vdots \\ \mathbf{z}^T(\mathbf{x}(k-I))\,\mathbf{z}(\mathbf{x}(k)) & \mathbf{z}^T(\mathbf{x}(k-I))\,\mathbf{z}(\mathbf{x}(k-1)) & \cdots & \mathbf{z}^T(\mathbf{x}(k-I))\,\mathbf{z}(\mathbf{x}(k-I)) \end{bmatrix},\end{aligned} \tag{3.38}$$

where, in this book, $I+1$ represents the number of input signal vectors retained in the data dictionary, a feature inherent to many kernel-based algorithms. Note that the inner products in the last equality entail $D+1$ multiplications. For larger D, the approximation to the kernel matrix becomes more accurate.

Example 3.1 Implement the kernel LMS algorithm and apply it to the problems following described [6] encompassing the decision feedback equalizer (DFE) (see Figure 3.2). A digital channel model can be represented by the following system of equations:

$$\begin{aligned} v(k) &= x(k) + 0.5x(k-1); \\ y'(k) &= \frac{v(k) + 0.2v^2(k) + 0.1v^3(k)}{\sigma_{y'}}; \\ y(k) &= y'(k) + n(k). \end{aligned}$$

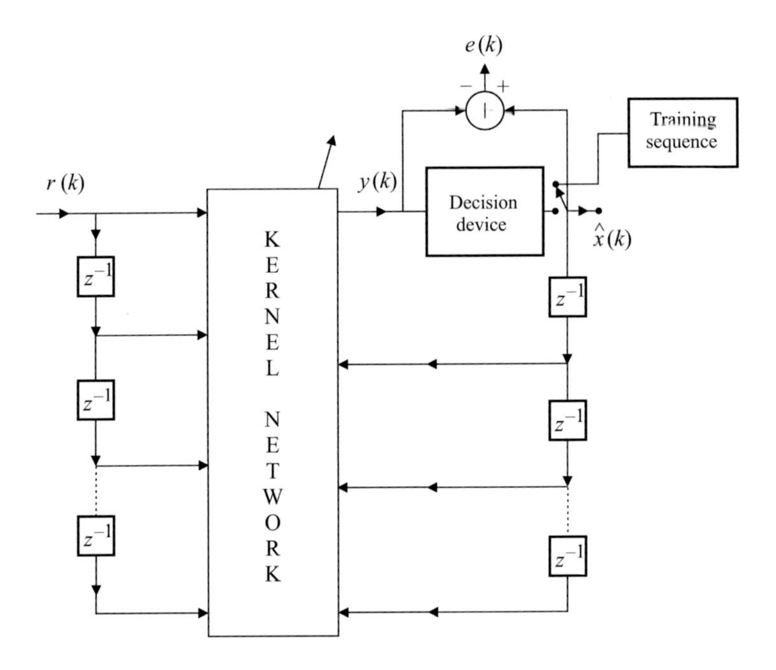

Figure 3.2 Decision feedback equalizer based on kernel.

The channel is corrupted by Gaussian white noise with variance $\sigma_n^2 = -10\,\text{dB}$. The training signal and the actual input signal consist of independent binary samples $\{-1, 1\}$. The training period depends on the algorithm, but our first attempt is to train the equalizer until no errors occur for 50 iterations, and after that one can start normal operation.

(a) Design an equalizer for this problem. Use a filter of appropriate order and plot the learning curves.

(b) Using the same number of adaptive-filter coefficients, implement a DFE equalizer with a normalized LMS algorithm in order to gain insight into the loss of performance of the linear-in-coefficients adaptive filters.

- Utilize the Gaussian kernel.
- Discuss the trade-off between the training and testing periods with respect to their MSEs.

Solution

(a) The results presented refer to the case where the following parameters are employed.

- The number of coefficients used in the feedforward and feedback structures were 8 and 2, respectively.
- The convergence factors in the feedforward and feedback structures were in both cases equal to 0.2.

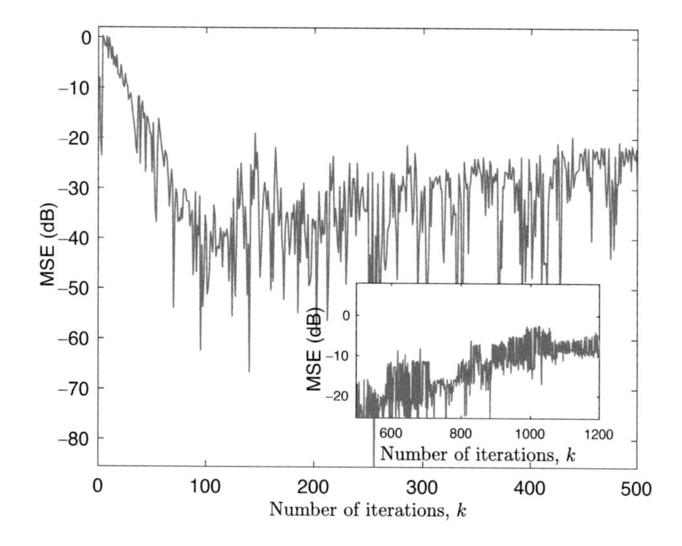

Figure 3.3 DFE results using the KLMS algorithms: The main plot shows the learning curve for the MSE, and the smaller plot depicts the blind period behavior.

- The Gaussian kernel variance was $\sigma_{\mathrm{k}} = 1$, $\gamma_{\mathrm{c}} = 0.99$, and the delay of the training signal was 2.

The dictionary was allowed to have a maximum of $I_{\max} = 200$ entries.

The simulation was set to allow a training period with a duration such that after 50 iterations without a decision error the algorithm would go blindly through a decision feedback equalizer. In a single run, the change from a training period to a DFE required 77 iterations. In the DFE period, the equalizer ran around 3200 iterations until the first decision error occurred. Figure 3.3 illustrates a typical early convergence of the equalization error followed by the period when the blind adaptation starts. Figure 3.3 also depicts the learning curve related to the MSE where we observe a fast convergence and a slow degradation in the equalization error in the blind period.

Figure 3.4 illustrates the decision errors and the accumulated error decisions[4] along the iterations, showing the behavior of the decisions during the training and blind periods. We have also reduced the dictionary size substantially, and it did not have any effect in the equalizer performance, even for $I = 10$, due to the simplicity of the problem at hand.

(b) The attempt to solve the equalization problem utilizing the normalized LMS algorithm failed since the channel is nonlinear. An alternative solution employing the Volterra normalized LMS algorithm can be observed in Figure 3.5 where bit error accumulates continuously much earlier than in the case of the kernel LMS algorithm.

[4] Normalized with respect to the overall number of iterations.

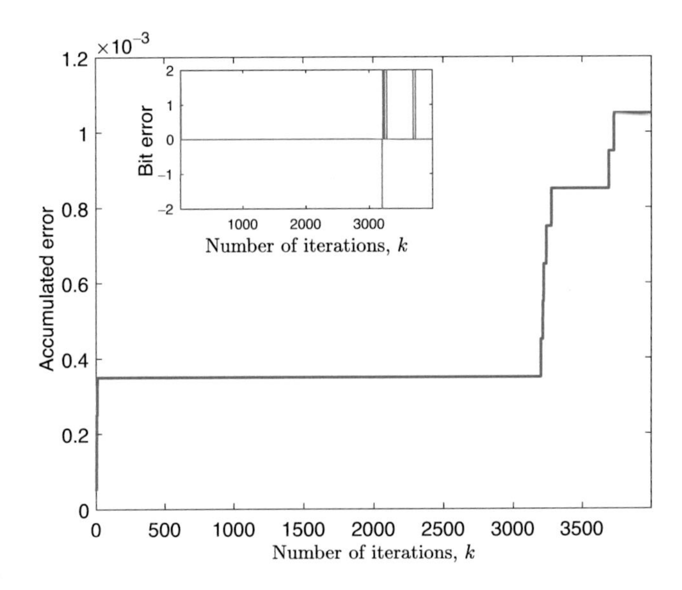

Figure 3.4 Bit error result, DFE example using the KLMS algorithm: The main plot shows the accumulated bit error, and the smaller plot depicts the bit error.

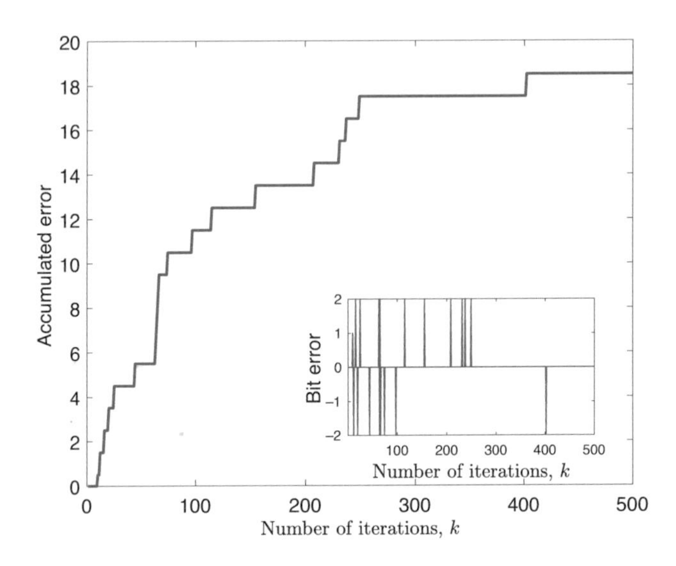

Figure 3.5 Bit error result, DFE example using the Volterra normalized LMS algorithm: accumulated bit error and individual bit error.

Example 3.2 An unknown system described by the model

$$\begin{aligned} d'(k) =& -0.3d'(k-1) + x(k) + 0.04x^2(k) + 0.1x^3(k) \\ d(k) =& d'(k) + n(k) \end{aligned}$$

should be identified utilizing the KLMS algorithm [6]. The additional noise is Gaussian white noise with variance $\sigma_n^2 = -10\,\text{dB}$, and the input signal is also a Gaussian white noise and the variance of $d'(k)$ is equal to one.

In a second experiment, verify the performance of the KLMS algorithm for the case where

$$d(k) = \alpha \cos(d'(k)) + n(k),$$

when the variance of $\alpha \cos(d'(k))$ is also 1.

In all cases utilize the polynomial, the Gaussian and the random Fourier kernels. In the latter case, choose $D = 9$.

Solution

The maximum number of entries in the dictionary was $I_{\max} = 200$. For the KLMS algorithm, we use 10 adaptive-filter coefficients and $\mu = 0.1$. As can be observed in Figure 3.6, the polynomial kernel presented a better performance for this particular example. By contrast, the random Fourier kernel closely follows the Gaussian kernel's performance, not shown here.

Figure 3.7 shows that for the case of the desired signal applying the cosine to $d'(k)$, all kernels have similar best steady-state behavior. However, the random Fourier and Gaussian kernels present the fastest convergence.

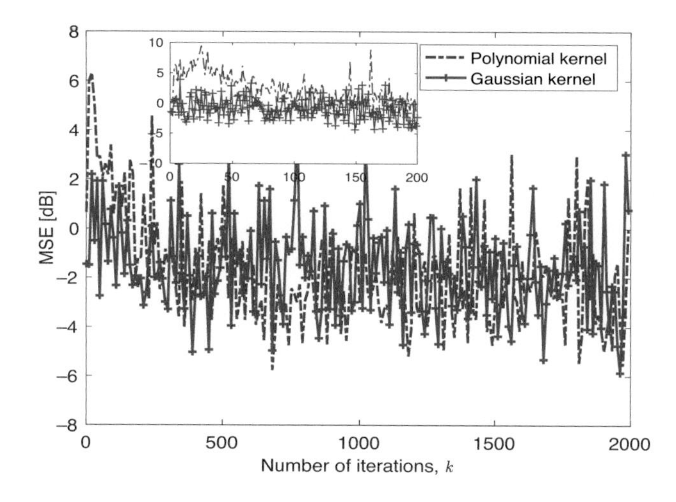

Figure 3.6 System identification example using KLMS algorithm: polynomial kernel parameters $a = n = 1$ and $b = 2$; Gaussian kernel parameter $\sigma^2 = 2$.

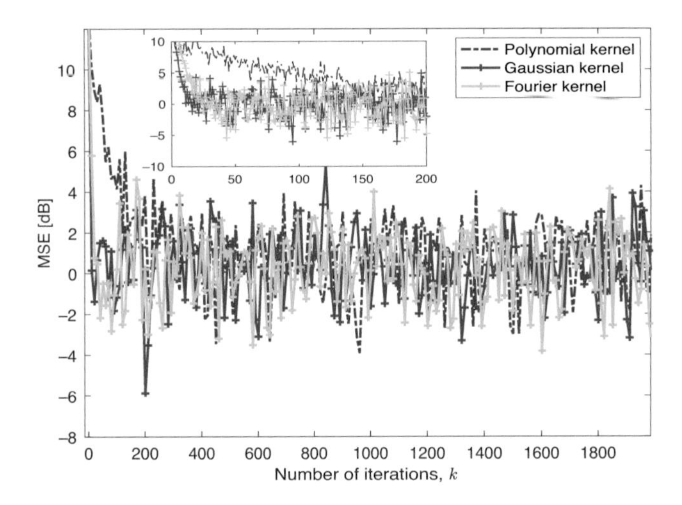

Figure 3.7 System identification example using KLMS algorithm, cosine function case.

3.4 The Kernel Affine Projection Algorithm

As in standard affine projection adaptive filtering algorithms, we assume the last $L+1$ input signal vectors are available and are assembled in their feature space to form the input information matrix as follows:

$$\mathbf{X}_{\mathrm{AP}}(k) = [\boldsymbol{\upsilon}(\mathbf{x}(k))\ \boldsymbol{\upsilon}(\mathbf{x}(k-1)) \ldots \boldsymbol{\upsilon}(\mathbf{x}(k-L))]. \tag{3.39}$$

Similarly, we can define the adaptive-filter output, the desired signal, and the error vectors at a given iteration k as:

$$\mathbf{y}_{\mathrm{AP}}(k) = \mathbf{X}_{\mathrm{AP}}^{\mathrm{T}}(k)\mathbf{w}(k) = \begin{bmatrix} y_{\mathrm{AP},0}(k) \\ y_{\mathrm{AP},1}(k) \\ \vdots \\ y_{\mathrm{AP},L}(k) \end{bmatrix}; \tag{3.40}$$

$$\mathbf{d}_{\mathrm{AP}}(k) = \begin{bmatrix} d(k) \\ d(k-1) \\ \vdots \\ d(k-L) \end{bmatrix}; \tag{3.41}$$

$$\mathbf{e}_{\mathrm{AP}}(k) = \begin{bmatrix} e_{\mathrm{AP},0}(k) \\ e_{\mathrm{AP},1}(k) \\ \vdots \\ e_{\mathrm{AP},L}(k) \end{bmatrix} = \begin{bmatrix} d(k) - y_{\mathrm{AP},0}(k) \\ d(k-1) - y_{\mathrm{AP},1}(k) \\ \vdots \\ d(k-L) - y_{\mathrm{AP},L}(k) \end{bmatrix} = \mathbf{d}_{\mathrm{AP}}(k) - \mathbf{y}_{\mathrm{AP}}(k), \tag{3.42}$$

respectively. Notice that, for the sake of simplicity, we are now using the notation $\boldsymbol{v}^{\mathrm{T}}(\mathbf{x}(k))\,\mathbf{w}(k)$ to also denote the inner product $\langle \boldsymbol{v}(\mathbf{x}(k)), \mathbf{w}(k)\rangle$ in the feature space $\mathcal{H}$.

An important ingredient of the KAP algorithm is the kernel matrix defined at the present discussion as:

$$\begin{aligned}\underline{\boldsymbol{K}}_{\mathrm{AP}}(k) =& \mathbf{X}_{\mathrm{AP}}^{\mathrm{T}}(k)\mathbf{X}_{\mathrm{AP}}(k)\\ =& [\boldsymbol{v}(\mathbf{x}(k))\ \boldsymbol{v}(\mathbf{x}(k-1))\ldots\boldsymbol{v}(\mathbf{x}(k-L))]^{\mathrm{T}}\\ & [\boldsymbol{v}(\mathbf{x}(k))\ \boldsymbol{v}(\mathbf{x}(k-1))\ldots\boldsymbol{v}(\mathbf{x}(k-L))]\\ =& \begin{bmatrix} \kappa(\mathbf{x}(k),\mathbf{x}(k)) & \kappa(\mathbf{x}(k),\mathbf{x}(k-1)) & \cdots & \kappa(\mathbf{x}(k),\mathbf{x}(k-L))\\ \kappa(\mathbf{x}(k-1),\mathbf{x}(k)) & \kappa(\mathbf{x}(k-1),\mathbf{x}(k-1)) & \cdots & \kappa(\mathbf{x}(k-1),\mathbf{x}(k-L))\\ \vdots & \vdots & \ddots & \vdots\\ \kappa(\mathbf{x}(k-L),\mathbf{x}(k)) & \kappa(\mathbf{x}(k-L),\mathbf{x}(k-1)) & \cdots & \kappa(\mathbf{x}(k-L),\mathbf{x}(k-L)) \end{bmatrix}.\end{aligned} \tag{3.43}$$

As in the standard affine projection algorithm, the objective here is to minimize

$$\begin{aligned}&\frac{1}{2}\|\mathbf{w}(k+1)-\mathbf{w}(k)\|^2\\ &\text{subject to :}\\ &\mathbf{d}_{\mathrm{AP}}(k)-\mathbf{X}_{\mathrm{AP}}^{\mathrm{T}}(k)\mathbf{w}(k+1)=\mathbf{0},\end{aligned} \tag{3.44}$$

with respect to $\mathbf{w}(k+1)$. The solution to this constrained minimization problem is derived through its transformation into an unconstrained minimization utilizing the method of Lagrange multipliers.

The resulting update equation resembles the conventional affine projection algorithm [6] with unity convergence factor. A trade-off between final misadjustment and convergence speed is achieved through the introduction of a convergence factor as follows:

$$\begin{aligned}\mathbf{w}(k+1) =& \mathbf{w}(k)+\mu\mathbf{X}_{\mathrm{AP}}(k)\left(\mathbf{X}_{\mathrm{AP}}^{\mathrm{T}}(k)\mathbf{X}_{\mathrm{AP}}(k)\right)^{-1}\mathbf{e}_{\mathrm{AP}}(k)\\ =& \mathbf{w}(k)+\mu\mathbf{X}_{\mathrm{AP}}(k)\underline{\boldsymbol{K}}_{\mathrm{AP}}^{-1}(k)\mathbf{e}_{\mathrm{AP}}(k).\end{aligned} \tag{3.45}$$

The distinct feature of the update Equation (3.45) is that the output signal is computed in the feature space as

$$y(k) = \boldsymbol{v}^{\mathrm{T}}(\mathbf{x}(k))\,\mathbf{w}(k), \tag{3.46}$$

and the information matrix $\mathbf{X}_{\mathrm{AP}}(k)$ is also computed in the feature space as per (3.39).

The obvious drawback with this formulation is that the output signal is calculated in a high-dimensional space where, in the case we represent the filter in vector form, the number of entries of the filter coefficients of $\mathbf{w}(k)$ might tend to infinite. A way around this problem is to define a representation for the filter coefficient vector as a linear combination of the feature vector computed at prescribed or pre-chosen set of $I+1$ vectors $\mathbf{z}(i)$ in the following form

$$\mathbf{w}(k)=\sum_{i=0}^{I}\beta_i(k)\boldsymbol{\upsilon}\left(\mathbf{z}(i)\right), \tag{3.47}$$

where the vectors $\mathbf{z}(i)$ represent a subset of the vectors acquired so far, that is, a subset of $\mathbf{x}(l)$ for $l=0,1,2,\ldots,k-1$. A matrix formed by the vectors $\mathbf{z}(i)$ forms a dictionary whose choice should be discussed separately and is denoted by

$$\mathbf{X}_{\text{dic}}(k)=[\mathbf{z}(0)\ \mathbf{z}(1)\ \mathbf{z}(2)\ldots\mathbf{z}(I)]. \tag{3.48}$$

By employing this formulation, we have

$$\begin{aligned}y(k)=&\sum_{i=0}^{I}\beta_i(k)\boldsymbol{\upsilon}^{\mathrm{T}}\left(\mathbf{x}(k)\right)\boldsymbol{\upsilon}\left(\mathbf{z}(i)\right)\\=&\sum_{i=0}^{I}\beta_i(k)\kappa\left(\mathbf{x}(k),\mathbf{z}(i)\right)\\=&\boldsymbol{\kappa}^{\mathrm{T}}(\mathbf{x}(k),\cdot)\boldsymbol{\beta}(k),\end{aligned} \tag{3.49}$$

where $\boldsymbol{\kappa}(\mathbf{x}(k),\cdot)$ is the vector of kernel elements, and $\boldsymbol{\beta}(k)$ is the vector of weights that are applied to the kernel elements.

It should be noticed that whenever $\mathbf{X}_{\text{dic}}(k)$ is updated, the vector $\boldsymbol{\kappa}(\mathbf{x}(k),\cdot)$ should also be updated. In general, the size of the dictionary must increase whenever the maximum value of $|\kappa\left(\mathbf{x}(k),\mathbf{z}(i)\right)|<\gamma_{\text{c}}$, where γ_{c} represents a threshold in the range $\gamma_{\text{c}}\in[0,1]$ related to the sparseness of the input signal, see Equation (3.28). The threshold parameter has some relation with the sparsity and coherence of the dictionary of kernel functions [14, 17].

But assembling the last $L+1$ error signals measured using the last $L+1$ adaptive filter output signals as in (3.49), we get

$$\begin{aligned}\mathbf{e}_{\text{AP}}(k)=&\mathbf{d}_{\text{AP}}(k)-\mathbf{y}_{\text{AP}}(k)\\=&\mathbf{d}_{\text{AP}}(k)-\boldsymbol{K}_{\text{AP}}^{\mathrm{T}}(k)\boldsymbol{\beta}(k),\end{aligned} \tag{3.50}$$

where in this case the kernel matrix of dimension $(I+1)\times(L+1)$ is defined as

$$\begin{aligned}\boldsymbol{K}_{\text{AP}}(k)&=\begin{bmatrix}\boldsymbol{\upsilon}^{\mathrm{T}}\left(\mathbf{z}(0)\right)\\\boldsymbol{\upsilon}^{\mathrm{T}}\left(\mathbf{z}(1)\right)\\\vdots\\\boldsymbol{\upsilon}^{\mathrm{T}}\left(\mathbf{z}(I)\right)\end{bmatrix}\left[\boldsymbol{\upsilon}\left(\mathbf{x}(k)\right)\ \boldsymbol{\upsilon}\left(\mathbf{x}(k-1)\right)\ldots\boldsymbol{\upsilon}\left(\mathbf{x}(k-L)\right)\right]\\&=\begin{bmatrix}\kappa\left(\mathbf{z}(0),\mathbf{x}(k)\right)&\kappa\left(\mathbf{z}(0),\mathbf{x}(k-1)\right)&\cdots&\kappa\left(\mathbf{z}(0),\mathbf{x}(k-L)\right)\\\kappa\left(\mathbf{z}(1),\mathbf{x}(k)\right)&\kappa\left(\mathbf{z}(1),\mathbf{x}(k-1)\right)&\cdots&\kappa\left(\mathbf{z}(1),\mathbf{x}(k-L)\right)\\\vdots&\vdots&\ddots&\vdots\\\kappa\left(\mathbf{z}(I),\mathbf{x}(k)\right)&\kappa\left(\mathbf{z}(I),\mathbf{x}(k-1)\right)&\cdots&\kappa\left(\mathbf{z}(I),\mathbf{x}(k-L)\right)\end{bmatrix}.\end{aligned} \tag{3.51}$$

In the notation above the vectors $\mathbf{z}(i)$, for $i = 0, 1, \ldots, I$, represent the elements of the dictionary that is built online, where $\mathbf{z}(0)$ represents the most recent one.

In the KAP algorithm, we replace this representation in the standard affine projection formulation of (3.44), where the coefficients to be updated are assembled in the vector $\boldsymbol{\beta}(k)$. This vector of the KAP algorithm is called kernel weight vector, whose update recursion has the following form:

$$\boldsymbol{\beta}(k+1) = \boldsymbol{\beta}(k) + \mu \boldsymbol{K}_{\mathrm{AP}}(k) \left(\boldsymbol{K}_{\mathrm{AP}}^{\mathrm{T}}(k) \boldsymbol{K}_{\mathrm{AP}}(k) + \gamma \mathbf{I} \right)^{-1} \mathbf{e}_{\mathrm{AP}}(k). \tag{3.52}$$

Note that μ represents the convergence factor for an affine projection algorithm. As usual, the matrix to be inverted includes a regularization factor to avoid numerical difficulties.

During the initialization process as well as during the normal operation, we need to invert

$$\mathbf{A}_{\mathrm{AP}}(k) = \left(\boldsymbol{K}_{\mathrm{AP}}^{\mathrm{T}}(k) \boldsymbol{K}_{\mathrm{AP}}(k) + \gamma \mathbf{I} \right), \tag{3.53}$$

with the knowledge of the inverse of $\mathbf{A}_{\mathrm{AP}}(k-1)$. It is straightforward to show that the $L \times L$ sub-matrix sitting on the upper-left corner of $\mathbf{A}_{\mathrm{AP}}(k-1)$ is related to the $L \times L$ sub-matrix sitting on the lower-right corner of $\mathbf{A}_{\mathrm{AP}}(k)$, allowing for some computational savings in the matrix inversion. This property originates from the fact that

$$\boldsymbol{K}_{\mathrm{AP}}(k) = \begin{bmatrix} \kappa(\mathbf{z}(0), \mathbf{x}(k)) & \boldsymbol{\kappa}_{\mathrm{AP}}^{\mathrm{T}}(\mathbf{z}(I), \mathbf{x}(k)) \\ \boldsymbol{\kappa}_{\mathrm{AP}}(\mathbf{z}(I), \mathbf{x}(k)) & \bar{\boldsymbol{K}}_{\mathrm{AP}}(k-1) \end{bmatrix}, \tag{3.54}$$

where $\boldsymbol{\kappa}_{\mathrm{AP}}(\mathbf{z}(I), \mathbf{x}(k)) = [\kappa(\mathbf{z}(1), \mathbf{x}(k)), \ldots, \kappa(\mathbf{z}(I-1), \mathbf{x}(k)), \kappa(\mathbf{z}(I), \mathbf{x}(k))]^{\mathrm{T}}$ and $\bar{\boldsymbol{K}}_{\mathrm{AP}}(k-1)$ represents the $L \times L$ upper-left corner of $\boldsymbol{K}_{\mathrm{AP}}(k-1)$[5].

The task now is to compute $\mathbf{A}_{\mathrm{AP}}^{-1}(k)$ assuming we know $\mathbf{A}_{\mathrm{AP}}^{-1}(k-1)$ utilizing an appropriate inversion formula. Let's define

$$\mathbf{A}_{\mathrm{AP}}(k) = \begin{bmatrix} a & \mathbf{b}^{\mathrm{T}} \\ \mathbf{b} & \tilde{\mathbf{C}} + \mathbf{b}\mathbf{b}^{\mathrm{T}} \end{bmatrix} = \begin{bmatrix} a & \mathbf{b}^{\mathrm{T}} \\ \mathbf{b} & \mathbf{C} \end{bmatrix}; \tag{3.55}$$

$$\mathbf{A}_{\mathrm{AP}}(k-1) = \begin{bmatrix} \bar{\mathbf{C}} & \bar{\mathbf{b}} \\ \bar{\mathbf{b}}^{\mathrm{T}} & \bar{a} \end{bmatrix}, \tag{3.56}$$

where in the vectors and matrices on the right-hand side we omit the time index.

Let the inverse of $\mathbf{A}_{\mathrm{AP}}(k-1)$ be given by

$$\mathbf{A}_{\mathrm{AP}}^{-1}(k-1) = \begin{bmatrix} \hat{\mathbf{A}} & \hat{\mathbf{b}} \\ \hat{\mathbf{b}}^{\mathrm{T}} & \hat{c} \end{bmatrix}, \tag{3.57}$$

[5] Here, we assume that the terms retained in $\bar{\boldsymbol{K}}_{\mathrm{AP}}(k-1)$ involve the selected elements $\mathbf{z}(\cdot)$ from the previous iteration whose the indexes are renumbered from 1 to I.

where (3.56) and (3.57) are inverse of each other if

$$\tilde{\mathbf{C}}\hat{\mathbf{A}} + \bar{\mathbf{b}}\hat{\mathbf{b}}^{\mathrm{T}} = \mathbf{I}; \tag{3.58}$$

$$\bar{\mathbf{b}}^{\mathrm{T}}\hat{\mathbf{b}} + \bar{a}\hat{c} = 1; \tag{3.59}$$

$$\tilde{\mathbf{C}}\hat{\mathbf{b}} + \bar{\mathbf{b}}\hat{c} = \mathbf{0}; \tag{3.60}$$

$$\bar{\mathbf{b}}^{\mathrm{T}}\hat{\mathbf{A}} + \bar{a}\hat{\mathbf{b}}^{\mathrm{T}} = \mathbf{0}. \tag{3.61}$$

From these equalities, it is possible to show that

$$\tilde{\mathbf{C}}^{-1} = \hat{\mathbf{A}} - \frac{\hat{\mathbf{b}}\hat{\mathbf{b}}^{\mathrm{T}}}{\hat{c}}. \tag{3.62}$$

However, we require the inverse of $\mathbf{C} = \tilde{\mathbf{C}} + \mathbf{b}\mathbf{b}^{\mathrm{T}}$ which can be calculated using the matrix inversion lemma as

$$\mathbf{C}^{-1} = \tilde{\mathbf{C}}^{-1} - \frac{1}{1 + \mathbf{b}^{\mathrm{T}}\tilde{\mathbf{C}}^{-1}\mathbf{b}}\left[\tilde{\mathbf{C}}^{-1} - \tilde{\mathbf{C}}^{-1}\mathbf{b}\mathbf{b}^{\mathrm{T}}\tilde{\mathbf{C}}^{-1}\right]. \tag{3.63}$$

After a few manipulations, one can conclude that

$$\mathbf{A}_{\mathrm{AP}}^{-1}(k) = \frac{1}{f}\begin{bmatrix} 1 & -\left(\mathbf{C}^{-1}\mathbf{b}\right)^{\mathrm{T}} \\ -\mathbf{C}^{-1}\mathbf{b} & \mathbf{C}^{-1}\left[\mathbf{I}f + \mathbf{b}\left(\mathbf{C}^{-1}\mathbf{b}\right)^{\mathrm{T}}\right] \end{bmatrix}, \tag{3.64}$$

where $f = a - \mathbf{b}^{\mathrm{T}}\mathbf{C}^{-1}\mathbf{b}$.

During initialization, we can define a growing kernel matrix denoted as $\boldsymbol{K}_{l,\mathrm{AP}}(l)$, where $l+1$ is the dimension of the growing matrix. For instance, when the first data pair is received, the following evaluations should be performed.

$$\boldsymbol{K}_{0,\mathrm{AP}}(0) = \kappa(\mathbf{x}(0), \mathbf{x}(0)); \tag{3.65}$$

$$\boldsymbol{K}_{0,\mathrm{AP}}^{-1}(0) = \frac{1}{\kappa(\mathbf{x}(0), \mathbf{x}(0))}; \tag{3.66}$$

$$\mathbf{A}_{\mathrm{AP}}^{-1}(0) = \frac{1}{\kappa^2(\mathbf{x}(0), \mathbf{x}(0))}. \tag{3.67}$$

Then, for the next L samples, we will complete the algorithm matrices and vectors, for $l = 1, \ldots, L$, as follows:

$$\boldsymbol{K}_{l,\mathrm{AP}}(l) = \begin{bmatrix} \kappa(\mathbf{x}(l), \mathbf{x}(l)) & \boldsymbol{\kappa}_{\mathrm{AP}}^{\mathrm{T}}(\mathbf{z}(l), \mathbf{x}(l)) \\ \boldsymbol{\kappa}_{\mathrm{AP}}(\mathbf{z}(l)), \mathbf{x}(l)) & \bar{\boldsymbol{K}}_{l-1,\mathrm{AP}}(l-1) \end{bmatrix}; \tag{3.68}$$

$$\mathbf{A}_{\mathrm{AP}}(l) = \left(\boldsymbol{K}_{l,\mathrm{AP}}^{\mathrm{T}}(k)\boldsymbol{K}_{l,\mathrm{AP}}(k) + \gamma\mathbf{I}\right); = \begin{bmatrix} a & \mathbf{b}^{\mathrm{T}} \\ \mathbf{b} & \mathbf{C} \end{bmatrix}; \tag{3.69}$$

$$\tilde{\mathbf{C}}^{-1} = \mathbf{A}_{\mathrm{AP}}^{-1}(l-1), \text{ and Equation (3.63)}; \tag{3.70}$$

$$\mathbf{A}_{\mathrm{AP}}^{-1}(l) = \frac{1}{f}\begin{bmatrix} 1 & -\left(\mathbf{C}^{-1}\mathbf{b}\right)^{\mathrm{T}} \\ -\mathbf{C}^{-1}\mathbf{b} & \mathbf{C}^{-1}\left[\mathbf{I}f + \mathbf{b}\left(\mathbf{C}^{-1}\mathbf{b}\right)^{\mathrm{T}}\right] \end{bmatrix}, \tag{3.71}$$

where all vectors and matrices during the initialization process are growing in dimension. In the case of $\boldsymbol{\beta}(l+1)$, the update equation originates from the solution of the following constrained minimization:

$$\begin{aligned} &\underset{\boldsymbol{\beta}(l+1)}{\text{minimize}} \frac{1}{2}\|\boldsymbol{\beta}(l+1) - \begin{bmatrix} 0 \\ \boldsymbol{\beta}(l) \end{bmatrix}\|^2 \\ &\text{subject to} \\ &\mathbf{d}_{\mathrm{AP}}(k) - \boldsymbol{K}_{l,\mathrm{AP}}^{\mathrm{T}}(l)\boldsymbol{\beta}(l+1) = \mathbf{0}, \end{aligned} \tag{3.72}$$

whose solution, after including a convergence factor, is

$$\boldsymbol{\beta}(l+1) = \begin{bmatrix} 0 \\ \boldsymbol{\beta}(l) \end{bmatrix} + \mu \boldsymbol{K}_{l,\mathrm{AP}}(k)\mathbf{A}_{\mathrm{AP}}^{-1}(l)\mathbf{e}_{\mathrm{AP}}(l), \tag{3.73}$$

where the error vector in this case is

$$\mathbf{e}_{\mathrm{AP}}(l) = \mathbf{d}_{\mathrm{AP}}(l) - \boldsymbol{K}_{l,\mathrm{AP}}^{\mathrm{T}}(l)\begin{bmatrix} 0 \\ \boldsymbol{\beta}(l) \end{bmatrix}. \tag{3.74}$$

Algorithm 3.13 outlines the KAP algorithm including a regularization factor γ as well as an input data selection method, also known as sparsification, in order to avoid that the number of input signal vectors information grows without bound. Typically whenever a new input data vector is acquired, we should verify if it brings about enough innovation in order to justify its inclusion in the set of input vectors we keep in the dictionary. If the newly acquired data

---

Algorithm 3.13 The kernel affine projection algorithm I

Initialization

$\mathbf{x}(-1) = \boldsymbol{\beta}(0) = [0\,0\ldots 0]^{\mathrm{T}}$

choose μ in the range $0 < \mu \leq 1$

choose γ_{d}, γ_{e}, and γ = small constant

$\boldsymbol{K}_{0,\mathrm{AP}}(0) = \kappa(\mathbf{x}(0),\mathbf{x}(0)) \quad \boldsymbol{K}_{0,\mathrm{AP}}^{-1}(0) = \frac{1}{\kappa(\mathbf{X}(0),\mathbf{X}(0))}$ (3.65) and (3.66)

$\mathbf{A}_{\mathrm{AP}}^{-1}(0) = \frac{1}{\kappa^2(\mathbf{X}(0),\mathbf{X}(0))}$ (3.67)

Do for $l = 1,\ldots,L$

$$\boldsymbol{K}_{l,\mathrm{AP}}(l) = \begin{bmatrix} \kappa(\mathbf{z}(0),\mathbf{x}(l)) & \boldsymbol{\kappa}_{\mathrm{AP}}^{\mathrm{T}}(\mathbf{z}(l),\mathbf{x}(l)) \\ \boldsymbol{\kappa}_{\mathrm{AP}}(\mathbf{z}(l),\mathbf{x}(l)) & \boldsymbol{K}_{l-1,\mathrm{AP}}(l-1) \end{bmatrix} \tag{3.68}$$

$$\mathbf{A}_{\mathrm{AP}}(l) = \left(\boldsymbol{K}_{l,\mathrm{AP}}^{\mathrm{T}}(l)\boldsymbol{K}_{l,\mathrm{AP}}(l) + \gamma\mathbf{I}\right) = \begin{bmatrix} a & \mathbf{b}^{\mathrm{T}} \\ \mathbf{b} & \mathbf{C} \end{bmatrix} \tag{3.69}$$

$$\tilde{\mathbf{C}}^{-1} = \mathbf{A}_{\mathrm{AP}}^{-1}(l-1) \tag{3.70}$$

$$\mathbf{C}^{-1} = \tilde{\mathbf{C}}^{-1} - \frac{1}{1+\mathbf{b}^{\mathrm{T}}\tilde{\mathbf{C}}^{-1}\mathbf{b}}\left[\tilde{\mathbf{C}}^{-1} - \tilde{\mathbf{C}}^{-1}\mathbf{b}\mathbf{b}^{\mathrm{T}}\tilde{\mathbf{C}}^{-1}\right] \tag{3.63}$$

$f = a - \mathbf{b}^{\mathrm{T}}\mathbf{C}^{-1}\mathbf{b}$

$$\mathbf{A}_{\mathrm{AP}}^{-1}(l) = \frac{1}{f}\begin{bmatrix} 1 & -\left(\mathbf{C}^{-1}\mathbf{b}\right)^{\mathrm{T}} \\ -\mathbf{C}^{-1}\mathbf{b} & \mathbf{C}^{-1}\left[\mathbf{I}f + \mathbf{b}\left(\mathbf{C}^{-1}\mathbf{b}\right)^{\mathrm{T}}\right] \end{bmatrix} \tag{3.71}$$

$$\mathbf{e}_{\mathrm{AP}}(l) = \mathbf{d}_{\mathrm{AP}}(l) - \boldsymbol{K}_{l,\mathrm{AP}}^{\mathrm{T}}(l)\begin{bmatrix} 0 \\ \boldsymbol{\beta}(l) \end{bmatrix} \tag{3.74}$$

$$\boldsymbol{\beta}(l+1) = \begin{bmatrix} 0 \\ \boldsymbol{\beta}(l) \end{bmatrix} + \mu\boldsymbol{K}_{l,\mathrm{AP}}(l)\mathbf{A}_{\mathrm{AP}}^{-1}(l)\mathbf{e}_{\mathrm{AP}}(l) \tag{3.73}$$

End do

$I = L$

For do $k \geq L+1$

If $\underset{l \in \mathcal{L}_I(k-1)}{\text{maximum}} |\kappa(\mathbf{x}(k), \mathbf{x}(k-l))| \leq \gamma_c$ and $I \leq I_{\max}$ (3.28)

$l_0(k+1) = k+1$, including $\mathbf{x}(k)$ in the dictionary

$$\boldsymbol{K}_{\text{AP}}(k) = \begin{bmatrix} \kappa(\mathbf{z}(0), \mathbf{x}(k)) & \boldsymbol{\kappa}_{\text{AP}}^{\text{T}}(\mathbf{z}(I)), \mathbf{x}(k)) \\ \boldsymbol{\kappa}_{\text{AP}}(\mathbf{z}(I), \mathbf{x}(k)) & \boldsymbol{K}_{\text{AP}}(k-1) \end{bmatrix} \quad (3.54)$$

$$\mathbf{e}_{\text{AP}}(k) = \mathbf{d}_{\text{AP}}(k) - \boldsymbol{K}_{\text{AP}}^{\text{T}}(k) \begin{bmatrix} 0 \\ \boldsymbol{\beta}(k) \end{bmatrix} \quad (3.74)$$

$$\mathbf{A}_{\text{AP}}(k) = \left(\boldsymbol{K}_{\text{AP}}^{\text{T}}(k)\boldsymbol{K}_{\text{AP}}(k) + \gamma\mathbf{I}\right) = \begin{bmatrix} a & \mathbf{b}^{\text{T}} \\ \mathbf{b} & \mathbf{C} \end{bmatrix}$$

$$\mathbf{A}_{\text{AP}}^{-1}(k-1) = \begin{bmatrix} \hat{\mathbf{A}} & \hat{\mathbf{b}} \\ \hat{\mathbf{b}}^{\text{T}} & \hat{c} \end{bmatrix} \quad (3.57)$$

$$\tilde{\mathbf{C}}^{-1} = \hat{\mathbf{A}} - \frac{\hat{\mathbf{b}}\hat{\mathbf{b}}^{\text{T}}}{\hat{c}} \quad (3.62)$$

$$\mathbf{C}^{-1} = \tilde{\mathbf{C}}^{-1} - \frac{1}{1+\mathbf{b}^{\text{T}}\tilde{\mathbf{C}}^{-1}\mathbf{b}} \left[\tilde{\mathbf{C}}^{-1} - \tilde{\mathbf{C}}^{-1}\mathbf{b}\mathbf{b}^{\text{T}}\tilde{\mathbf{C}}^{-1}\right] \quad (3.63)$$

$$f = a - \mathbf{b}^{\text{T}}\mathbf{C}^{-1}\mathbf{b}$$

$$\mathbf{A}_{\text{AP}}^{-1}(k) = \frac{1}{f} \begin{bmatrix} 1 & -\left(\mathbf{C}^{-1}\mathbf{b}\right)^{\text{T}} \\ -\mathbf{C}^{-1}\mathbf{b} & \mathbf{C}^{-1} + \mathbf{C}^{-1}\left[\mathbf{I}f + \mathbf{b}\left(\mathbf{C}^{-1}\mathbf{b}\right)^{\text{T}}\right] \end{bmatrix} \quad (3.64)$$

$$\boldsymbol{\beta}(k+1) = \begin{bmatrix} 0 \\ \boldsymbol{\beta}(k) \end{bmatrix} + \mu \boldsymbol{K}_{\text{AP}}(k)\mathbf{A}_{\text{AP}}^{-1}(k)\mathbf{e}_{\text{AP}}(k) \quad (3.73)$$

$I = I + 1$

Else

$\boldsymbol{K}_{\text{AP}}(k) = \boldsymbol{K}_{\text{AP}}(k-1)$, $\mathbf{A}_{\text{AP}}^{-1}(k) = \mathbf{A}_{\text{AP}}^{-1}(k-1)$

$$\mathbf{e}_{\text{AP}}(k) = \mathbf{d}_{\text{AP}}(k) - \boldsymbol{K}_{\text{AP}}^{\text{T}}(k)\boldsymbol{\beta}(k) \quad (3.50)$$

$$\boldsymbol{\beta}(k+1) = \boldsymbol{\beta}(k) + \mu \boldsymbol{K}_{\text{AP}}(k)\mathbf{A}_{\text{AP}}^{-1}(k)\mathbf{e}_{\text{AP}}(k) \quad (3.52)$$

End if

End do

---

is considered important, it should be added to the current dictionary whose number of members should be increased to $I+1$. Algorithm 3.13 applies the sparsification through the coherence approach of (3.28). For a dictionary with a prescribed maximum number of entries, denoted as $I_{\max}$, the strategy described in Algorithm 3.13 is to keep the most innovative input data observed up to the iteration where $I_{\max}$ is reached.

Algorithm 3.14 describes an alternative implementation where the sparsification is performed utilizing the strategies based on Equations (3.26) and (3.27), respectively. Also for this algorithm, we increase the dictionary size up to a prescribed value $I_{\max}$.

After the dictionary size has reached its maximum prescribed value, we have the option to replace an old entry of the dictionary by the new one, see the discussion in Subsection 3.5.3. The procedure entails the inclusion of the newly acquired data if it is considered important, by removing another less important input signal vector from the dictionary in order to keep the number of members limited. Two criteria are usually employed: one is how distant from each of the current dictionary members is the newly incoming data; if the new data is not

Algorithm 3.14 The kernel affine projection algorithm II

Initialization

$\mathbf{x}(-1) = \boldsymbol{\beta}(0) = [0\,0\ldots 0]^{\mathrm{T}}$

choose μ in the range $0 < \mu \leq 1$

choose γ_{d}, γ_{e}, and $\gamma =$ small constant

$\boldsymbol{K}_{0,\mathrm{AP}}(0) = \kappa(\mathbf{x}(0), \mathbf{x}(0)) \quad \boldsymbol{K}^{-1}_{0,\mathrm{AP}}(0) = \frac{1}{\kappa(\mathbf{X}(0),\mathbf{X}(0))}$ (3.65) and (3.66)

$\mathbf{A}^{-1}_{\mathrm{AP}}(0) = \frac{1}{\kappa^2(\mathbf{X}(0),\mathbf{X}(0))}$ (3.67)

Do for $l = 1, \ldots, L$

$$\boldsymbol{K}_{l,\mathrm{AP}}(l) = \begin{bmatrix} \kappa(\mathbf{z}(0), \mathbf{x}(l)) & \boldsymbol{\kappa}^{\mathrm{T}}_{\mathrm{AP}}(\mathbf{z}(l), \mathbf{x}(l)) \\ \boldsymbol{\kappa}_{\mathrm{AP}}(\mathbf{z}(l), \mathbf{x}(l)) & \boldsymbol{K}_{l-1,\mathrm{AP}}(l-1) \end{bmatrix} \quad (3.68)$$

$$\mathbf{A}_{\mathrm{AP}}(l) = \left(\boldsymbol{K}^{\mathrm{T}}_{l,\mathrm{AP}}(l)\boldsymbol{K}_{l,\mathrm{AP}}(l) + \gamma\mathbf{I}\right) = \begin{bmatrix} a & \mathbf{b}^{\mathrm{T}} \\ \mathbf{b} & \mathbf{C} \end{bmatrix} \quad (3.69)$$

$$\tilde{\mathbf{C}}^{-1} = \mathbf{A}^{-1}_{\mathrm{AP}}(l-1) \quad (3.70)$$

$$\mathbf{C}^{-1} = \tilde{\mathbf{C}}^{-1} - \frac{1}{1+\mathbf{b}^{\mathrm{T}}\tilde{\mathbf{C}}^{-1}\mathbf{b}}\left[\tilde{\mathbf{C}}^{-1} - \tilde{\mathbf{C}}^{-1}\mathbf{b}\mathbf{b}^{\mathrm{T}}\tilde{\mathbf{C}}^{-1}\right] \quad (3.63)$$

$$f = a - \mathbf{b}^{\mathrm{T}}\mathbf{C}^{-1}\mathbf{b}$$

$$\mathbf{A}^{-1}_{\mathrm{AP}}(l) = \frac{1}{f}\begin{bmatrix} 1 & -\left(\mathbf{C}^{-1}\mathbf{b}\right)^{\mathrm{T}} \\ -\mathbf{C}^{-1}\mathbf{b} & \mathbf{C}^{-1}\left[\mathbf{I}f + \mathbf{b}\left(\mathbf{C}^{-1}\mathbf{b}\right)^{\mathrm{T}}\right] \end{bmatrix} \quad (3.71)$$

$$\mathbf{e}_{\mathrm{AP}}(l) = \mathbf{d}_{\mathrm{AP}}(l) - \boldsymbol{K}^{\mathrm{T}}_{l,\mathrm{AP}}(l)\begin{bmatrix} 0 \\ \boldsymbol{\beta}(l) \end{bmatrix} \quad (3.74)$$

$$\boldsymbol{\beta}(l+1) = \begin{bmatrix} 0 \\ \boldsymbol{\beta}(l) \end{bmatrix} + \mu\boldsymbol{K}_{l,\mathrm{AP}}(l)\mathbf{A}^{-1}_{\mathrm{AP}}(l)\mathbf{e}_{\mathrm{AP}}(l) \quad (3.73)$$

End do

$I = L$

For do $k \geq L+1$

If $\|\mathbf{x}(k) - \mathbf{x}(l)\| > \gamma_{\mathrm{d}}$ for $l \in \mathcal{L}_I(k-1)$ and $I \leq I_{\max}$

$$\boldsymbol{K}_{\mathrm{AP}}(k) = \begin{bmatrix} \kappa(\mathbf{z}(0), \mathbf{x}(k)) & \boldsymbol{\kappa}^{\mathrm{T}}_{\mathrm{AP}}(\mathbf{z}(I)), \mathbf{x}(k)) \\ \boldsymbol{\kappa}_{\mathrm{AP}}(\mathbf{z}(I), \mathbf{x}(k)) & \boldsymbol{K}_{\mathrm{AP}}(k-1) \end{bmatrix} \quad (3.54)$$

$$\mathbf{e}_{\mathrm{AP}}(k) = \mathbf{d}_{\mathrm{AP}}(k) - \boldsymbol{K}^{\mathrm{T}}_{\mathrm{AP}}(k)\begin{bmatrix} 0 \\ \boldsymbol{\beta}(k) \end{bmatrix} \quad (3.74)$$

If $\|\mathbf{e}_{\mathrm{AP}}(k)\| > \gamma_{\mathrm{e}}$

$l_0(k+1) = k+1$, including $\mathbf{x}(k)$ in the dictionary

$$\mathbf{A}_{\mathrm{AP}}(k) = \left(\boldsymbol{K}^{\mathrm{T}}_{\mathrm{AP}}(k)\boldsymbol{K}_{\mathrm{AP}}(k) + \gamma\mathbf{I}\right) = \begin{bmatrix} a & \mathbf{b}^{\mathrm{T}} \\ \mathbf{b} & \mathbf{C} \end{bmatrix}$$

$$\mathbf{A}^{-1}_{\mathrm{AP}}(k-1) = \begin{bmatrix} \hat{\mathbf{A}} & \hat{\mathbf{b}} \\ \hat{\mathbf{b}}^{\mathrm{T}} & \hat{c} \end{bmatrix} \quad (3.57)$$

$$\tilde{\mathbf{C}}^{-1} = \hat{\mathbf{A}} - \frac{\hat{\mathbf{b}}\hat{\mathbf{b}}^{\mathrm{T}}}{\hat{c}} \quad (3.62)$$

$$\mathbf{C}^{-1} = \tilde{\mathbf{C}}^{-1} - \frac{1}{1+\mathbf{b}^{\mathrm{T}}\tilde{\mathbf{C}}^{-1}\mathbf{b}}\left[\tilde{\mathbf{C}}^{-1} - \tilde{\mathbf{C}}^{-1}\mathbf{b}\mathbf{b}^{\mathrm{T}}\tilde{\mathbf{C}}^{-1}\right] \quad (3.63)$$

$$f = a - \mathbf{b}^{\mathrm{T}}\mathbf{C}^{-1}\mathbf{b}$$

$$\mathbf{A}^{-1}_{\mathrm{AP}}(k) = \frac{1}{f}\begin{bmatrix} 1 & -\left(\mathbf{C}^{-1}\mathbf{b}\right)^{\mathrm{T}} \\ -\mathbf{C}^{-1}\mathbf{b} & \mathbf{C}^{-1}\left[\mathbf{I}f + \mathbf{b}\left(\mathbf{C}^{-1}\mathbf{b}\right)^{\mathrm{T}}\right] \end{bmatrix} \quad (3.64)$$

$$\boldsymbol{\beta}(k+1) = \begin{bmatrix} 0 \\ \boldsymbol{\beta}(k) \end{bmatrix} + \mu\boldsymbol{K}_{\mathrm{AP}}(k)\mathbf{A}^{-1}_{\mathrm{AP}}(k)\mathbf{e}_{\mathrm{AP}}(k) \quad (3.73)$$

$I = I + 1$

Else

$\boldsymbol{K}_{\mathrm{AP}}(k) = \boldsymbol{K}_{\mathrm{AP}}(k-1)$, $\mathbf{A}_{\mathrm{AP}}^{-1}(k) = \mathbf{A}_{\mathrm{AP}}^{-1}(k-1)$

$\mathbf{e}_{\mathrm{AP}}(k) = \mathbf{d}_{\mathrm{AP}}(k) - \boldsymbol{K}_{\mathrm{AP}}^{\mathrm{T}}(k)\boldsymbol{\beta}(k)$ (3.50)

$\boldsymbol{\beta}(k+1) = \boldsymbol{\beta}(k) + \mu\boldsymbol{K}_{\mathrm{AP}}(k)\mathbf{A}_{\mathrm{AP}}^{-1}(k)\mathbf{e}_{\mathrm{AP}}(k)$ (3.52)

End if

Else

$\boldsymbol{K}_{\mathrm{AP}}(k) = \boldsymbol{K}_{\mathrm{AP}}(k-1)$, $\mathbf{A}_{\mathrm{AP}}^{-1}(k) = \mathbf{A}_{\mathrm{AP}}^{-1}(k-1)$

$\mathbf{e}_{\mathrm{AP}}(k) = \mathbf{d}_{\mathrm{AP}}(k) - \boldsymbol{K}_{\mathrm{AP}}^{\mathrm{T}}(k)\boldsymbol{\beta}(k)$ (3.50)

$\boldsymbol{\beta}(k+1) = \boldsymbol{\beta}(k) + \mu\boldsymbol{K}_{\mathrm{AP}}(k)\mathbf{A}_{\mathrm{AP}}^{-1}(k)\mathbf{e}_{\mathrm{AP}}(k)$ (3.52)

End if

End do

close enough to previous one we test if the corresponding error signal is small enough, a case in which the data is also discarded. Therefore, in the case the dictionary is full, two options might be considered when the new incoming data should be incorporated into the dictionary such as: remove the oldest data entry $\mathbf{x}(I)$; or the one leading to the minimum value of $\kappa(\mathbf{x}(k), \mathbf{z}(l))$ for $l \in \mathcal{L}_I(k-1)$.

In summary, there are some options to configure the sparsification strategy for the KAP algorithm such as:

- Utilize the coherence approach based on Equation (3.28) without input data replacement when the dictionary is full, as in Algorithm 3.13.
- Utilize the coherence approach based on Equation (3.28) with data replacement when the dictionary is full, employing data permutation to be explained in Section 3.5.3.
- Utilize the strategy based on Equations (3.26) and (3.27) without dictionary replacement when the dictionary is full, as in Algorithm 3.14.
- Utilize the strategy based on Equations (3.26) and (3.27) with dictionary replacement when the dictionary is full, also requiring data permutation.
- Utilize the approximately linear dependent (ALD) based on Equation (3.85) to be discussed.

It should be remarked that in the KAP algorithms presented, there are many computational savings to exploit in order to turn the implementation more efficient, although the details are not included here. Take, for example, the formation of $\mathbf{A}_{\mathrm{AP}}^{-1}(k)$ in Equation (3.64), where the term $\mathbf{C}^{-1}\mathbf{b}$ appears several times and should be computed only once. In addition, the combination of ways to select the data to be included in the dictionary along with the decision to prescribe or not the dictionary size may lead to several alternatives for the algorithm implementation. The computational budget, as well as required performance, will dictate the best choice for each given application.

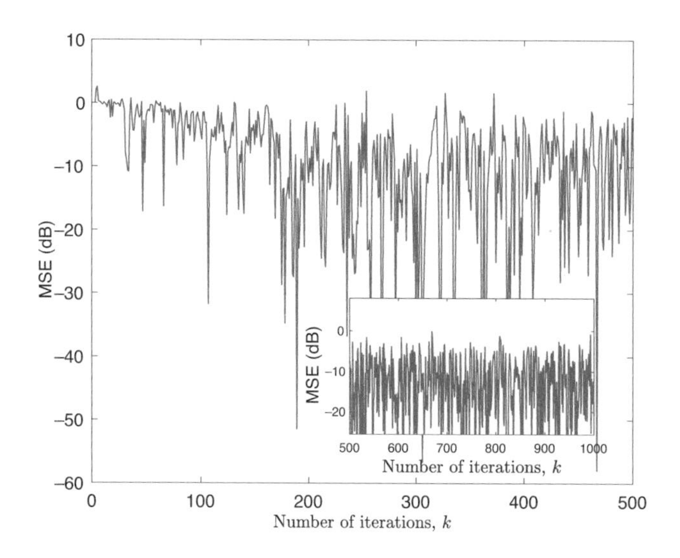

Figure 3.8 Equalization results using the KAP algorithm.

Example 3.3 Utilizing the channel model of Example 3.1 and employing the KAP algorithm perform the channel equalization considering the channel noise variance equal to -20 dB in comparison to the channel filtered transmitted signal at the receiver end. The choice of the kernel is Gaussian, and it utilizes a decision-directed (DD) blind equalization solution after training.

Solution

For the dictionary size, we opted to use $I_{\max} = 2000$, the same number of iterations the algorithm was tested, whereas the value of the convergence factor was $\mu = 0.4$. In a single run with a nonlinear kernel adaptive filter with six coefficients, no errors were noticed after 180 iterations. Figure 3.8 depicts the MSE learning curve with the equalization error growing slowly in the blind period, when it is expected degradation in the learning process.

In Figure 3.9, the decision errors and the accumulated error decisions are shown, illustrating the equalizer behavior during the iterations.

3.5 The Kernel Recursive Least-Squares Algorithm

This section addresses the derivation of the recursive least-squares algorithm incorporating the concept of kernel learning. As in the conventional RLS algorithm, the idea is to calculate the coefficients of the adaptive filter such that the output signal $y(k)$, during the period of observation, will match the desired

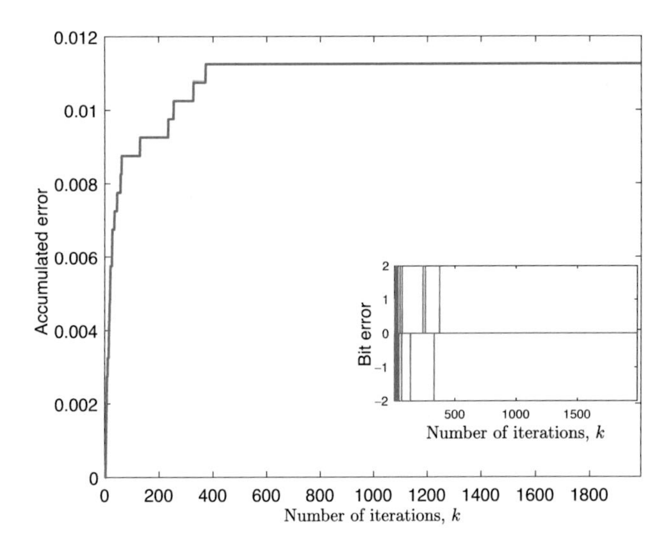

Figure 3.9 Accumulated bit error and bit error results, equalizer example using the KAP algorithm.

signal in the least-squares sense. As happens with all kernel-based algorithms the adaptive filter coefficients, as we know them, cannot be updated directly since they entail operations in the feature space $\mathcal{H}$, which are not accessible and would typically require infinite dimensions. Also, in the KRLS algorithm, the minimization process requires the information of the input signal available so far and the cost function is deterministic.

The basic idea of a generic kernel FIR adaptive filter realized by the direct mapping of the input-signal vector in the feature domain is depicted in Figure 3.10. The input signal information vector at a given instant k is mapped into the feature space as follows:

$$\boldsymbol{\upsilon}\left(\mathbf{x}(k)\right) = \boldsymbol{\upsilon}\left(\left[x(k)\; x(k-1) \ldots \; x(k-I)\right]^{\mathrm{T}}\right), \tag{3.75}$$

where $I+1$ is the dimension of the input signal vector. The adaptive filter functions in the feature space are denoted as $\mathbf{w}(k)$ and adapted aiming at the minimization of the weighted least-squares the objective function given by

$$\begin{aligned}\xi^{d}(k) &= \sum_{i=0}^{k} \lambda^{k-i} \varepsilon^{2}(i) \\ &= \sum_{i=0}^{k} \lambda^{k-i} \left[d(i) - \langle \mathbf{w}(k), \boldsymbol{\upsilon}(\mathbf{x}(k)) \rangle\right]^{2},\end{aligned} \tag{3.76}$$

where $\mathbf{w}(k)$ is the adaptive-filter function in the feature space, and $\varepsilon(i)$ is the *a posteriori* output error at instant i. The parameter λ is the exponential

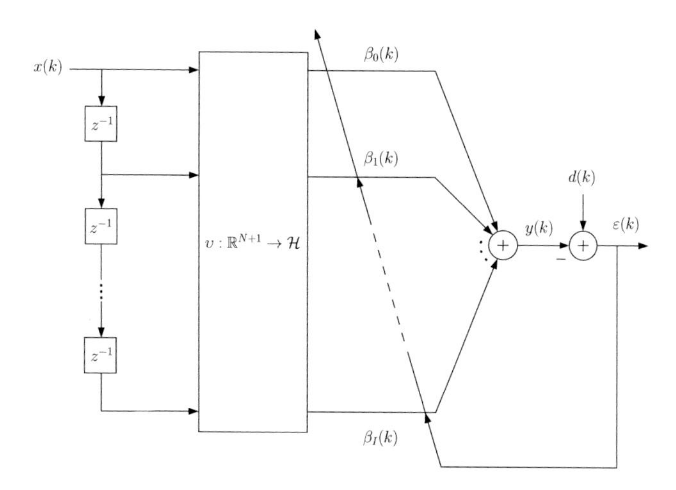

Figure 3.10 Kernel adaptive FIR filter.

weighting factor of the weighted least-squares objective function, which should be chosen in the range $0 \ll \lambda \leq 1$.The parameter λ is also known as forgetting factor given that the information of the distant past has an increasingly negligible effect on the coefficient updating.

As can be noted, each error consists of the difference between the desired signal and the filter output, using the most recent coefficients $\mathbf{w}(k)$. By differentiating $\xi^d(k)$ with respect to $\mathbf{w}(k)$, it follows that:

$$\frac{\partial \xi^d(k)}{\partial \mathbf{w}(k)} = -2 \sum_{i=0}^{k} \lambda^{k-i} \boldsymbol{\upsilon}(\mathbf{x}(i)) \left[d(i) - \langle \mathbf{w}(k), \boldsymbol{\upsilon}(\mathbf{x}(k)) \rangle \right]. \tag{3.77}$$

By equating the result to 0, it is possible to find the optimal function $\mathbf{w}(k)$ that minimizes the least-squares error, through the following relation:

$$-\sum_{i=0}^{k} \lambda^{k-i} \boldsymbol{\upsilon}(\mathbf{x}(i)) \langle \mathbf{w}(k), \boldsymbol{\upsilon}(\mathbf{x}(k)) \rangle + \sum_{i=0}^{k} \lambda^{k-i} \boldsymbol{\upsilon}(\mathbf{x}(i))\, d(i) = \mathbf{0}. \tag{3.78}$$

The resulting expression for the optimal coefficient function $\mathbf{w}(k)$ is given by

$$\begin{aligned} \mathbf{w}(k) &= \left[\sum_{i=0}^{k} \lambda^{k-i} \boldsymbol{\upsilon}(\mathbf{x}(i))\, \boldsymbol{\upsilon}^{\mathrm{T}}(\mathbf{x}(i)) \right]^{-1} \sum_{i=0}^{k} \lambda^{k-i} \boldsymbol{\upsilon}(\mathbf{x}(i))\, d(i) \\ &= \mathbf{R}_{\mathrm{D}}^{-1}(k) \mathbf{p}_{\mathrm{D}}(k), \end{aligned} \tag{3.79}$$

where $\mathbf{R}_{\mathrm{D}}(k)$ and $\mathbf{p}_{\mathrm{D}}(k)$ are called the deterministic correlation matrix in the feature space and the deterministic cross-correlation vector between the information vector in the feature space and desired signals, respectively. We assumed that $\mathbf{R}_{\mathrm{D}}(k)$ is nonsingular. The issues to be addressed are the data dimensions related to the growing dictionary and the high dimensions of the adaptive filter function.

3.5.1 Dictionary Evaluation

Given a new incoming data, we want to verify if it brings about enough information to be included in the dictionary or not. If all previous data is available, not practical for online implementations, this can be verified by minimizing the following cost function with respect to the coefficients $\alpha_i(k)$.

$$\delta_{\rm d}(k) = \|\boldsymbol{\upsilon}\left(\mathbf{x}(k)\right) - \sum_{i=0}^{k-1} \alpha_i(k)\boldsymbol{\upsilon}\left(\mathbf{x}(i)\right)\|^2. \tag{3.80}$$

A solution for an online implementation is to verify if the new incoming data is well represented by the previous selected data as follows:

$$\delta_{\rm d}(k) = \|\boldsymbol{\upsilon}\left(\mathbf{x}(k)\right) - \sum_{i\in\mathcal{L}_I(k-1))} \alpha_i(k)\boldsymbol{\upsilon}\left(\mathbf{x}(l_i(k-1))\right)\|^2 \leq \gamma_d, \tag{3.81}$$

where γ_d is a prescribed threshold aiming at keeping the number of members in the data dictionary limited to $I+1$. It is also possible not to prescribe the number of members in the dictionary, and in this case, γ_d would determine the degree of sparsity of the data set acquired so far. This test determines if $\boldsymbol{\upsilon}(\mathbf{x}(k))$ is ALD of the current dictionary entries. For an online implementation, it would be desirable that if the new incoming data is not ALD, it replaces the oldest data represented by $\boldsymbol{\upsilon}(\mathbf{x}(l_I(k-1))$. In [19], their proposed solution does not discard elements of the dictionary, whereas here, we also discuss a solution for the cases where some sort of nonstationary behavior in the environment turns the oldest data in the dictionary less relevant. Here, we discuss the possibility of either discard or increase the dictionary size.

The minimization of (3.81) can be rewritten as

$$\begin{aligned}\delta_{\rm d}(k) =& \underset{\boldsymbol{\alpha}(k)}{\text{minimize}} \left\{ \sum_{i,j\in\mathcal{L}_I(k-1)} \alpha_i(k)\alpha_j(k)\langle\boldsymbol{\upsilon}\left(\mathbf{x}(i)\right), \boldsymbol{\upsilon}\left(\mathbf{x}(j)\right)\rangle \right.\\ &\left. -2\sum_{i\in\mathcal{L}_I(k-1)} \alpha_i(k)\langle\boldsymbol{\upsilon}\left(\mathbf{x}(i)\right), \boldsymbol{\upsilon}\left(\mathbf{x}(k)\right)\rangle + \langle\boldsymbol{\upsilon}\left(\mathbf{x}(k)\right), \boldsymbol{\upsilon}\left(\mathbf{x}(k)\right)\rangle \right\}.\end{aligned} \tag{3.82}$$

Recall that $\kappa(\mathbf{x}(i), \mathbf{x}(j)) = \langle\boldsymbol{\upsilon}\left(\mathbf{x}(i)\right), \boldsymbol{\upsilon}\left(\mathbf{x}(j)\right)\rangle$, where in the present discussion, $i, j \in \mathcal{L}_I(k-1)$. Therefore, Equation (3.82) can be rewritten as

$$\begin{aligned}\delta_{\rm d}(k) =& \underset{\boldsymbol{\alpha}(k)}{\text{minimize}} \left\{ \sum_{i,j\in\mathcal{L}_I(k-1)} \alpha_i(k)\alpha_j(k)\kappa(\mathbf{x}(i), \mathbf{x}(j)) \right.\\ &\left. -2\sum_{i\in\mathcal{L}_I(k-1)} \alpha_i(k)\kappa(\mathbf{x}(i), \mathbf{x}(k)) + \kappa(\mathbf{x}(k), \mathbf{x}(k)) \right\}\\ =& \underset{\boldsymbol{\alpha}(k)}{\text{minimize}} \left\{\boldsymbol{\alpha}^{\rm T}(k)\boldsymbol{K}(k-1)\boldsymbol{\alpha}(k) - 2\boldsymbol{\kappa}^{\rm T}(\mathbf{x}(k), \cdot)\boldsymbol{\alpha}(k) + \kappa(\mathbf{x}(k), \mathbf{x}(k))\right\},\end{aligned} \tag{3.83}$$

where the i-th element of $\boldsymbol{\kappa}(\mathbf{x}(k), \cdot)$ is $\kappa(\mathbf{x}(k), \mathbf{x}(i))$ and the entries of vector $\boldsymbol{\alpha}(k)$ are $\alpha_i(k)$ for $i \in \mathcal{L}_I(k-1)$.

The solution of this minimization is given by

$$\boldsymbol{\alpha}(k) = \boldsymbol{K}^{-1}(k-1)\boldsymbol{\kappa}(\mathbf{x}(k), \cdot), \tag{3.84}$$

so that the approximation error is

$$\delta_{\mathrm{d}}(k) = \kappa(\mathbf{x}(k), \mathbf{x}(k)) - \boldsymbol{\alpha}^{\mathrm{T}}(k)\boldsymbol{\kappa}(\mathbf{x}(k), \cdot). \tag{3.85}$$

In case $\delta_{\mathrm{d}}(k) > \gamma_{\mathrm{d}}$, we should include $\mathbf{x}(k)$ in the dictionary and eventually remove or not the oldest one, corresponding to the actual data $\mathbf{x}(k-l)$ for $l = l_I(k-1)$ since $\mathcal{L}_I(k-1) = \{l_0(k-1) \ \ldots \ l_I(k-1)\}$. If $\mathbf{x}(k)$ is included in the dictionary and no data is removed, then we need to add a new element to the set of selected data by increasing I by 1.

3.5.2 General Formulation of the KRLS Algorithm

The following discussion addresses how to minimize a general RLS-type cost function, relying on the selected incoming data utilizing ALD or another criterion for selection. Let's consider that

$$\begin{aligned}\mathbf{w}(k) &= \sum_{i=0}^{I} \beta_i(k)\boldsymbol{\upsilon}(\mathbf{z}(i)) \\ &= \sum_{i=0}^{I} \beta_{I-i}(k)\boldsymbol{\upsilon}(\mathbf{x}(l_{I-i}(k))) \\ &= \boldsymbol{\Upsilon}(k)\boldsymbol{\beta}(k),\end{aligned} \tag{3.86}$$

where $\mathbf{z}(i)$ represents the i-th selected dictionary element,

$$\boldsymbol{\beta}(k) = [\beta_0(k), \beta_1(k), \ldots, \beta_I(k)]^{\mathrm{T}}$$

and

$$\begin{aligned}\boldsymbol{\Upsilon}(k) &= [\boldsymbol{\upsilon}(\mathbf{z}(0)), \boldsymbol{\upsilon}(\mathbf{z}(1)), \ldots, \boldsymbol{\upsilon}(\mathbf{z}(I))] \\ &= [\boldsymbol{\upsilon}(\mathbf{x}(l_I(k))), \boldsymbol{\upsilon}(\mathbf{x}(l_{I-1}(k))), \ldots, \boldsymbol{\upsilon}(\mathbf{x}(l_1(k))), \boldsymbol{\upsilon}(\mathbf{x}(k))].\end{aligned}$$

Note that the number of data $I+1$ in the summation represents the most recent data selected at instant k, including the most recent data $\mathbf{x}(k)$, so that $\mathbf{x}(k-l)$ for $l \in \mathcal{L}_I(k)$, where $\mathcal{L}_I(k) = \{k, \ l_1(k), \ \ldots, \ l_I(k)\}$ represents the data set available at the instant k.

As per Equation (3.86), the RLS objective function can be rewritten as

$$\begin{aligned}
\xi^d(k) =& \sum_{l=0}^{k} \lambda^{k-l} \varepsilon^2(l), \text{ for } l \in \mathcal{L}_I(k) \\
=& \sum_{l=0}^{k} \lambda^{k-l} \left[\left(d(l) - \boldsymbol{\upsilon}^{\mathrm{T}}(\mathbf{x}(l)) \sum_{i=0}^{I} \beta_i(k) \boldsymbol{\upsilon}(\mathbf{z}(i)) \right) \right]^2 && (3.87) \\
=& \| \boldsymbol{\Lambda}_{\mathrm{KRLS}}^{\frac{1}{2}}(k) \left(\mathbf{d}_{\mathrm{KRLS}}(k) - \boldsymbol{\Upsilon}^{\mathrm{T}}(k) \boldsymbol{\Upsilon}(k) \boldsymbol{\beta}(k) \right) \|^2 && (3.88) \\
=& \| \boldsymbol{\Lambda}_{\mathrm{KRLS}}^{\frac{1}{2}}(k) \left(\mathbf{d}_{\mathrm{KRLS}}(k) - \boldsymbol{K}(k) \boldsymbol{\beta}(k) \right) \|^2, && (3.89)
\end{aligned}$$

where

$$\boldsymbol{\Lambda}_{\mathrm{KRLS}}(k) = \begin{bmatrix} \lambda^{l_I(k)} & & & & \\ & \lambda^{l_{I-1}(k)} & & \mathbf{0} & \\ & \mathbf{0} & \ddots & & \\ & & & \lambda^{l_1(k)} & \\ & & & & 1 \end{bmatrix}$$

and

$$\mathbf{d}_{\mathrm{KRLS}}(k) = \begin{bmatrix} d\left(k - l_I(k-1)\right) \\ \vdots \\ d\left(k - l_1(k)\right) \\ d\left(k\right) \end{bmatrix}.$$

The solution for $\boldsymbol{\beta}(k)$ is expressed as

$$\boldsymbol{\beta}(k) = \boldsymbol{K}^{\dagger}(k) \mathbf{d}_{\mathrm{KRLS}}(k) = \left(\boldsymbol{K}^{\mathrm{T}}(k) \boldsymbol{K}(k) \right)^{-1} \boldsymbol{K}^{\mathrm{T}}(k) \mathbf{d}_{\mathrm{KRLS}}(k), \tag{3.90}$$

where $(\cdot)^{\dagger}$ represents the pseudo inverse of $(\cdot)$.Assuming the kernel matrix is invertible, the solution becomes

$$\boldsymbol{\beta}(k) = \boldsymbol{K}^{-1}(k) \mathbf{d}_{\mathrm{KRLS}}(k). \tag{3.91}$$

The problem with this solution lies in the fact that Equation (3.89) might have an infinite number of solutions given that the dimension of feature space is, in some cases, much higher than the number of data points.[6]

A solution to the problem above is to include a regularization factor to the objective function such as

$$\xi^d(k) = \| \boldsymbol{\Lambda}_{\mathrm{KRLS}}^{\frac{1}{2}}(k) \left(\boldsymbol{K}(k) \boldsymbol{\beta}(k) - \mathbf{d}_{\mathrm{KRLS}}(k) \right) \|^2 + \gamma \boldsymbol{\beta}^{\mathrm{T}}(k) \boldsymbol{K}(k) \boldsymbol{\beta}(k). \tag{3.92}$$

[6] This represents an overfitting problem where in the case the learning algorithm has more degrees of freedom than required, so that the extra degrees of freedom fit the noise in the data, generating an inaccurate model. Overfitting phenomenon is not fully described here, see [13] for details.

In this case, the optimal solution to $\boldsymbol{\beta}(k)$ is given by

$$\boldsymbol{\beta}(k) = \left(\boldsymbol{K}(k) + \gamma \boldsymbol{\Lambda}_{\mathbf{KRLS}}^{-1}(k)\mathbf{I}\right)^{-1} \mathbf{d}_{\mathrm{KRLS}}(k) = \boldsymbol{K}_{\mathrm{RLS}}^{-1}(k)\mathbf{d}_{\mathrm{KRLS}}(k), \tag{3.93}$$

where γ represents the regularization constant, like the one present in the standard RLS algorithms, and $\mathbf{I}$ is the identity matrix.[7]

In the general formulation, it is possible to increase the number of dictionary elements $I+1$ by verifying if the most recent vector acquired is not ALD, see for example Problem 3.15. In the following, we consider a fixed amount of dictionary entries.

3.5.3 The Sliding-Window Kernel RLS Algorithm

It is possible to generate a kernel-based RLS algorithm without growing dimension matrices and vectors through the use of the sliding-window strategy.

For the sliding-window algorithm, we consider $\lambda = 1$ so that the objection function becomes

$$\xi^d(k) = \| (\boldsymbol{K}(k)\boldsymbol{\beta}(k) - \mathbf{d}_{\mathrm{KRLS}}(k)) \|^2 + \gamma \boldsymbol{\beta}^{\mathrm{T}}(k)\boldsymbol{K}(k)\boldsymbol{\beta}(k). \tag{3.94}$$

As before the optimal solution to $\boldsymbol{\beta}(k)$ is given by

$$\boldsymbol{\beta}(k) = \boldsymbol{K}_{\mathrm{RLS}}^{-1}(k)\mathbf{d}_{\mathrm{KRLS}}(k) \tag{3.95}$$

The challenge now is to invert $\boldsymbol{K}_{\mathrm{RLS}}(k)$ with the knowledge of the inverse of $\boldsymbol{K}_{\mathrm{RLS}}(k-1)$ in the case the new incoming data is accepted in the dictionary. Assume at instant $k-1$ that the kernel matrix $\boldsymbol{K}_{\mathrm{RLS}}(k-1)$ of dimension $M \times M$ is reduced by eliminating its first row and column generating $\bar{\boldsymbol{K}}_{\mathrm{RLS}}(k-1)$ of dimension $(M-1) \times (M-1)$, removing this way the information related to the oldest input data vector. This action brings about two benefits: avoids a growing dimension inherent to the classical kernel-based algorithms and provides room to account for the new incoming data pair. With the new input data vector, the kernel matrix should be resized as

$$\begin{aligned}
\boldsymbol{K}_{\mathrm{RLS}}(k) &= \begin{bmatrix} \bar{\boldsymbol{K}}_{\mathrm{RLS}}(k-1) & \boldsymbol{\kappa}_{\mathrm{RLS}}(\mathbf{x}(k-1),\mathbf{x}(k)) \\ \boldsymbol{\kappa}_{\mathrm{RLS}}^{\mathrm{T}}(\mathbf{x}(k-1),\mathbf{x}(k)) & \kappa(\mathbf{x}(k),\mathbf{x}(k)) + \gamma \end{bmatrix} \\
&= \begin{bmatrix} \mathbf{C} & \boldsymbol{\kappa}_{\mathrm{RLS}}(\mathbf{x}(k-1),\mathbf{x}(k)) \\ \boldsymbol{\kappa}_{\mathrm{RLS}}^{\mathrm{T}}(\mathbf{x}(k-1),\mathbf{x}(k)) & \kappa(\mathbf{x}(k),\mathbf{x}(k)) + \gamma \end{bmatrix},
\end{aligned} \tag{3.96}$$

where $\boldsymbol{\kappa}_{\mathrm{RLS}}(\mathbf{x}(k-1),\mathbf{x}(k)) = [\kappa(\mathbf{x}(k-M),\mathbf{x}(k)), \kappa(\mathbf{x}(k-M+1),\mathbf{x}(k)), \ldots, \kappa(\mathbf{x}(k-1),\mathbf{x}(k))]$, and matrix $\mathbf{C}$ is equal to $\bar{\boldsymbol{K}}_{\mathrm{RLS}}(k-1)$. It should be noticed

[7] Note that the forgetting factor λ will basically modify the effect of the regularization factor related to the matrix $\boldsymbol{K}(k)$. That means that in the long run the actual regularization factor is controlled by $\gamma\lambda^{-l_1(k)}$.

that the *a posteriori* error from the solution of (3.94) related to the most recent data is calculated by

$$\varepsilon(k) = d(k) - \left[\boldsymbol{\kappa}_{\rm RLS}^{\rm T}(\mathbf{x}(k-1),\mathbf{x}(k)) \;\; \kappa(\mathbf{x}(k),\mathbf{x}(k)) + \gamma\right] \boldsymbol{\beta}(k). \tag{3.97}$$

Note that this error will be close to 0 and if one wishes to compare the KRLS algorithm with the competing algorithms, the *a priori* error should be computed using $\boldsymbol{\beta}(k-1)$.

The next step is to compute the inverse of the kernel matrix in an efficient way. Specifically we want to obtain $\boldsymbol{K}_{\rm RLS}^{-1}(k)$ assuming we know $\boldsymbol{K}_{\rm RLS}(k)$ and $\bar{\boldsymbol{K}}_{\rm RLS}^{-1}(k-1)$, utilizing an inversion formula as following discussed.

Let's first describe a form to express matrix $\boldsymbol{K}_{\rm RLS}(k-1)$ in a general form ready to be reduced in dimension, where the iteration index is omitted on the right-hand side since it is irrelevant for this discussion. Defining

$$\boldsymbol{K}_{\rm RLS}(k-1) = \begin{bmatrix} a & \mathbf{b}^{\rm T} \\ \mathbf{b} & \mathbf{C} \end{bmatrix} \text{ and}$$

$$\boldsymbol{K}_{\rm RLS}^{-1}(k-1) = \begin{bmatrix} \bar{a} & \bar{\mathbf{b}}^{\rm T} \\ \bar{\mathbf{b}} & \bar{\mathbf{C}} \end{bmatrix}, \tag{3.98}$$

we note that matrix $\mathbf{C}$ corresponds to $\bar{\boldsymbol{K}}_{\rm RLS}(k-1)$ obtained in the process of reducing the dimension of $\boldsymbol{K}_{\rm RLS}(k-1)$. The matrices above are inverse of each other if

$$\begin{aligned} \bar{a}\mathbf{b} + \mathbf{C}\bar{\mathbf{b}} &= \mathbf{0} \text{ and} \\ \mathbf{b}\bar{\mathbf{b}}^{\rm T} + \bar{\mathbf{C}}\mathbf{C} &= \mathbf{I}, \end{aligned} \tag{3.99}$$

leading to

$$\mathbf{C}^{-1} = \bar{\mathbf{C}} - \frac{\bar{\mathbf{b}}\bar{\mathbf{b}}^{\rm T}}{\bar{a}}, \tag{3.100}$$

where, as a result, $\mathbf{C}^{-1}$ corresponds to $\bar{\boldsymbol{K}}_{\rm RLS}^{-1}(k-1)$.

In the process of adding a column and a row to generate $\boldsymbol{K}_{\rm RLS}(k)$, the matrix inversion formulas are

$$\boldsymbol{K}_{\rm RLS}(k) = \begin{bmatrix} \mathbf{C} & \tilde{\mathbf{b}} \\ \tilde{\mathbf{b}}^{\rm T} & c \end{bmatrix} \text{ and}$$

$$\boldsymbol{K}_{\rm RLS}^{-1}(k) = \begin{bmatrix} \hat{\mathbf{A}} & \hat{\mathbf{b}} \\ \hat{\mathbf{b}}^{\rm T} & \hat{c} \end{bmatrix}, \tag{3.101}$$

where these matrices are inverse of each other if

$$\begin{aligned} \mathbf{C}\hat{\mathbf{A}} + \tilde{\mathbf{b}}\hat{\mathbf{b}}^{\rm T} &= \mathbf{I}; \\ \mathbf{C}\hat{\mathbf{b}} + \tilde{\mathbf{b}}\hat{c} &= \mathbf{0}; \text{ and} \\ \tilde{\mathbf{b}}^{\rm T}\hat{\mathbf{b}} + \hat{c}c &= 1, \end{aligned} \tag{3.102}$$

leading, after a few manipulations, to

$$\boldsymbol{K}_{\mathrm{RLS}}^{-1}(k) = \begin{bmatrix} \mathbf{C}^{-1}\left(\mathbf{I} - \tilde{\mathbf{b}}\tilde{\mathbf{b}}^{\mathrm{T}}\mathbf{C}^{-1}\hat{c}\right) & -\mathbf{C}^{-1}\tilde{\mathbf{b}}\hat{c} \\ -\left(\mathbf{C}^{-1}\tilde{\mathbf{b}}\right)^{\mathrm{T}}\hat{c} & \hat{c} \end{bmatrix}, \tag{3.103}$$

where $\hat{c} = (c - \tilde{\mathbf{b}}^{\mathrm{T}}\mathbf{C}^{-1}\tilde{\mathbf{b}})^{-1}$.

The initialization of the sliding-window KRLS algorithm can be performed as [21]:

$$\boldsymbol{K}_{\mathrm{RLS}}(-1) = (1+\gamma)\mathbf{I}. \tag{3.104}$$

The complete online sliding-window KRLS algorithm is described in Algorithm 3.15.

An alternative solution for the selection of the entry to discard in the sliding-window KRLS algorithm is to choose the least significant data point among those included in the dictionary. There are many ways to prune the least significant data point; one is to swap the column and row with the lowest value of $|\beta_i(k)|$, where $\beta_i(k)$ corresponds to the i-th entry of $\boldsymbol{\beta}(k)$ in (3.93). Here, we present the so-called fixed-budget KRLS algorithm [22]. This lower value of $|\beta_i(k)|$ indicates a less critical influence of the i-th data pair to the nonlinear mapping. The algorithm procedure is to exchange the position of the

---

Algorithm 3.15 Sliding-window kernel RLS algorithm

$\gamma = 0.01$ (prescribed small constant)

Initialization

$\boldsymbol{K}_{\mathrm{RLS}}^{-1}(-1) = \frac{\mathbf{I}}{1+\gamma}$ and $\boldsymbol{K}_{\mathrm{RLS}}(-1) = (1+\gamma)\mathbf{I}$ (3.104)

$\mathbf{x}(-1) = \mathbf{0}$

For do $k = 0, 1, \ldots$

Remove the first column and row of $\boldsymbol{K}_{\mathrm{RLS}}(k-1)$ to obtain $\bar{\boldsymbol{K}}_{\mathrm{RLS}}(k-1)$

Since $\boldsymbol{K}_{\mathrm{RLS}}^{-1}(k-1) = \begin{bmatrix} \bar{a} & \bar{\mathbf{b}}^{\mathrm{T}} \\ \bar{\mathbf{b}} & \bar{\mathbf{C}} \end{bmatrix}$ (3.98) and (3.99)

$\mathbf{C}^{-1} = \bar{\mathbf{C}} - \frac{\bar{\mathbf{b}}\bar{\mathbf{b}}^{\mathrm{T}}}{\bar{a}}$ (3.100)

$\boldsymbol{K}_{\mathrm{RLS}}(k) = \begin{bmatrix} \bar{\boldsymbol{K}}_{\mathrm{RLS}}(k-1) & \boldsymbol{\kappa}_{\mathrm{RLS}}(\mathbf{x}(k-1),\mathbf{x}(k)) \\ \boldsymbol{\kappa}_{\mathrm{RLS}}^{\mathrm{T}}(\mathbf{x}(k-1),\mathbf{x}(k)) & \kappa(\mathbf{x}(k),\mathbf{x}(k)) + \gamma \end{bmatrix} = \begin{bmatrix} \mathbf{C} & \tilde{\mathbf{b}} \\ \tilde{\mathbf{b}}^{\mathrm{T}} & c \end{bmatrix}$ (3.96)

$\hat{c} = (c - \tilde{\mathbf{b}}^{\mathrm{T}}\mathbf{C}^{-1}\tilde{\mathbf{b}})^{-1}$ (3.103)

$\boldsymbol{K}_{\mathrm{RLS}}^{-1}(k) = \begin{bmatrix} \mathbf{C}^{-1}\left(\mathbf{I} - \tilde{\mathbf{b}}\tilde{\mathbf{b}}^{\mathrm{T}}\mathbf{C}^{-1}\hat{c}\right) & -\mathbf{C}^{-1}\tilde{\mathbf{b}}\hat{c} \\ -\left(\mathbf{C}^{-1}\tilde{\mathbf{b}}\right)^{\mathrm{T}}\hat{c} & \hat{c} \end{bmatrix}$ (3.101), (3.102), and (3.103)

Remove the oldest entry (first entry) of $\mathbf{d}_{\mathrm{KRLS}}(k-1)$ and add the newest data sample $d(k)$ (last entry)

$\boldsymbol{\beta}(k) = \boldsymbol{K}_{\mathrm{RLS}}^{-1}(k)\mathbf{d}_{\mathrm{KRLS}}(k)$ (3.93)

$\varepsilon(k) = d(k) - \left[\boldsymbol{\kappa}_{\mathrm{RLS}}^{\mathrm{T}}(\mathbf{x}(k-1),\mathbf{x}(k)) \;\; \kappa(\mathbf{x}(k),\mathbf{x}(k)) + \gamma\right]\boldsymbol{\beta}(k)$ (3.97)

End do

---

i-th column and row with a permutation matrix according to the situation. Two distinct permutations are required such as: the exchange of the i-th column and row with the first column and row of a matrix can be performed by pre-multiplying and post-multiplying the matrix by

$$\mathbf{P}_i = \begin{bmatrix} 0 & \mathbf{0} & 1 & \mathbf{0} \\ \mathbf{0} & \mathbf{I}_{i-2} & \mathbf{0} & \mathbf{0} \\ 1 & \mathbf{0} & 0 & \mathbf{0} \\ \mathbf{0} & \mathbf{0} & \mathbf{0} & \mathbf{I}_{M-i} \end{bmatrix}. \tag{3.105}$$

Similarly, by pre-multiplying and post-multiplying a given matrix by the following matrix

$$\hat{\mathbf{P}}_i = \begin{bmatrix} \mathbf{0} & \mathbf{I}_{i-1} & \mathbf{0} \\ 1 & \mathbf{0} & \mathbf{0} \\ \mathbf{0} & \mathbf{0} & \mathbf{I}_{M-i} \end{bmatrix}, \tag{3.106}$$

we are able to place i-th column and row to its first column and row. Note that the permutation matrices are orthonormal, that is, $\mathbf{P}_i^{\mathrm{T}}\mathbf{P}_i = \hat{\mathbf{P}}_i^{\mathrm{T}}\hat{\mathbf{P}}_i = \mathbf{I}$.

The sliding-window KRLS algorithm including the choice of the least significant data pair, called fixed-budget kernel RLS algorithm, is described in Algorithm 3.16.

Algorithm 3.16 Fixed-budget kernel RLS algorithm

$\gamma = 0.01$ (prescribed small constant)

Initialization

$\boldsymbol{K}_{\mathrm{RLS}}^{-1}(-1) = \frac{\mathbf{I}}{1+\gamma}$ and $\boldsymbol{K}_{\mathrm{RLS}}(-1) = (1+\gamma)\mathbf{I}$ (3.104)

$\mathbf{x}(-1) = \mathbf{0}$

For do $k = 0, 1, \ldots$

Choose i corresponding to the minimum value of $|\beta_i(k)|$

Assign: $\boldsymbol{K}_{\mathrm{RLS}}(k-1) \Leftarrow \mathbf{P}_i \boldsymbol{K}_{\mathrm{RLS}}(k-1)\mathbf{P}_i$

Remove the first column and row of $\boldsymbol{K}_{\mathrm{RLS}}(k-1)$ to obtain $\bar{\boldsymbol{K}}_{\mathrm{RLS}}(k-1)$

Since $\boldsymbol{K}_{\mathrm{RLS}}^{-1}(k-1) = \begin{bmatrix} \bar{a} & \bar{\mathbf{b}}^{\mathrm{T}} \\ \bar{\mathbf{b}} & \bar{\mathbf{C}} \end{bmatrix}$ (3.98) and (3.99)

Assign: $\boldsymbol{K}_{\mathrm{RLS}}^{-1}(k-1) \Leftarrow \mathbf{P}_i \boldsymbol{K}_{\mathrm{RLS}}^{-1}(k-1)\mathbf{P}_i$

$\bar{\boldsymbol{K}}_{\mathrm{RLS}}^{-1}(k-1) = \mathbf{C}^{-1} = \bar{\mathbf{C}} - \frac{\bar{\mathbf{b}}\bar{\mathbf{b}}^{\mathrm{T}}}{\bar{a}}$ (3.100)

Assign: $\bar{\boldsymbol{K}}_{\mathrm{RLS}}^{-1}(k-1) \Leftarrow \hat{\mathbf{P}}_i \bar{\boldsymbol{K}}_{\mathrm{RLS}}^{-1}(k-1)\hat{\mathbf{P}}_i$

$\boldsymbol{K}_{\mathrm{RLS}}(k) = \begin{bmatrix} \bar{\boldsymbol{K}}_{\mathrm{RLS}}(k-1) & \boldsymbol{\kappa}_{\mathrm{RLS}}(\mathbf{x}(k-1), \mathbf{x}(k)) \\ \boldsymbol{\kappa}_{\mathrm{RLS}}^{\mathrm{T}}(\mathbf{x}(k-1), \mathbf{x}(k)) & \kappa(\mathbf{x}(k), \mathbf{x}(k)) + \gamma \end{bmatrix} = \begin{bmatrix} \mathbf{C} & \tilde{\mathbf{b}} \\ \tilde{\mathbf{b}}^{\mathrm{T}} & c \end{bmatrix}$

(3.96)

$\hat{c} = (c - \tilde{\mathbf{b}}^{\mathrm{T}}\mathbf{C}^{-1}\tilde{\mathbf{b}})^{-1}$ (3.103)

$$\boldsymbol{K}_{\text{RLS}}^{-1}(k) = \begin{bmatrix} \mathbf{C}^{-1}\left(\mathbf{I} - \tilde{\mathbf{b}}\tilde{\mathbf{b}}^{\text{T}}\mathbf{C}^{-1}\hat{c}\right) & -\mathbf{C}^{-1}\tilde{\mathbf{b}}\hat{c} \\ -\left(\mathbf{C}^{-1}\tilde{\mathbf{b}}\right)^{\text{T}}\hat{c} & \hat{c} \end{bmatrix} \quad (3.101),\ (3.102),\ \text{and}$$

(3.103)

Remove the i-th of $\mathbf{d}_{\text{KRLS}}(k-1)$ and add the newest data sample $d(k)$ (last entry)

$\boldsymbol{\beta}(k) = \boldsymbol{K}_{\text{RLS}}^{-1}(k)\mathbf{d}_{\text{KRLS}}(k)$ (3.93)

$\varepsilon(k) = d(k) - \left[\boldsymbol{\kappa}_{\text{RLS}}^{\text{T}}(\mathbf{x}(k-1), \mathbf{x}(k)) \quad \kappa(\mathbf{x}(k), \mathbf{x}(k)) + \gamma\right]\boldsymbol{\beta}(k)$ (3.97)

End do

Example 3.4 Repeat the Example 3.3 employing the KRLS algorithm and utilizing a DD equalization after training and with additional noise with variance $\sigma_n = -10\,\text{dB}$.

Solution

Like in the previous examples, $I_{\text{max}} = 1000$ and the training period lasted for all iterations. In this case, we utilized four coefficients for the KRLS adaptive filter, with the Gaussian kernel bandwidth chosen in this experiment equal to 1. In a single run, the equalizer remained error-free up to 180 iterations and presented a few errors in the first 1000 iterations. One should note that this particular setup is not the most favorable for learning algorithms including matrix inversion, such as KRLS and KAP algorithms, if one considers that the transmitted signal is binary. The MSE is shown in Figure 3.11, where it is possible to verify the learning curve of the equalizer error.

Figures 3.12 shows the decision errors and the accumulated error decisions to indicate two essential aspects of the performance.

The next example explores the use of the polynomial kernel.

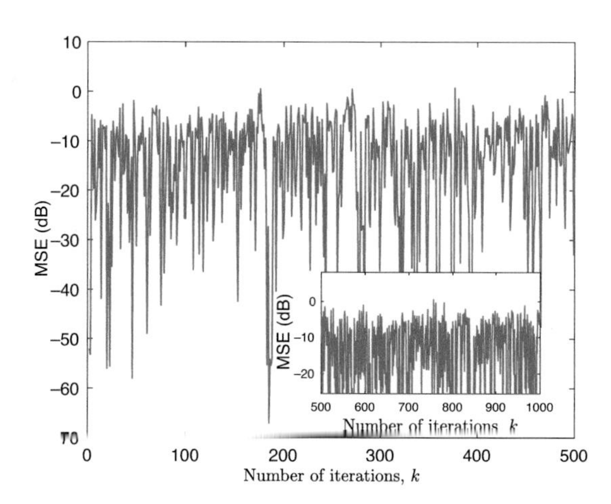

Figure 3.11 Equalizer results, KRLS algorithm.

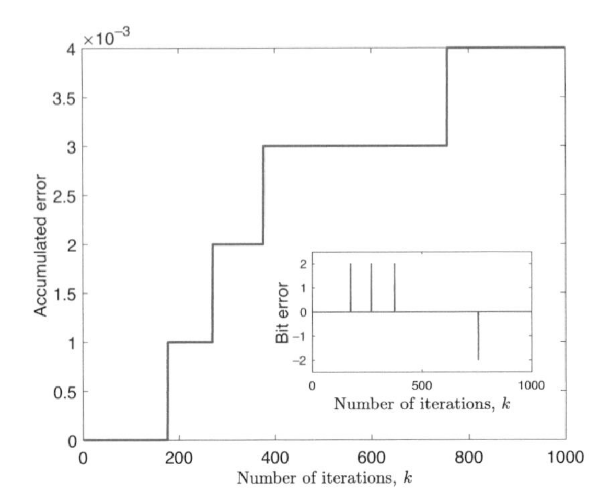

Figure 3.12 Accumulated bit error and bit error results, equalizer example using the KRLS algorithm.

Example 3.5 Repeat Example 3.2 using the KRLS algorithm using polynomial kernel.

Solution

The parameters of the polynomial kernel are $a = b = 1$ and $n = 2$. Figures 3.13.a and 3.13.b depict the learning curves for the two sets of reference signals. As can be seen, the KRLS algorithm converges quite fast in this example employing 10 adaptive coefficients.

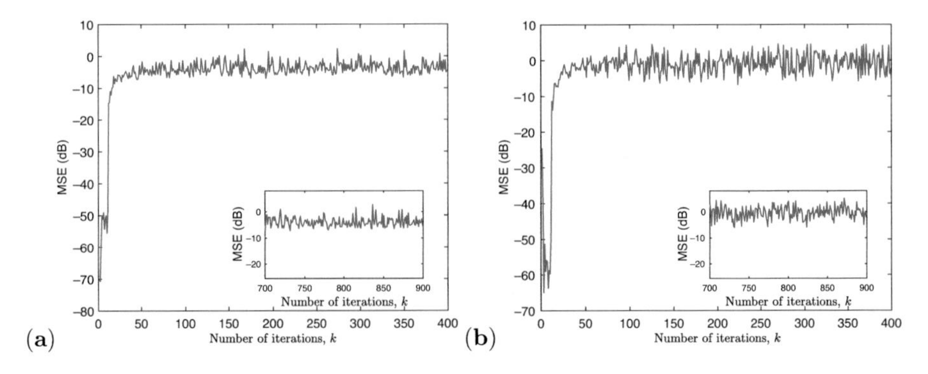

Figure 3.13 System identification example using KRLS algorithm whose results are averages of 1000 independent runs: (a) learning curve for the MSE and (b) learning curve for the MSE, the cosine function case.

3.6 The Set-Membership Kernel Affine Projection Algorithm

The derivation of the family of SM-KAP algorithms follows closely the ones utilized for the original SM-AP algorithms in [6]. It is worth mentioning that initially, the kernel normalized LMS algorithm was proposed in [23] and later extended in [24]. The work [25] describes a comprehensive extension of the nonlinear kernel-based adaptive filtering algorithms, including the set-membership strategy to the NLMS and the Bounding Ellipsoidal Adaptive CONstrained (BEACON) algorithms. The BEACON algorithm contains expressions somewhat related to the RLS algorithm.

In particular, we will present the simplified version of the SM-KAP algorithm whose update equation is given by

$$\boldsymbol{\beta}(k+1) = \boldsymbol{\beta}(k) + \boldsymbol{K}_{\mathrm{AP}}(k)\left(\boldsymbol{K}_{\mathrm{AP}}^{\mathrm{T}}(k)\boldsymbol{K}_{\mathrm{AP}}(k) + \gamma\mathbf{I}\right)^{-1}\mu(k)e(k)\mathbf{u}_1, \tag{3.107}$$

where $\mathbf{u}_1^{\mathrm{T}} = [1\,0\ldots0]$ and

$$\mathbf{e}_{\mathrm{AP}}(k) = \mathbf{d}_{\mathrm{AP}}(k) - \boldsymbol{K}_{\mathrm{AP}}^{\mathrm{T}}(k)\boldsymbol{\beta}(k); \tag{3.108}$$

$$\mu(k) = \begin{cases} 1 - \frac{\bar{\gamma}}{|e(k)|} & \text{if } |e(k)| > \bar{\gamma}, \\ 0 & \text{otherwise.} \end{cases} \tag{3.109}$$

As in the original SM-AP algorithm, this algorithm minimizes the Euclidean distance $\|\boldsymbol{\beta}(k+1) - \boldsymbol{\beta}(k)\|^2$ subject to the constraint $\boldsymbol{\beta}(k+1) \in \psi^{L+1}(k)$ such that the *a posteriori* errors at iteration $k-i$, $\varepsilon(k-i)$, are kept constant for $i = 2, \ldots, L+1$. Therefore, the simplified SM-KAP algorithm given by Equation (3.107) will perform an update if and only if $\boldsymbol{\beta}(k) \notin \mathcal{H}(k)$, or $e(k) > \bar{\gamma}$. The step-by-step description of the simplified SM-AP algorithm is presented in Algorithm 3.17.

Example 3.6 (a) The channel model of Example 3.1 is utilized in an equalization problem with the SM-KAP algorithm, where the channel noise variance is equal to -20 dB in comparison to the channel filtered transmitted signal at the receiver end.

(b) Test the SM-KAP for a polynomial kernel with parameters $a = n = 1$ and $b = 4$, and in this case, perform training to the equalizer for 200 iterations and after that turn the equalizer to a DD equalizer.

Solution

(a) We utilized $I_{\max} = 2000$, the same number of iterations the algorithm was tested. A nonlinear kernel adaptive filter with 10 coefficients showed no error decisions in a single run after 100 iterations. In this simulation, the adaptive

Algorithm 3.17 The set-membership kernel affine projection algorithm

Initialization

$\mathbf{x}(-1) = \boldsymbol{\beta}(0) = [0\,0\ldots0]^{\mathrm{T}}$

choose μ in the range $0 < \mu \leq 1$

choose $\bar{\gamma} = \sqrt{5}\sigma_n$

choose γ_{d} and γ_{e}

$\gamma =$ small constant

$\boldsymbol{K}_{0,\mathrm{AP}}(0) = \kappa(\mathbf{x}(0), \mathbf{x}(0)) \quad \boldsymbol{K}_{0,\mathrm{AP}}^{-1}(0) = \frac{1}{\kappa(\mathbf{X}(0),\mathbf{X}(0))}$ (3.65) and (3.66)

$\mathbf{A}_{\mathrm{AP}}^{-1}(0) = \frac{1}{\kappa^2(\mathbf{X}(0),\mathbf{X}(0))}$ (3.67)

Do for $l = 1, \ldots, L$

$$\boldsymbol{K}_{l,\mathrm{AP}}(l) = \begin{bmatrix} \kappa(\mathbf{z}(0), \mathbf{x}(l)) & \boldsymbol{\kappa}_{\mathrm{AP}}^{\mathrm{T}}(\mathbf{z}(l), \mathbf{x}(l)) \\ \boldsymbol{\kappa}_{\mathrm{AP}}(\mathbf{z}(l), \mathbf{x}(l)) & \bar{\boldsymbol{K}}_{l-1,\mathrm{AP}}(l-1) \end{bmatrix} \quad (3.68)$$

$$\mathbf{A}_{\mathrm{AP}}(l) = \left(\boldsymbol{K}_{l,\mathrm{AP}}^{\mathrm{T}}(l)\boldsymbol{K}_{l,\mathrm{AP}}(l) + \gamma\mathbf{I}\right) = \begin{bmatrix} a & \mathbf{b}^{\mathrm{T}} \\ \mathbf{b} & \mathbf{C} \end{bmatrix} \quad (3.69)$$

$$\tilde{\mathbf{C}}^{-1} = \mathbf{A}_{\mathrm{AP}}^{-1}(l-1) \quad (3.70)$$

$$\mathbf{C}^{-1} = \tilde{\mathbf{C}}^{-1} - \frac{1}{1+\mathbf{b}^{\mathrm{T}}\tilde{\mathbf{C}}^{-1}\mathbf{b}}\left[\tilde{\mathbf{C}}^{-1} - \tilde{\mathbf{C}}^{-1}\mathbf{b}\mathbf{b}^{\mathrm{T}}\tilde{\mathbf{C}}^{-1}\right] \quad (3.63)$$

$$f = a - \mathbf{b}^{\mathrm{T}}\mathbf{C}^{-1}\mathbf{b}$$

$$\mathbf{A}_{\mathrm{AP}}^{-1}(l) = \frac{1}{f}\begin{bmatrix} 1 & -\left(\mathbf{C}^{-1}\mathbf{b}\right)^{\mathrm{T}} \\ -\mathbf{C}^{-1}\mathbf{b} & \mathbf{C}^{-1}\left[\mathbf{I}f + \mathbf{b}\left(\mathbf{C}^{-1}\mathbf{b}\right)^{\mathrm{T}}\right] \end{bmatrix} \quad (3.71)$$

$$\mathbf{e}_{\mathrm{AP}}(l) = \mathbf{d}_{\mathrm{AP}}(l) - \boldsymbol{K}_{l,\mathrm{AP}}^{\mathrm{T}}(l)\begin{bmatrix} 0 \\ \boldsymbol{\beta}(l) \end{bmatrix} \quad (3.74)$$

$$\boldsymbol{\beta}(l+1) = \begin{bmatrix} 0 \\ \boldsymbol{\beta}(l) \end{bmatrix} + \mu\boldsymbol{K}_{l,\mathrm{AP}}(l)\mathbf{A}_{\mathrm{AP}}^{-1}(l)\mathbf{e}_{\mathrm{AP}}(l) \quad (3.73)$$

End do

$I = L$

For do $k \geq L + 1$

If $\underset{l \in \mathcal{L}_I(k-1)}{\mathrm{maximum}}|\kappa\left(\mathbf{x}(k), \mathbf{x}(k-l)\right)| \leq \gamma_{\mathrm{c}}$ and $I \leq I_{\max}$ (3.28)

$l_0(k+1) = k+1$, including $\mathbf{x}(k)$ in the dictionary

$$\boldsymbol{K}_{\mathrm{AP}}(k) = \begin{bmatrix} \kappa(\mathbf{z}(0), \mathbf{x}(k)) & \boldsymbol{\kappa}_{\mathrm{AP}}^{\mathrm{T}}(\mathbf{z}(I)), \mathbf{x}(k)) \\ \boldsymbol{\kappa}_{\mathrm{AP}}(\mathbf{z}(I), \mathbf{x}(k)) & \bar{\boldsymbol{K}}_{\mathrm{AP}}(k-1) \end{bmatrix} \quad (3.54)$$

$$\mathbf{e}_{\mathrm{AP}}(k) = \mathbf{d}_{\mathrm{AP}}(k) - \boldsymbol{K}_{\mathrm{AP}}^{\mathrm{T}}(k)\begin{bmatrix} 0 \\ \boldsymbol{\beta}(k) \end{bmatrix} \quad (3.74)$$

$$\mathbf{A}_{\mathrm{AP}}(k) = \left(\boldsymbol{K}_{\mathrm{AP}}^{\mathrm{T}}(k)\boldsymbol{K}_{\mathrm{AP}}(k) + \gamma\mathbf{I}\right) = \begin{bmatrix} a & \mathbf{b}^{\mathrm{T}} \\ \mathbf{b} & \mathbf{C} \end{bmatrix}$$

$$\mathbf{A}_{\mathrm{AP}}^{-1}(k-1) = \begin{bmatrix} \hat{\mathbf{A}} & \hat{\mathbf{b}} \\ \hat{\mathbf{b}}^{\mathrm{T}} & \hat{c} \end{bmatrix} \quad (3.57)$$

$$\tilde{\mathbf{C}}^{-1} = \hat{\mathbf{A}} - \frac{\hat{\mathbf{b}}\hat{\mathbf{b}}^{\mathrm{T}}}{\hat{c}} \quad (3.62)$$

$$\mathbf{C}^{-1} = \tilde{\mathbf{C}}^{-1} - \frac{1}{1+\mathbf{b}^{\mathrm{T}}\tilde{\mathbf{C}}^{-1}\mathbf{b}}\left[\tilde{\mathbf{C}}^{-1} - \tilde{\mathbf{C}}^{-1}\mathbf{b}\mathbf{b}^{\mathrm{T}}\tilde{\mathbf{C}}^{-1}\right] \quad (3.63)$$

$$f = a - \mathbf{b}^{\mathrm{T}}\mathbf{C}^{-1}\mathbf{b}$$

$$\mathbf{A}_{\mathrm{AP}}^{-1}(k) = \frac{1}{f}\begin{bmatrix} 1 & -\left(\mathbf{C}^{-1}\mathbf{b}\right)^{\mathrm{T}} \\ -\mathbf{C}^{-1}\mathbf{b} & \mathbf{C}^{-1}\left[\mathbf{I}f + \mathbf{b}\left(\mathbf{C}^{-1}\mathbf{b}\right)^{\mathrm{T}}\right] \end{bmatrix} \quad (3.64)$$

$$\mu(k) = \begin{cases} 1 - \frac{\bar{\gamma}}{|e(k)|} & \text{if } |e(k)| > \bar{\gamma} \\ 0 & \text{otherwise} \end{cases}$$

$$\boldsymbol{\beta}(k+1) = \begin{bmatrix} 0 \\ \boldsymbol{\beta}(k) \end{bmatrix} + \boldsymbol{K}_{\mathrm{AP}}(k)\mathbf{A}_{\mathrm{AP}}^{-1}(k)\mu(k)e(k)\mathbf{u}_1 \quad (3.109)$$

$I = I + 1$

Else

$\boldsymbol{K}_{\mathrm{AP}}(k) = \boldsymbol{K}_{\mathrm{AP}}(k-1)$, $\mathbf{A}_{\mathrm{AP}}^{-1}(k) = \mathbf{A}_{\mathrm{AP}}^{-1}(k-1)$

$$\mathbf{e}_{\mathrm{AP}}(k) = \mathbf{d}_{\mathrm{AP}}(k) - \boldsymbol{K}_{\mathrm{AP}}^{\mathrm{T}}(k)\boldsymbol{\beta}(k) \quad (3.50)$$

$$\mu(k) = \begin{cases} 1 - \frac{\bar{\gamma}}{|e(k)|} & \text{if } |e(k)| > \bar{\gamma} \\ 0 & \text{otherwise} \end{cases}$$

$$\boldsymbol{\beta}(k+1) = \boldsymbol{\beta}(k) + \boldsymbol{K}_{\mathrm{AP}}(k)\mathbf{A}_{\mathrm{AP}}^{-1}(k)\mu(k)e(k)\mathbf{u}_1$$

End if

End do

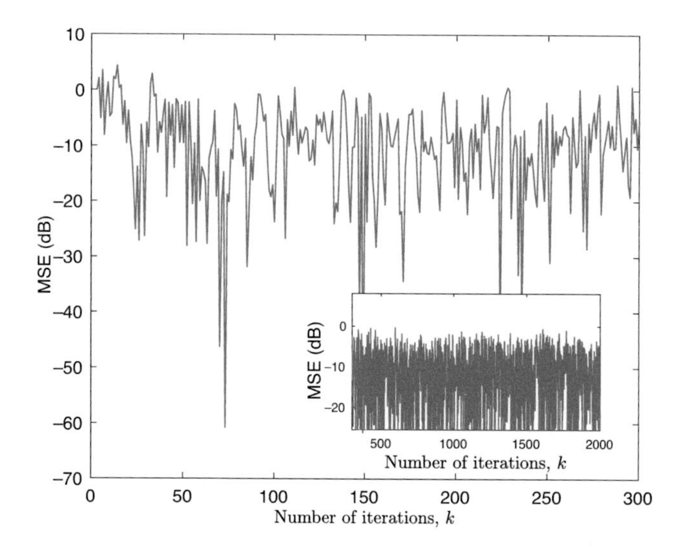

Figure 3.14 Equalization results using the SM-KAP algorithm with the Gaussian kernel.

filter has eight coefficients. Figure 3.14 depicts the MSE learning curve with the equalization error reducing slowly in the training period.

Figure 3.15 depicts the decision errors and the accumulated error decision, showing that the equalization was effective after approximately 400 iterations. Its distinct feature is the fact that the SM-KAP algorithm does not update every iteration. In this particular example it updated 15.1% of the time, using a $\bar{\gamma} = 0.3\sigma_{\mathrm{k}}^2$, with $\sigma_{\mathrm{k}}^2 = 1$ representing the variance of the Gaussian kernel.

(b) The SM-KAP algorithm utilized a polynomial kernel and $\bar{\gamma} = 0.5\sigma_{\mathrm{k}}^2$, utilized 10 coefficients, and updated 4.65% of the time. We employed $I_{\max} = 2000$, for a nonlinear kernel adaptive filter with 10 coefficients. Figure 3.16 depicts the MSE learning curve for a single run of the equalization. The equalization error remains low in the blind period.

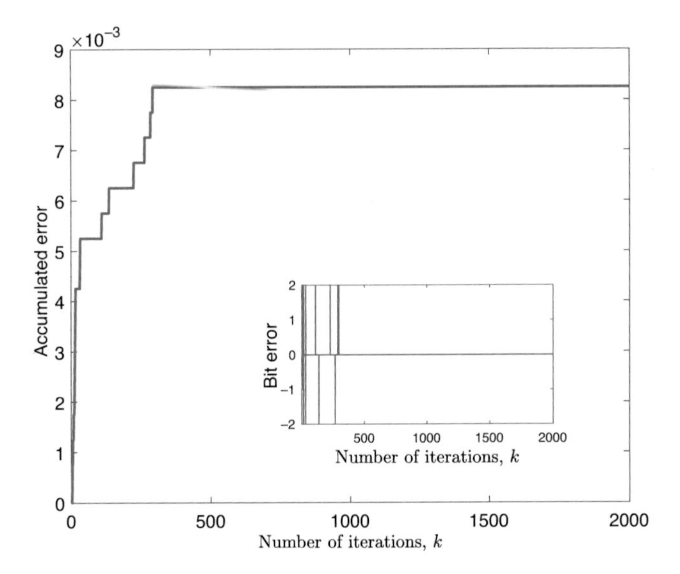

Figure 3.15 Accumulated bit error and bit error results, equalization example using the SM-KAP algorithm.

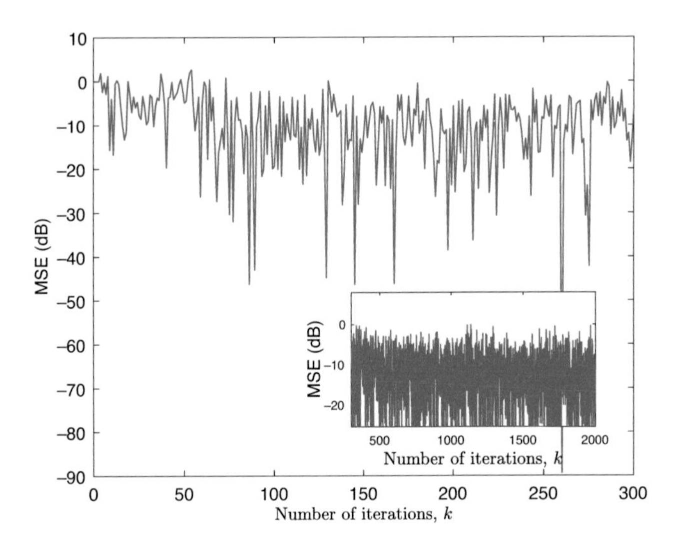

Figure 3.16 Equalization results using the SM-KAP algorithm.

Figures 3.17 shows that no decision errors occurred for over 1000 iterations after training. It is worth mentioning that the results for any particular single run can be quite different; however, the simulation presented is typical for this particular example. A single run simulation illustrates a situation closer to real-life implementations.

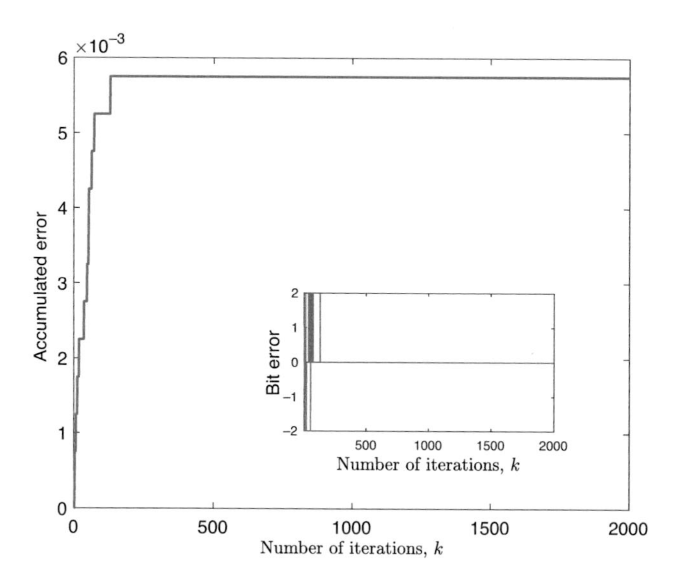

Figure 3.17 Accumulated bit error and bit error results, equalization example using the SM-KAP algorithm.

It is worth mentioning that the set-membership approach is not the only method to implement data selection in adaptive filtering algorithms. An alternative way is to prescribe a probability of coefficient updating based on measured statistical approximations of the error signal, such as discussed in [31, 32]. The same ideas apply to kernel adaptive filtering, as proposed in [33] and can be successfully embedded in machine learning algorithms [34].

The principal component analysis (PCA) also employs the kernel concept, and in this area, there is a growing interest in proposing online algorithms for the implementation. However, these solutions are more computationally intensive in comparison to the kernel-based adaptive filtering algorithms discussed here and fall in the class of blind signal processing learning algorithms, see [35–37], and the references therein.

3.7 Conclusion

This chapter shows how to apply the concepts of kernel-based learning to the typical adaptive filtering environment, where in most cases the model updates are performed online. As in the traditional kernel framework, the idea is to perform linear learning in a high-dimensional feature space through the Mercer kernel and the use of the kernel trick to avoid dealing with the computationally complex feature space vectors.

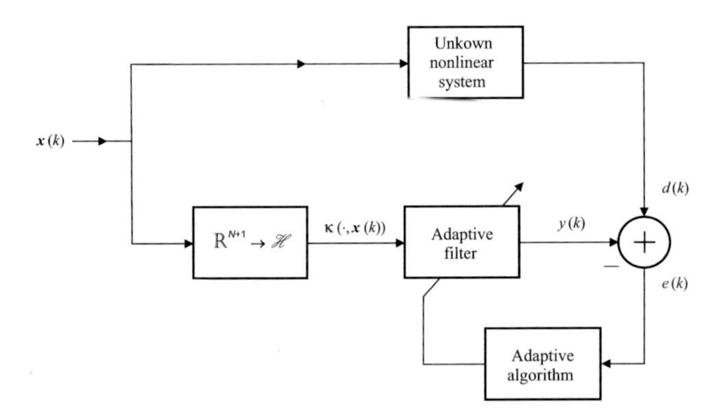

Figure 3.18 Kernel-based system identification.

Most classical kernel-based learning algorithms increase in complexity when the number of data samples also increases, hindering their widespread use in online learning applications. In this chapter, the main adaptive filtering algorithms are derived using the kernel concept that is meant to be applied in online time-varying environments. The key ingredient of the online kernel adaptive filtering algorithms is the inclusion of some sort of sparsification strategy to choose a subset of data samples that represent well the problem at hand. The algorithms presented, namely the KLMS, KAP, KRLS, and the SM-AP algorithms, consist of an alternative set of nonlinear adaptive filtering algorithms.

Problems

3.1 By employing a Gaussian kernel with variance 0.2, compare the performances of kernel LMS, AP, and RLS algorithms in the identification, see Figure 3.18, of the following system.

$$\begin{aligned} d(k) = & -0.76x(k) - 1.0x(k-1) + 1.0x(k-2) + 0.5x^2(k) \\ & + 2.0x(k)x(k-2) - 1.6x^2(k-1) + 1.2x^2(k-2) \\ & + 0.8x(k-1)x(k-2) + n(k). \end{aligned}$$

The input signal is a uniformly distributed white noise with variance $\sigma_{n_x}^2 = 0.1$, filtered by all-pole filter given by

$$H(z) = \frac{z}{z - 0.95}.$$

An additional Gaussian white noise with variance 10^{-2} is present at unknown system output.

3.2 Identify a system with the following nonlinear input to output relation:

$$\begin{aligned} d(k) = & -0.08x(k) - 0.15x(k-1) + 0.14x(k-2) + 0.055x^2(k) \\ & + 0.30x(k)x(k-2) - 0.16x^2(k-1) + 0.14x^2(k-2) + n(k) \end{aligned}$$

utilizing the kernel LMS, AP, and RLS algorithms. The input signal is Gaussian white noise with variance $\sigma_x^2 = 0.7$, and the measurement noise is also Gaussian white noise independent of the input signal with variance $\sigma_n^2 = 0.01$. Use the Gaussian kernel.

For the KAP algorithm, use $L = 1$.

3.3 For Problem 3.1, compare the performances of kernel LMS algorithm for the cases where the sparsity is implemented with the coherence approach and with the selections based on both Equations (3.26) and (3.27). Verify the number of dictionary members that would be required to guarantee a good performance.

3.4 In a digital channel, the input signal $x(k)$ traverses the linear transfer function $H(z)$ generating signal $u(k)$, which is then applied to a memoryless nonlinear function giving rise to $y(k) = \frac{2}{1+e^{-u(k)}} - 1$. The received signal is given by $r(k) = y(k) + n(k)$, where the variance σ_n^2 of the noise is equal to -20 dB. The training signal as well as the transmitted signal consist of independent binary sample $(-1, 1)$. Define an appropriate training period such that the following symbol are properly detected after equalization for

$$H(z) = \frac{1}{4} + \frac{1}{4}z^{-1} + \frac{1}{4}z^{-2} + \frac{1}{4}z^{-3}.$$

Utilize the KLMS, KAP, and the KRLS algorithms by plotting their learning curves as well as their bit error rate. Start with 10 adaptive coefficients and verify if this number is sufficient for good performance. Utilize the Gaussian kernel.

3.5 Verify the validity of Equations (3.62) and (3.64).

3.6 Repeat the Problem 3.4 assuming the given linear transfer function $H(z)$ changes to

$$H(z) = \frac{1}{4} - \frac{1}{4}z^{-1} + \frac{1}{4}z^{-2} - \frac{1}{4}z^{-3}$$

after 1500 iterations.

3.7 In Problem 3.6, compare the performances of the KAP and SM-KAP algorithms.

3.8 Identify the nonlinear system following described using kernel-based algorithms utilizing at least five adaptive coefficients:

$$d(k) = \left(e^{-|d(k-1)|} + x(k)\right)^2 + n(k),$$

where $d(0) = 0.25$, $x(k)$ is a zero-mean white Gaussian noise with unitary variance and uncorrelated with the noise $n(k)$, which is also a zero-mean white Gaussian noise with variance 0.15. Test the following algorithms:

- KLMS
- KAP
- KRLS
- SM-KAP

Utilize the Gaussian kernel.

3.9 Predict the next outcome of Mackey–Glass time-delay differential equation given by [18]:

$$\frac{dx(t)}{dt} = -\alpha_1 x(t) + \frac{\alpha_2 x(t-\tau)}{1 + x^{10}(t-\tau)}$$

using eight previous sample of the series, using $\alpha_1 = 0.1$, $\alpha_2 = 0.2$, and $\tau = 17$. Utilize the polynomial kernel of order 4 and the KLMS and KRLS algorithms.

The following script from MATLAB can be utilized to load the Mackey–Glass samples.

```
load mgdata.dat
a = mgdata;
time = a(:, 1);
x_t = a(:, 2);
plot(time, x_t);
xlabel('Time (sec)','fontsize',10); ylabel('x(t)','fontsize',10);
title('Mackey-Glass Chaotic Time Series','fontsize',10);
```

3.10 Identify the following nonlinear channel using the KAP and SM-KAP algorithms utilizing a third-order polynomial kernel.

$$\begin{aligned} u(k) =& x(k) - x(k-1); \\ d(k) =& u(k) + u^3(k) + n(k), \end{aligned}$$

where $x(k)$ is a uniformly distributed noise with unitary variance, and the noise $n(k)$ is a zero-mean Gaussian noise with variance 0.25. Choose the best parameters for the given algorithms.

3.11 Verify the validity of (3.99) and (3.100).

3.12 Verify the validity of (3.102) and (3.103).

3.13 In a signal enhancement problem, the reference signal is

$$d(k) = \sin(0.2\pi k) + n(k),$$

where $n(k)$ is zero-mean Gaussian white noise with variance $\sigma_n^2 = 10$. The input signal to the signal enhancer is given by $n(k)$ passed through a nonlinear filter described by

$$x(k) = n(k) - x(k-1)n(k-1) - 0.3x(k-1) - 0.5x(k-2).$$

Using the KAP and SM-KAP algorithms, implement the signal enhancer and plot how the error signal evolves. Choose the appropriate parameters for good convergence and proper signal enhancement.

3.14 Show the permutation properties of (3.105) and (3.106).

3.15 *Kernel RLS Algorithm with ALD*: One of the earliest KRLS algorithm is described in Algorithm 3.18 [19]. This algorithm performs the ALD test to decide if the new income data should be included in the dictionary set. It can be shown that if the chosen kernel is a continuous set and $\mathbb{R}^{I+1}$ is a compact subset of the Banach space, the number of entries in the dictionary will be finite whenever $\delta_{\mathrm{d}} > 0$ and for an infinite number of incoming data [19].

Algorithm 3.18 Kernel RLS algorithm including ALD

Initialization

$\gamma_{\mathrm{d}} = 0.01$ (prescribed small constant)

$\boldsymbol{K}_0(0) = \kappa(\mathbf{x}(0), \mathbf{x}(0))$

$\boldsymbol{K}_0^{-1}(0) = \frac{1}{\kappa(\mathbf{X}(0),\mathbf{X}(0))}$

$\boldsymbol{\beta}_0(0) = \frac{d(0)}{\kappa(\mathbf{X}(0),\mathbf{X}(0))}$

$l = 0$, and $\mathbf{B}(0) = [1]$

For do $k \geq 1$ (at this point $\boldsymbol{K}_l(k)$ is a growing matrix of dimension $l+1$)

$\boldsymbol{\kappa}_l(k) = [\kappa(\mathbf{z}(0), \mathbf{x}(k)), \kappa(\mathbf{z}(1), \mathbf{x}(k)), \ldots, \kappa(\mathbf{z}(I), \mathbf{x}(k))]$ ($\mathbf{z}(i)$ represents previously included dictionary entry)

$\mathbf{a}_l(k) = \boldsymbol{K}_l^{-1}(k-1)\boldsymbol{\kappa}_l(k)$

$\delta_{\mathrm{d}}(k) = \kappa(\mathbf{x}(k), \mathbf{x}(k)) - \mathbf{a}_l^{\mathrm{T}}(k)\boldsymbol{\kappa}_l(k)$

If $\delta_{\mathrm{d}}(k) > \gamma_{\mathrm{d}}$, add $\mathbf{x}(k)$ to the dictionary

$$\boldsymbol{K}_{l+1}^{-1}(k) = \frac{1}{\delta_{\mathrm{d}}(l)} \begin{bmatrix} \delta_{\mathrm{d}}(l)\boldsymbol{K}_l^{-1}(k-1) + \mathbf{a}_l(k)\mathbf{a}_l^{\mathrm{T}}(k) & -\mathbf{a}_l(k) \\ -\mathbf{a}_l^{\mathrm{T}}(k) & 1 \end{bmatrix}$$

$$\mathbf{B}_{l+1}(k) = \begin{bmatrix} \mathbf{B}_l(k-1) & \mathbf{0} \\ \mathbf{0} & 1 \end{bmatrix},$$ auxiliary matrix to cope with dimension increase.

$$\boldsymbol{\beta}_{l+1}(k) = \begin{bmatrix} \boldsymbol{\beta}_l(k-1) - \frac{\mathbf{a}_l(k)}{\delta_{\mathrm{d}}(k)}\left(d(k) - \boldsymbol{\kappa}_l^{\mathrm{T}}(k)\boldsymbol{\beta}_l(k-1)\right) \\ \frac{1}{\delta_{\mathrm{d}}(k)}\left(d(k) - \boldsymbol{\kappa}_l^{\mathrm{T}}(k)\boldsymbol{\beta}_l(k-1)\right) \end{bmatrix}$$

$l = l + 1$

Else

$\boldsymbol{K}_l^{-1}(k) = \boldsymbol{K}_l^{-1}(k-1)$

$\mathbf{B}_l(k) = \mathbf{B}_l(k-1) - \frac{\mathbf{B}_l(k-1)\mathbf{a}_l(k)\mathbf{a}_l^{\mathrm{T}}(k)\mathbf{B}_l(k-1)}{1+\mathbf{a}_l(k)^{\mathrm{T}}\mathbf{B}_l(k-1)\mathbf{a}_l(k)}$

$\boldsymbol{\beta}_l(k) = \boldsymbol{\beta}_l(k-1) - \boldsymbol{K}_l^{-1}(k)\frac{\mathbf{B}_l(k-1)\mathbf{a}_l(k)}{1+\mathbf{a}_l^{\mathrm{T}}(k)\mathbf{B}_l(k-1)\mathbf{a}_l(k)}\left(d(k) - \boldsymbol{\kappa}_l^{\mathrm{T}}(k)\boldsymbol{\beta}_l(k-1)\right)$

End if

End do

Repeat Problem 3.9 employing the KRLS algorithm including ALD and comment on the data dimensions after convergence.

3.16 Discuss the validity of Equation (3.36).

3.17 Prove Equation (3.37).

References

[1] N. Aronszajn, Theory of reproducing kernels. Transactions of the American Mathematical Society **63**, pp. 337–404 (1950).

[2] M. A. Aizerman, E. M. Braverman, and L. I. Rozoner, Theoretical foundations of the potential function method in pattern recognition learning. Automation and Remote Control **25**, pp. 821–837 (1964).

[3] V. Vapnik, *The Nature of Machine Learning*, 2nd ed. (Springer, New York, 1999).

[4] V. Vapnik, *Statistical Learning Theory* (Wiley Interscience, New York, 1998).

[5] S. Mallat, *A Wavelet Tour of Signal Processing* (Academic Press, Burlington, 2009).

[6] P. S. R. Diniz, *Adaptive Filtering: Algorithms and Practical Implementations*, 5th ed. (Springer, Cham, 2020).

[7] W. Liu, P.P. Pokharel, and J.C. Príncipe, The kernel least-mean-square algorithm. IEEE Transactions on Signal Processing. **56**, pp. 543–554 (2008).

[8] W. Liu, J.C. Príncipe, S. Haykin, *Kernel Adaptive Filtering: A Comprehensive Introduction* (Wiley, Hoboken, 2010).

[9] C. Bishop, *Pattern Recognition and Machine Learning* (Springer, New York, 2007).

[10] B. Schölkopf and A. L. Smola, *Learning with Kernels: Support Vector Machine, Regularization, Optimization and Beyond* (The MIT Press, Cambridge, 2001).

[11] K. P. Murphy, *Machine Learning: A Probabilistic Perspective* (The MIT Press, Cambridge, 2012).

[12] S. Theodoridis, *Machine Learning: A Bayesian and Optimization Perspective* (Academic Press, Oxford, 2015).

[13] Y. S. Abu-Mostafa, M. Magdon-Ismail, and H.-T. Lin, *Learning from Data* (AMLbook.com, 2012).

[14] C. Richard, J. C. M. Bermudez, and P. Honeine, Online prediction of time series data with kernels. IEEE Transactions on Signal Processing **57**, pp. 1058–1067 (2009).

[15] W. D. Parreira, J. C. M. Bermudez, C. Richard, and J.-Y. Tourneret, Stochastic behavior analysis of the Gaussian kernel least-mean-square algorithm. IEEE Transactions on Signal Processing **60**, pp. 2208–2222 (2012).

[16] K. Ozeki, *Theory of Affine Projection Algorithms for Adaptive Filtering* (Springer, New York, 2015).

[17] F. Albu, D. Coltuc, M. Rotaru, and K. Nishikawa, An efficient implementation of the kernel affine projection algorithm. 8th International Symposium on Image and Signal Processing and Analysis (ISPA 2013), Trieste, Italy, 2013, pp. 349–353.

[18] P. Honeine, Approximation errors of online sparsification criteria. IEEE Transactions on Signal Processing **63**, pp. 4700–4709 (2015).

[19] Y. Engel, S. Mannor, and R. Meir, The kernel recursive least-squares algorithm. IEEE Transactions on Signal Processing **52**, pp. 2275–2285 (2004).

[20] S. Van Vaerenbergh, J. Vía, and I. Santamaría, A sliding-window kernel RLS and its application to nonlinear channel identification. Proceedings of the IEEE International Conference on Acoustics, Speech, and Signal Processing, Toulouse, France, May 2006, pp. V-789–V-792.

[21] S. Van Vaerenbergh, J. Vía, and I. Santamaría, Nonlinear system identification using new sliding-window kernel RLS algorithm. Journal of Communications **2**, pp. 1–8 (2007).

[22] S. Van Vaerenbergh, I. Santamaría, W. Liu, and J. C. Príncipe, Fixed-budget kernel recursive least-squares. Proceedings of the IEEE International Conference on Acoustics, Speech, and Signal Processing, Toulouse, France, May 2006, pp. V-789–V-792.

[23] A. V. Malipatil, Y.-F. Huang, S. Andra, and K. Bennett, Kernelized set-membership approach to nonlinear adaptive filtering. Proceedings of the IEEE International Conference on Acoustics, Speech, and Signal Processing, Philadelphia, USA, May 2005, pp. IV-149–IV-152.

[24] A. Flores and R. C. de Lamare, Set-membership kernel adaptive algorithms. Proceedings of the IEEE International Conference on Acoustics, Speech, and Signal Processing, New Orleans, USA, May 2017, pp. 2676–2680.

[25] K. Chen, S. Werner, A. Kuh, and Y.-F. Huang, Nonlinear adaptive filtering with kernel set-membership approach. IEEE Transactions on Signal Processing **68**, pp. 1515–1528 (2020).

[26] A. Rahimi and B. Recht, Random features for large-scale kernel machines. NIPS'07: Proceedings of the 20th International Conference on Neural Information Processing Systems, Curran Associates, Inc., 2008, pp. 1–8.

[27] B. K. Sriperumbudur and Z. Szabó, Optimal rates for random Fourier features. NIPS'15: Proceedings of the 28th International Conference on Neural Information Processing Systems, pp. 1144–1152.

[28] D. J. Sutherland and J. Schneider, On the error of random Fourier features. arXiv:1506.02785, 2015, pp. 1–10.

[29] K. Muandet, K. Fukumizu, B. Sriperumbudur, and B. Schölkopf, *Kernel Mean Embedding of Distributions: A Review and Beyond* (NOW Publishers, Delft, 2017).

[30] W. Rudin, *Fourier Analysis on Groups* (Dover Publications, New York, 2017).

[31] P. S. R. Diniz, On data-selective adaptive filtering. IEEE Transactions on Signal Processing **66**, pp. 4239–4252 (2018).

[32] M. O. K. Mendonça, J. O. Ferreira, C. G. Tsinos, P. S. R. Diniz, and T. N. Ferreira, On fast converging data-selective adaptive filtering. Algorithms **12**, pp. 1–15 (2019).

[33] P. S. R. Diniz, J. O. Ferreira, M. O. K. Mendonça, and T. N. Ferreira, Data selection kernel conjugate gradient algorithm. Proceedings of the IEEE International Conference on Acoustics, Speech, and Signal Processing, Barcelona, Spain, May 2020, pp. 1–5.

[34] J. O. Ferreira, M. O. K. Mendonça, and P. S. R. Diniz, Data selection in neural networks. IEEE Open Journal of Signal Processing **2**, pp. 1–15 (2021).

[35] J. B. Souza Filho and P. S. R. Diniz, A recursive least square algorithm for online kernel principal component extraction. Neuralcomputing **237**, pp. 255–264 (2017).

[36] J. B. Souza Filho, P. S. R. Diniz, J. B. Souza Filho, and P. S. R. Diniz, Fixed-point online kernel principal component extraction algorithm. IEEE Transactions on Signal Processing **65**, pp. 6244–6259 (2017).

[37] J. B. Souza Filho and P. S. R. Diniz, Improving KPCA online extraction by orthonormalization in the feature space. IEEE Transactions on Neural Networks and Learning Systems **29**, pp. 2162–2388 (2018).

4 Distributed Adaptive Filters

4.1 Introduction

This chapter describes distributed learning methods that may be utilized when nodes of a network have the necessary computational capacity to process data locally. Collecting data from multiple sensors may improve network performance and robustness through diversity and cooperation. However, the use of resources to transmit all the data to a fusion center for processing can be prohibitive and inefficient, particularly for large and complex networks. Decentralized data processing offers the advantage of diversity together with reduced communication requirements. In addition to performance gains, distributed networks offer increased reliability in the event of faulty nodes, or broken links.

Similar to forming an opinion pool among experts, distributed adaptive filters may involve four distinct steps: sharing data with peers, making a local assessment of the problem, sharing the result of the assessment with peers, and making an informed decision. In the context of the algorithms discussed in this chapter, the peers are the nodes in the neighborhood, the assessment is the local estimate of the parameters, and the informed final decision is the move toward a consensus. The different ways these steps can be implemented result in different algorithms, with different storage and bandwidth requirements, and also different computational complexities.

The chapter starts with a discussion on equilibrium and consensus through examples drawn from the pari-mutuel betting system. It presents the expert opinion pool as a key concept to improved estimation and data modeling and uses DeGroot's algorithm to formalize equilibrium at consensus with the aid of Markov chains. The process of sharing, adapting, and aggregating is introduced with a token-ring strategy for the incremental buildup of the estimate, whereby the LMS and the RLS algorithms are developed. The more generic network topology is considered in diffusion update strategies, which include Combine-then-Adapt (CtA) and Adapt-then-Combine (AtC) LMS implementations. A whole section is dedicated to distributed adaptive filters with selective updating due to its importance for computational and transmission savings provided by the set-membership (SM) theory. The last section outlines adaptation of parameters for distributed detection.

4.2 Equilibrium and Consensus Strategies

Consensus may present itself in different ways, as many are the possibilities to aggregate contributions of experts. According to dictionaries, consensus means agreement, but not necessarily unanimity. In a decision-making system that shall take into account the contributions of a cohort of different experts, the question becomes how to build consensus weighting individual strengths and levels of confidence. The methods may require action from a decision maker, as a fusion center, not only to decide how to combine but also to perform the combination itself. However, a distributed processing system may want to do without the central node, and this is the arrangement of interest in this chapter. Furthermore, reaching consensus means that the algorithm underneath converges and remains meaningful, and this equilibrium point will also be a concern herein.

4.2.1 A Day at the Races: The Pari-Mutuel Betting System

Before we delve into consensus and distributed adaptation algorithms, let us examine the strategy suggested by Eisenberg and Gale [1] for a well-known betting system in which individuals autonomously maximize a utility function that dynamically changes along the process. An equilibrium point does exist, as we shall see, although consensus may not be particularly interesting in wagering.

The pari-mutuel betting system is very common for regulating the horse-race rewarding process. Let B_i, where $i = 1, \ldots, M$, represent the bettors and H_j, where $j = 1, \ldots, N$, represent the horses. We may construct a matrix $\mathbf{P}$ whose entries, $p_{i,j}$, are the prior probabilities, in the eyes of the i-th bettor, that the jth horse will win the race. These prior probabilities may reflect, for example, historical performances, gathered statistics from previous races, and possibly information regarding the well-being of animals and jockeys. Without loss of generality, let us assume that all horses deserve some credit from at least one bettor, which translates in a matrix $\mathbf{P}$ with entries in every column. Each bettor has a budget, denoted by b_i, and $\beta_{i,j}$ is the amount bettor i places on horse j. For simplicity, let us assume the total amount of bets adds up to \$1,

$$\sum_{i=1}^{M} b_i = 1 \tag{4.1}$$

and, naturally, we must have

$$\sum_{j=1}^{N} \beta_{i,j} = b_i. \tag{4.2}$$

The rules state that after all bets are placed and the betting system is closed, house-takes and taxes are deducted, and the payouts are calculated. An amount

$\sum_i \beta_{i,j}$ is totaled for each horse H_j as

$$\pi_j = \sum_{i=1}^{M} \beta_{i,j}. \tag{4.3}$$

For the purpose of this discussion and for the sake of simplicity, let us assume that deductions are waived, and only simple bets are allowed rewarding solely the winning horse. The payout for each horse H_j is calculated as the total amount of bets, $\sum_i b_i$, divided by the amount totaled for H_j, that is, π_j. As we have normalized $\sum_i b_i = 1$, the payout for horse H_j, if the winner, will be $1/\pi_j$ for every \$1 wagered on it. π_j also reflects the expectation, aggregated by the cohort of bettors and developed throughout the betting process, that the jth horse will win. Combining the previous three equations, we obtain, as expected,

$$\sum_{j=1}^{N} \pi_j = 1. \tag{4.4}$$

In the pari-mutuel system, betting on the horse with the highest probability of winning the race may be the safest, but not the most rewarding strategy. As one possible alternative [1], B_i may want to maximize the subjective expectation that will balance wining probability and payout odds. This changes the focus off betting the whole budget b_i on the horse H_j for which $p_{i,j}$ is the largest for all j. Instead, B_i shall examine and maximize the ratio $p_{i,j}/\pi_j$. B_i wants to bet only on the horse H_j for which $p_{i,j}/\pi_j$ is maximized, that is, for bettor B_i, $\beta_{i,j} > 0$ only if j is such that $p_{i,j}/\pi_j$ is maximal.

The difficulty is that B_i's bet changes H_j's payout on the process, therefore influencing π_j! This sets the pari-mutuel apart from other betting systems, for the offered payout is not known at the moment B_i is placing the bet. Although one may argue that strict distributed implementation of the strategy presented above has implicit difficulties and perhaps unparalleled obstacles at the races, there is an equilibrium point which adheres to the bettors' strategies and also respects the pari-mutuel principle [1, 2].

Let $\phi : \mathbb{R}^{M\times N} \to \mathbb{R}$ be a function defined as

$$\phi(\xi_{1,1}, \ \ldots, \ \xi_{M,N}) = \sum_{i=1}^{M} b_i \log \sum_{j=1}^{N} p_{i,j}\xi_{i,j} \tag{4.5}$$

with domain defined as the set of all $\xi_{i,j}$ such that

$$\xi_{i,j} \geq 0, \tag{4.6}$$

for all i, j, and

$$\sum_{i=1}^{M} \xi_{i,j} = 1, \tag{4.7}$$

for all j.

The equilibrium point turns out to be unique [1] for the set of probabilities

$$\pi_j = \max_i \frac{\partial \phi}{\partial \xi_{i,j}} = \max_i \frac{b_i p_{i,j}}{\sum_l p_{i,l}\xi_{i,l}} = \frac{b_i p_{i,j}}{\sum_l p_{i,l}\bar{\xi}_{i,l}}. \tag{4.8}$$

The corresponding amount bettor i places on horse j can be

$$\beta_{i,j} = \pi_j \bar{\xi}_{i,j}, \tag{4.9}$$

which also satisfies equilibrium.

As suggested by Eisenberg and Gale [1], ϕ is the weighted sum of the logarithms of the subjective expectation that B_i has when betting $\beta_{i,j}$ on horse H_j. The weights are the respective budgets, or how much each bettor is able to influence the total budget. Therefore, if ϕ is maximized at equilibrium, then the mindset of bettors, as a group, is to maximize the weighted subjective expectations.

The following example illustrates the system and the equilibrium obtained. Let there be two bettors, $M = 2$, and two horses, $N = 2$. The budgets of the bettors are divided as $b_1 = 0.2$ and $b_2 = 0.8$, and their prior assessment of the horses' chances is:

$$\mathbf{P} = \begin{bmatrix} 0.3 & 0.7 \\ 0.4 & 0.6 \end{bmatrix}. \tag{4.10}$$

If B_1 and B_2 want to put their money for maximum ratio $p_{i,j}/\pi_j$, the equilibrium formulas point out to $\pi_1 = 0.4$ and $\pi_2 = 0.6$. This leads to $p_{1,1}/\pi_1 = 3/4$ and $p_{1,2}/\pi_2 = 7/6$, which suggests that B_1 should choose H_2 over H_1, whereas $p_{2,1}/\pi_1 = p_{2,2}/\pi_2 = 1$, which suggests that B_2 should split the budget between H_1 and H_2. Therefore, B_1's money should not go to H_1, that is,

$$\beta_{1,1} = 0 \Rightarrow \bar{\xi}_{1,1} = 0 \tag{4.11}$$

and the whole budget should go to H_2,

$$\beta_{1,2} = b_1 = 0.2 \Rightarrow \bar{\xi}_{1,2} = 0.2/\pi_2 = 1/3. \tag{4.12}$$

For B_2, we can use the fact that $\bar{\xi}_{2,1} = 1 - \bar{\xi}_{1,1} = 1$ and $\bar{\xi}_{2,2} = 1 - \bar{\xi}_{1,2} = 2/3$, and obtain

$$\beta_{2,1} = \pi_1 \bar{\xi}_{2,1} = 0.4 \tag{4.13}$$

and

$$\beta_{2,2} = \pi_2 \bar{\xi}_{2,2} = 0.4, \tag{4.14}$$

which is the equal split of B_2's budget between H_1 and H_2, as expected. The reader is invited to verify that no other feasible equilibrium points π_1 and π_2 exist that maximizes the ratios $p_{i,j}/\pi_j$ for each bettor B_i over all horses H_j.

A beautiful and simple algorithm that reaches the equilibrium solution of the Eisenberg–Gale betting strategy for the pari-mutuel system was proposed by Norvig [3] and is summarized in Algorithm 4.19.

Algorithm 4.19 A dynamic model for the Eisenberg–Gale betting strategy for the pari-mutuel system [3]

Initialization
$\mathbf{P} \in \mathbb{R}^{M \times N}$ a matrix with the prior probabilities
$\mathbf{b} \in \mathbb{R}^{M}$ a vector with the individual budgets, $\|\mathbf{b}\|_1 = 1$
$\boldsymbol{\pi} \in \mathbb{R}^{N}$ a random vector with $\|\boldsymbol{\pi}\|_1 = 1$ and $0 \leq \pi_j \leq 1$
For $k = 1, \ldots, K$ **do**
  For $i = 1, \ldots, M$ **do**
    ind = index of max.($\mathbf{P}[i,:]./\boldsymbol{\pi}$) # first index of max. ratio $\mathbf{P}_{i,j}/\boldsymbol{\pi}_j$
    $\boldsymbol{\pi}[\text{ind}] = \boldsymbol{\pi}[\text{ind}] + \mathbf{b}[i]$
  end
end
$\boldsymbol{\pi} = \boldsymbol{\pi}/\|\boldsymbol{\pi}\|_1$
For $i = 1, \ldots, M$ **do**
  ind = index of max.($\mathbf{P}[i,:]./\boldsymbol{\pi}$) # index(es) of max. ratio $\mathbf{P}_{i,j}/\boldsymbol{\pi}_j$
  len = length(ind)
  $\beta_{i,\text{ind}} = \mathbf{b}[i]/\text{len}$ # otherwise $\beta_{i,j} = 0$
end

4.2.2 Forming an Opinion Pool

In yet another possible strategy, let the M bettors be replaced by experts to be consulted on a particular subject. Let $f(\boldsymbol{\theta}|\mathrm{E}_i)$, where $i = 1, \ldots, M$, be the contribution, or opinion, of the i-th expert, in the form of a probability density function of some process outcome, parameterized by a vector $\boldsymbol{\theta}$. For example, f can be the probability of a successful outcome of an entrepreneurship. Perhaps, the most obvious method to obtain $f(\boldsymbol{\theta}|\mathrm{E}_1, \ldots, \mathrm{E}_M)$, the single, consensus-based, decision-making system, is to take a weighted average of the M individual contributions, often referred to as the result of an opinion pool [4]:

$$f(\boldsymbol{\theta}|\mathrm{E}_1, \ldots, \mathrm{E}_M) = \sum_{i=1}^{M} w_i f(\boldsymbol{\theta}|\mathrm{E}_i), \tag{4.15}$$

where usually $w_i \geq 0$ and $\sum_i w_i = 1$ [5].

Another possibility is to combine the parameters themselves, as

$$\boldsymbol{\theta} = \sum_{i=1}^{M} w_i \boldsymbol{\theta}_i, \tag{4.16}$$

where $\boldsymbol{\theta}_i$ is the set of parameters provided, or used, by the i-th expert. Other possibilities do exist, as the following example will illustrate. As usually there is no correct answer, caution and experience should not be spared.

As an example, let us assume we want to endeavor some new enterprise, and that we are completely ignorant about our chance of success, θ, except that it is some number between 0 and 1. With no prior knowledge of how θ is distributed,

Table 4.1 Data D_1 and D_2 provided by experts E_1 and E_2

E_1		E_2	
#successes	#failures	#successes	#failures
30	70	80	120

we will model $f(\theta)$, the probability density function of the parameter that indicates our chance of success, as uniformly distributed between 0 and 1, that is, $f(\theta) = 1,\ 0 \leq \theta \leq 1$.

Before we make a decision, we decide to consult with experts in the field, E_1 and E_2. They provide us data that will help us build posterior distributions of parameter θ, hopefully guiding us toward a more educated decision. Table 4.1 summarizes the information provided by the experts, indicating us the rates of success and failure they have observed in situations similar to ours. The first data set shows that in 100 cases, E_1 observed only 30 cases of success, whereas a different data set, provided by E_2, shows a success rate of 80 success cases in 200.

From Bayes' theorem, we know that the posterior distribution is proportional to the likelihood function times the prior distribution. Therefore, the respective posterior density functions are

$$f(\theta|D_1) \propto \theta^{30}(1-\theta)^{70} \tag{4.17}$$

and

$$f(\theta|D_2) \propto \theta^{80}(1-\theta)^{120}, \tag{4.18}$$

respectively. For better or worse, after using data provided by experts, we were able to move from the uniform distribution of θ to beta posterior distributions with parameters $(31, 71)$ and $(81, 121)$. Figure 4.1(a) shows the beta probability density functions obtained after taking into account the data provided by each expert separately, whereas the dashed line in Figure 4.1(b) shows an attempt to achieve consensus with a convex combination of the posteriors, although from a statistical point of view it may make no sense at all:

$$f(\theta|D_1, D_2) = w_1 f(\theta|D_1) + w_2 f(\theta|D_2). \tag{4.19}$$

We could, arguably, make the convex combination of the estimates θ_1 and θ_2 that individually maximize the posteriors, or else, propose a consensus distribution using the aggregated data sets, therefore obtaining *a posterior* distribution which is beta $(111, 191)$ [5]. The solid line in Figure 4.1(b) shows the latter strategy.

The experts may be given the opportunity to reassess their judgements based on the feedback they receive from other experts' opinions, a process that can be repeated a number of times. Unless their contributions are very different and the experts only trust their own data and assessment, the feedback–reassessment

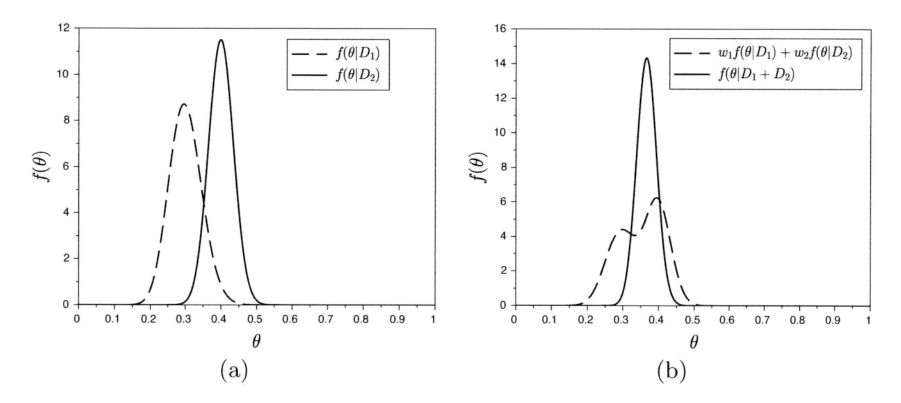

Figure 4.1 Posterior density functions: (a) each expert information considered separately and (b) convex combination of $f(\theta|D_1)$ and $f(\theta|D_2)$ with $w_1 = w_2 = 0.5$ versus the posterior density function using the aggregated data.

repeated routine can lead to consensus. In the next section, we will investigate the situation where the experts are apprised of their peers' assessments and are invited to review their own.

4.2.3 DeGroot's Algorithm

Let $p_{i,j}$ be the weight expert E_i assigns to the opinion of expert E_j, $p_{i,j} \geq 0$ and $\sum_{j=1}^{M} p_{i,j} = 1$, for all values of $i = 1, \ldots, M$. A reasonable reassessment of E_i's opinion may be a weighted sum of the previous assessments made by his peers, as follows:

$$f_2(\boldsymbol{\theta}|\mathrm{E}_i) = \sum_{j=1}^{M} p_{i,j} f_1(\boldsymbol{\theta}|\mathrm{E}_j), \tag{4.20}$$

where the sub-index indicates the iteration and each individual i is responsible for choosing the appropriate weights $p_{i,j}$, $j = 1, \ldots, M$.

Let $\mathbf{P}$ be the stochastic matrix whose entries are $p_{i,j}$, and let

$$\mathbf{f}_k(\boldsymbol{\theta}|\mathbf{E}) = \begin{bmatrix} f_k(\boldsymbol{\theta}|\mathrm{E}_1) & f_k(\boldsymbol{\theta}|\mathrm{E}_2) & \cdots & f_k(\boldsymbol{\theta}|\mathrm{E}_M) \end{bmatrix}^{\mathrm{T}} \tag{4.21}$$

be the vector with all individual contributions in iteration k. Repeated feedback and reassessment yields

$$\mathbf{f}_k(\boldsymbol{\theta}|\mathbf{E}) = \mathbf{P}^{k-1}\mathbf{f}_1(\boldsymbol{\theta}|\mathbf{E}). \tag{4.22}$$

The routine can be repeated until an eventual equilibrium is reached, defined as the condition $\mathbf{f}_{k+1}(\boldsymbol{\theta}|\mathbf{E}) = \mathbf{f}_k(\boldsymbol{\theta}|\mathbf{E})$. Consensus at equilibrium, however, is reached if $f_k(\boldsymbol{\theta}|\mathrm{E}_i) = f_k(\boldsymbol{\theta}|\mathrm{E}_j) = \bar{f}(\boldsymbol{\theta}|\mathbf{E})$ for all $i, j = 1, \ldots, M$, or, equivalently,

$$\mathbf{P}^k\mathbf{f}_1(\boldsymbol{\theta}|\mathbf{E}) = \bar{f}(\boldsymbol{\theta}|\mathbf{E})\mathbf{1}, \tag{4.23}$$

where $\mathbf{1} = \begin{bmatrix} 1 & 1 & \cdots & 1 \end{bmatrix}^{\mathrm{T}}$. Equation (4.23) has a solution if, and only if, all rows of matrix $\mathbf{P}^k$ are equal for some k sufficiently large to guarantee equilibrium. Let us define this row, at equilibrium, as

$$\boldsymbol{\pi}^{\mathrm{T}} = \begin{bmatrix} \pi_1 & \pi_2 & \cdots & \pi_M \end{bmatrix}, \tag{4.24}$$

where $\pi_j \geq 0$ and $\sum_{j=1}^{M} \pi_j = 1$. This solution, if exists, is unique [6].

P is a stochastic matrix in an one-step Markov chain [7] with M states and stationary probabilities. According to DeGroot [6], consensus, as defined above, is reached if there is at least one column of matrix **P** with all entries strictly positive. Analogously, we may say that consensus is reached if all the recurrent states in the equivalent Markov chain communicate with each other and are aperiodic [6]. The following examples illustrate equilibrium and consensus, as applicable.

For a first example, let there be $M = 2$ experts, who make a first assessment of the probability density functions $f_1(\boldsymbol{\theta}|\mathrm{E}_1)$ and $f_1(\boldsymbol{\theta}|\mathrm{E}_2)$, respectively. Let the weighting matrix, which indicates the confidence the experts have on each other and on themselves, be

$$\mathbf{P} = \begin{bmatrix} 0.3 & 0.7 \\ 0.4 & 0.6 \end{bmatrix} \tag{4.25}$$

which gives, at equilibrium,[1]

$$\lim_{k\to\infty} \mathbf{P}^k = \begin{bmatrix} 4/11 & 7/11 \\ 4/11 & 7/11 \end{bmatrix}. \tag{4.26}$$

Therefore, $\pi_1 = 4/11$ and $\pi_2 = 7/11$, and the consensus probability density function is

$$\bar{f}(\boldsymbol{\theta}|\mathbf{E}) = \pi_1 f_1(\boldsymbol{\theta}|\mathrm{E}_1) + \pi_2 f_1(\boldsymbol{\theta}|\mathrm{E}_2). \tag{4.27}$$

It is clear, in the example, that all the recurrent states communicate with each other and are aperiodic. In a different situation, in which the experts were insecure about their assessments, they might consider each other's opinions in lieu of their own, which would be reflected as the following matrix **P**:

$$\mathbf{P} = \begin{bmatrix} 0 & 1 \\ 1 & 0 \end{bmatrix}. \tag{4.28}$$

Although the states are recurrent, in this case they are periodic, and consensus is not reached.

Now suppose a different example in which a third expert is included in the pool, unbeknown to the first two, such that the weighting matrix becomes

$$\mathbf{P} = \begin{bmatrix} 0.3 & 0.7 & 0 \\ 0.4 & 0.6 & 0 \\ 0.3 & 0.3 & 0.4 \end{bmatrix}. \tag{4.29}$$

[1] For a discussion of powers of stochastic matrices, see, for example, [8, 9].

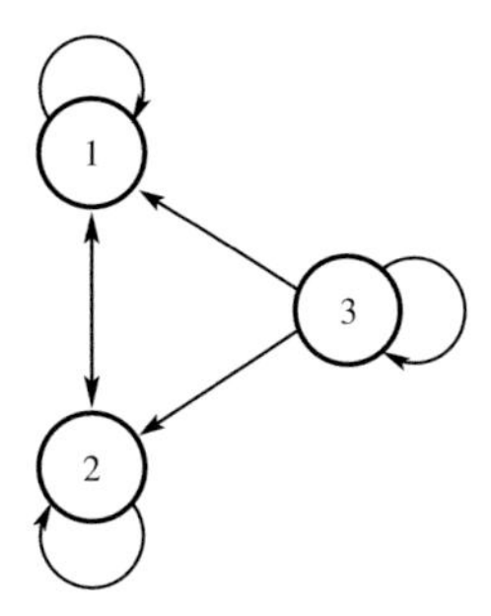

Figure 4.2 Markov chain with one transient state.

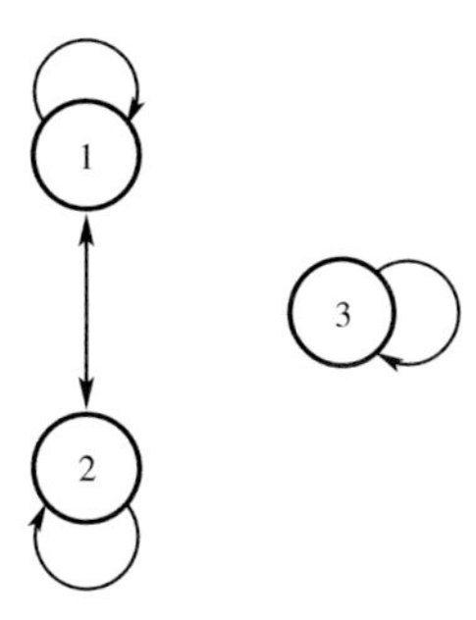

Figure 4.3 Markov chain with two closed irreducible sets of states.

The consensus is achieved at equilibrium with $\pi_1 = 4/11$, $\pi_2 = 7/11$, and $\pi_3 = 0$, that is, $f_1(\boldsymbol{\theta}|\mathrm{E}_3)$ receives no consideration. Figure 4.2 shows that state 3 in the Markov chain, represented by the third expert, is transient and vanishes at consensus.

As a last example, consider a pool of the same three experts as above, but in this case, E_1 and E_2 are also unbeknown to E_3:

$$\mathbf{P} = \begin{bmatrix} 0.3 & 0.7 & 0 \\ 0.4 & 0.6 & 0 \\ 0 & 0 & 1 \end{bmatrix}. \tag{4.30}$$

Equilibrium is reached, but no consensus:

$$\lim_{k\to\infty} \mathbf{P}^k = \begin{bmatrix} 4/11 & 7/11 & 0 \\ 4/11 & 7/11 & 0 \\ 0 & 0 & 1 \end{bmatrix}. \tag{4.31}$$

As can be seen in Figure 4.3, in this case all states are recurrent, but not all of them communicate with each other. Therefore, we may not expect consensus.

4.2.4 An Incremental Strategy for Reaching Consensus

A particular and interesting case of DeGroot's algorithm arises when the pool of experts acts like a token ring, receiving feedback from only one neighbor

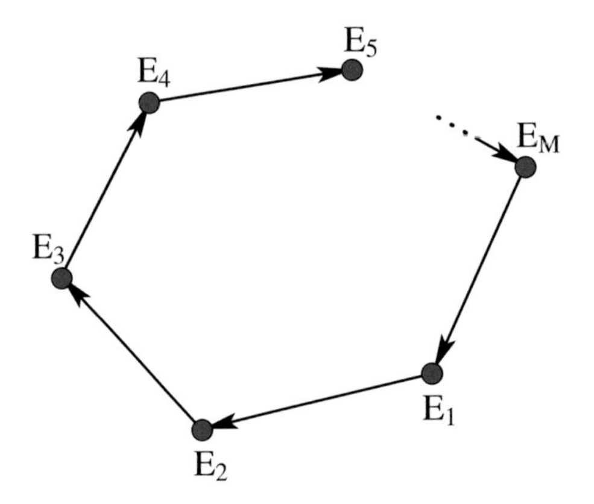

Figure 4.4 A token ring for incremental updates.

in order to reassess his own estimate, and sharing the new estimate with the next in line. One may argue that message passing may delay consensus, for reassessment is done in series, instead of in parallel. However, the incremental update requires limited knowledge of the network, improving security, and needed communication resources are minimal.

Let $\{\mathrm{E_i}\}$, where $i = 1, \ldots, M$, be a list of experts ordered as they are allowed to communicate: $\mathrm{E_1}$ shares his knowledge with $\mathrm{E_2}$, who shares his knowledge with $\mathrm{E_3}$, and so on. $\mathrm{E_M}$ receives information from $\mathrm{E_{M-1}}$ and shares his own reassessment with $\mathrm{E_1}$, closing the ring, as depicted in Figure 4.4. Whoever has the token must have received the latest update from the previous expert in the list and will use this information to reassess his own estimate. Afterwards, he shall pass the token with his updated estimate to the next in the list.

We may need to modify the notation in Equation (4.21) in order to take into account the token passing and the complete cycles. Let

$$\mathbf{f}_{k,i}(\boldsymbol{\theta}|\mathbf{E}) = \left[f_{k,i}(\boldsymbol{\theta}|\mathrm{E}_1) \quad f_{k,i}(\boldsymbol{\theta}|\mathrm{E}_2) \quad \cdots \quad f_{k,i}(\boldsymbol{\theta}|\mathrm{E}_M)\right]^{\mathrm{T}} \tag{4.32}$$

be the vector with the individual contributions of the experts at cycle, or iteration, k when expert i has the token. We may write

$$\mathbf{f}_{k,2}(\boldsymbol{\theta}|\mathbf{E}) = \begin{bmatrix} 1 & 0 & 0 & \cdots & 0 \\ p_{2,1} & p_{2,2} & 0 & \cdots & 0 \\ 0 & 0 & 1 & \cdots & 0 \\ \vdots & & & \ddots & \vdots \\ 0 & & & \cdots & 1 \end{bmatrix} \mathbf{f}_{k,1}(\boldsymbol{\theta}|\mathbf{E}) \tag{4.33}$$

for the reassessment of the opinion of expert E_2,

$$\mathbf{f}_{k,3}(\boldsymbol{\theta}|\mathbf{E}) = \begin{bmatrix} 1 & 0 & 0 & \cdots & 0 \\ 0 & 1 & 0 & \cdots & 0 \\ 0 & p_{3,2} & p_{3,3} & \cdots & 0 \\ \vdots & & & \ddots & \vdots \\ 0 & & & \cdots & 1 \end{bmatrix} \mathbf{f}_{k,2}(\boldsymbol{\theta}|\mathbf{E}) \tag{4.34}$$

for the reassessment of the opinion of expert E_3. Each cycle completes with expert E_1, as

$$\mathbf{f}_{k+1,1}(\boldsymbol{\theta}|\mathbf{E}) = \begin{bmatrix} p_{1,1} & 0 & 0 & \cdots & p_{1,M} \\ 0 & 1 & 0 & \cdots & 0 \\ 0 & 0 & 1 & \cdots & 0 \\ \vdots & & & \ddots & \vdots \\ 0 & & & \cdots & 1 \end{bmatrix} \mathbf{f}_{k,M}(\boldsymbol{\theta}|\mathbf{E}). \tag{4.35}$$

Each of the stochastic matrices represented above indicates the trust a particular expert has on his own opinion and on the opinion of the adjacent neighbor.

A simple example illustrates how the incremental updating unfolds and how it compares with the strategies examined in the previous section. Let the pool be formed with three experts, E_1, E_2, and E_3. The composite stochastic matrix $\mathbf{P}$ which induces the transformation performed onto $\mathbf{f}_k(\boldsymbol{\theta}|\mathbf{E})$ after one cycle is, in this case,

$$\mathbf{P} = \begin{bmatrix} p_{1,1} + p_{1,3}p_{3,2}p_{2,1} & p_{1,3}p_{3,2}p_{2,2} & p_{1,3}p_{3,3} \\ p_{2,1} & p_{2,2} & 0 \\ p_{3,2}p_{2,1} & p_{3,2}p_{2,2} & p_{3,3} \end{bmatrix}. \tag{4.36}$$

Notice that the reassessment made by E_3 takes into account the opinion of E_1, $f_k(\boldsymbol{\theta}|E_1)$, albeit indirectly and weighted by the trust E_2 has on E_1, denoted by $p_{2,1}$, and also the trust E_3 has on E_2, denoted by $p_{3,2}$. Notice also that the trust E_3 puts on E_2's opinion is itself affected by the trust E_2 has on himself, denoted by $p_{2,2}$, multiplied by the trust E_3 has on E_2. Consensus occurs provided matrix $\mathbf{P}$ in Equation (4.36) has at least one column with all entries strictly positive. Figure 4.5 illustrates how the information is passed from one expert to the next.

We have chosen E_1 arbitrarily for starting the cycle sharing his opinion $f_{k,1}(\theta|E_1)$ with E_2, as seen in Figure 4.5(a). Therefore, the cycle ends also with E_1, who is responsible for receiving the information from E_3 and forming the vector of individual contributions for the next iteration, $\mathbf{f}_{k+1,1}(\theta|\mathbf{E})$, as seen in Figure 4.5(c).

4.3 Distributed Adaptive Strategies

In the previous section, we investigated equilibrium and consensus strategies that involved distributed decision-making for maximizing some utility function.

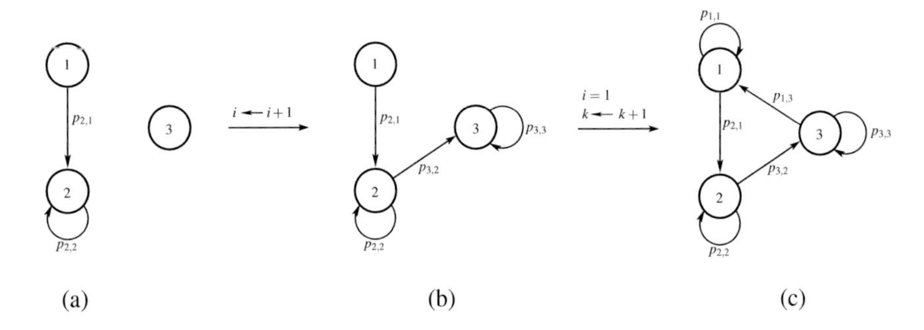

Figure 4.5 Three experts share incrementally their information: (a) $(k,i)=(k',2)$, (b) $(k,i)=(k',3)$, and (c) $(k,i)=(k'+1,1)$.

Common sense motivated the feedback–reassessment routine, which may be able to provide a more robust and meaningful solution to the task at hand. We were not concerned with the local assessment made by each expert, or, more appropriately from now on, how each node makes local estimates of the parameters. Based on how much information is available locally, and also how and to what extent this information shall be used, different algorithms result. As another important part of the process, we also assumed that each individual could judge how much trust his neighbors deserved. In a scenario where this information cannot be assumed available, adaptation may be required for the local parameter estimates, and the opinion pool may be formed taking into account the quality of data each individual handles, resulting in different combination strategies for pursuing equilibrium and consensus.

Application of distributed estimation has been fueled by the increased availability of nodes, or sensors, with enough processing capacity to render local updates more advantageous than transmitting data to a central processing unit. Advantage may be in terms of reduced communication costs, or increased security, reliability, and coverage range. In addition, spatial diversity comes irrespective of explicit knowledge of network topology.

Before we go into truly distributed algorithms, let us first formulate the global minimum mean squared error solution of a problem which has distributed data, collected in time and space. For the purposes of the current exposition, consider a network with M nodes, possibly geographically distributed, but fully synchronized, and let each node m have access to a data pair $\{\mathbf{x}_m(k),\ d_m(k)\}$ in iteration k, where $\mathbf{x}_m(k) \in \mathbb{C}^N$ and $d_m(k) \in \mathbb{C}$. In iteration k, the network collectively assesses data that can be gathered in a vector of complex reference signals,

$$\mathbf{d}(k) = [d_1(k)\ d_2(k)\ \ldots\ d_M(k)]^{\mathrm{T}}, \tag{4.37}$$

and an $N \times M$ matrix of complex input signal vectors,

$$\mathbf{X}(k) = [\mathbf{x}_1(k)\ \mathbf{x}_2(k)\ \ldots\ \mathbf{x}_M(k)]. \tag{4.38}$$

We wish to find $\mathbf{w}_o \in \mathbb{C}^N$, which is the global solution that minimizes

$$\xi(k) = \mathbb{E}\{\|\mathbf{d}(k) - \mathbf{X}^{\mathrm{T}}(k)\mathbf{w}^*\|^2\}. \tag{4.39}$$

4.3.1 Incremental Update Strategy for Distributing the Load

For minimizing Equation (4.39), we may use the steepest descent method at each iteration k:

$$\begin{aligned}\mathbf{w}(k+1) &= \mathbf{w}(k) - \mu\nabla_{\mathbf{w}(k)}\xi(k)\\ &= \mathbf{w}(k) - \mu\sum_{m=1}^{M}\nabla_{\mathbf{w}(k)}\xi_m(k),\end{aligned} \tag{4.40}$$

where $\nabla_{\mathbf{w}(k)}\xi(k)$ stands for the gradient vector of the cost function $\xi(k)$ with respect to $\mathbf{w}(k)$.

The solution to the minimization problem stated above could be calculated by a central node, but that is precisely what shall be avoided. Instead, we may realize that Equation (4.39) can be rewritten as

$$\begin{aligned}\xi(k) &= \sum_{m=1}^{M}\mathbb{E}\{|d_m(k) - \mathbf{w}^{\mathrm{H}}(k)\mathbf{x}_m(k)|^2\}\\ &= \sum_{m=1}^{M}\xi_m(k),\end{aligned} \tag{4.41}$$

in order to distribute the updating load among the nodes [10–17].

Each node may run its own steepest descent algorithm and make a contribution toward the global solution:

$$\mathbf{w}_m(k) = \mathbf{w}_{m-1}(k) - \mu\nabla_{\mathbf{w}(k)}\xi_m(k). \tag{4.42}$$

If we make $\mathbf{w}_0(k) = \mathbf{w}(k)$, the global solution obtained in iteration k, then the last cycle yields $\mathbf{w}(k+1) = \mathbf{w}_M(k)$. However, the equation above requires local knowledge of the gradient calculated at $\mathbf{w}(k)$, $\nabla_{\mathbf{w}(k)}\xi_m(k)$, for $\xi_m(k) = \mathbb{E}\{\|d_m(k) - \mathbf{w}^{\mathrm{H}}(k)\mathbf{x}_m(k)\|^2\}$, which may be unavailable. This can be remedied with the approximation

$$\mathbf{w}_{m+1}(k) = \mathbf{w}_m(k) - \mu\nabla_{\mathbf{w}_m(k)}\xi_m(k). \tag{4.43}$$

In either Equation (4.42) or in Equation (4.43), the solution is built incrementally: Each node aggregates its contribution founded on its own data and on its immediate neighbor's local estimate. After one complete cycle, when all nodes have contributed once, the network will have taken a step toward the minimum mean squared error solution of Equation (4.39).

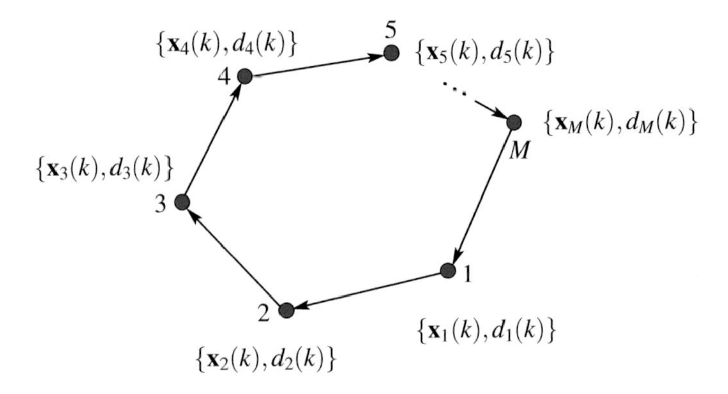

Figure 4.6 Network for incremental adaptation.

Adaptive Incremental Strategies: The LMS Algorithm

Let the set of M nodes forming the network be connected such that information can be passed from node to node only once within one cycle, as in a Hamiltonian path. An example of such network is depicted in Figure 4.6.

In the pursuit of the solution which minimizes Equation (4.41), we may formulate the cost function of the incremental LMS algorithm as follows [18]:

$$\xi_{\mathrm{ILMS}}(k) = \sum_{m=1}^{M} \xi_{\mathrm{ILMS},m}(k), \tag{4.44}$$

where

$$\xi_{\mathrm{ILMS},m}(k) = \|\mathbf{w}(k+1) - \mathbf{w}(k)\|^2_{\mathbf{P}_m(k)} + \mu|\bar{\varepsilon}_m(k)|^2, \tag{4.45}$$

$$\bar{\varepsilon}_m(k) = d_m(k) - \mathbf{w}^{\mathrm{H}}(k+1)\mathbf{x}_m(k), \tag{4.46}$$

and[2]

$$\mathbf{P}_m(k) - \mathbf{I} \quad \mu\mathbf{x}_m(k)\mathbf{x}_m^{\mathrm{H}}(k). \tag{4.47}$$

Calculating the gradient vector of $\xi_{\mathrm{ILMS}}(k)$ with respect to $\mathbf{w}(k+1)$, we obtain

$$\begin{aligned}\nabla_{\mathbf{w}(k+1)}\xi_{\mathrm{ILMS}}(k) = \sum_{m=1}^{M} [\, & 2\mathbf{P}_m(k)\mathbf{w}(k+1) - 2\mathbf{P}_m(k)\mathbf{w}(k) \\ & - 2\mu d_m^*(k)\mathbf{x}_m(k) + 2\mu\mathbf{x}_m(k)\mathbf{x}_m^{\mathrm{H}}(k)\mathbf{w}(k+1)\,],\end{aligned} \tag{4.48}$$

which vanishes for $\xi_{\mathrm{ILMS}}(k)$ minimum. Therefore, the solution that minimizes Equation (4.44) is

$$\mathbf{w}(k+1) = \mathbf{w}(k) + \sum_{m=1}^{M} \frac{\mu}{M}\bar{e}_m^*(k)\mathbf{x}_m(k), \tag{4.49}$$

[2] Matrix $\mathbf{P}_m(k)$ defined herein for the vector norm bears no relation with the stochastic matrix $\mathbf{P}$.

where

$$\bar{e}_m(k) = d_m(k) - \mathbf{w}^{\mathrm{H}}(k)\mathbf{x}_m(k). \tag{4.50}$$

In order to make the algorithm incremental, where each node receives from its neighbor a partial solution and makes a contribution using local data before passing the updated solution forward in the network, we must follow the same procedure already described for the distributed steepest descent algorithm in Equation (4.42). If we make $\mathbf{w}_0(k+1) = \mathbf{w}(k)$ and update $\mathbf{w}_m(k)$ as

$$\mathbf{w}_m(k+1) = \mathbf{w}_{m-1}(k+1) + \frac{\mu}{M}\bar{e}_m^*(k)\mathbf{x}_m(k), \tag{4.51}$$

then

$$\mathbf{w}(k+1) = \mathbf{w}_M(k+1), \tag{4.52}$$

which is the updated global solution that shall be passed on to node 1 for the next iteration.

Notice that the knowledge of the global gradient needed in Equation (4.42) reveals itself also here, as expected, for the calculation of $\bar{e}_m(k)$, which depends upon the global solution from the previous iteration, $\mathbf{w}(k)$. However, we may want to substitute $\mathbf{w}(k)$ by $\mathbf{w}_{m-1}(k+1)$, which may be available. In fact, given that $\mathbf{w}(k) = \mathbf{w}_0(k+1)$, by using $\mathbf{w}_{m-1}(k+1)$ instead, node m employs a better estimate of the gradient, which is known to improve performance. In this version of the algorithm,

$$\mathbf{w}_m(k+1) = \mathbf{w}_{m-1}(k+1) + \frac{\mu}{M}e_m^*(k)\mathbf{x}_m(k), \tag{4.53}$$

where

$$e_m(k) = d_m(k) - \mathbf{w}_{m-1}^{\mathrm{H}}(k+1)\mathbf{x}_m(k) \tag{4.54}$$

and

$$\mathbf{w}(k+1) = \mathbf{w}_M(k+1). \tag{4.55}$$

In this case, the incremental LMS algorithm can be written as summarized in Algorithm 4.20.

Adaptive Incremental Strategies: The RLS Algorithm

As an alternative to the LMS algorithm, the RLS algorithm can also be derived in a distributed cooperative setup. But before we delve into the adaptive incremental solution to the distributed problem, let us first formulate the problem from a centralized point of view. Let the network have M nodes, which at any iteration k can access data pairs $\{\mathbf{x}_m(k),\ d_m(k)\}$, $m = 1, \ldots, M$, gathered as vector $\mathbf{d}(k)$ and matrix $\mathbf{X}(k)$, as defined in Equations (4.37) and (4.38), respectively. The least squares counterpart of the objective function of Equation (4.39) can be defined as [19]:

Algorithm 4.20 The incremental LMS algorithm

Initialization

For a network with M nodes arranged as a Hamiltonian path, set:

$\mu > 0$

$\mathbf{w}(1) = \mathbf{0}$

For $k > 0$ **do**

  $\mathbf{w}_0(k+1) = \mathbf{w}(k)$

  For $m = 1, \ldots, M$ **do**

    $e_m(k) = d_m(k) - \mathbf{w}_{m-1}^{\mathrm{H}}(k+1)\mathbf{x}_m(k)$

    $\mathbf{w}_m(k+1) = \mathbf{w}_{m-1}(k+1) + \frac{\mu}{M}e_m^*(k)\mathbf{x}_m(k)$

    If $m < M$

      Share $\mathbf{w}_m(k+1)$ with node $m+1$

    end

  end

  $\mathbf{w}(k+1) = \mathbf{w}_M(k+1)$

end

$$
\begin{aligned}
\xi_{\mathrm{IRLS}}(k) &= \lambda^k M\|\mathbf{w}(k+1)\|^2 + \sum_{i=1}^{k}\lambda^{k-i}\|\mathbf{d}(i) - \mathbf{X}^{\mathrm{T}}(i)\mathbf{w}^*(k+1)\|^2 \\
&= \lambda^k M\|\mathbf{w}(k+1)\|^2 + \sum_{i=1}^{k}\lambda^{k-i}\sum_{m=1}^{M}|d_m(i) - \mathbf{w}^{\mathrm{H}}(k+1)\mathbf{x}_m(i)|^2 \\
&= \lambda^k M\|\mathbf{w}(k+1)\|^2 + \sum_{m=1}^{M}\sum_{i=1}^{k}\lambda^{k-i}|\bar{\varepsilon}_m(i)|^2,
\end{aligned} \tag{4.56}
$$

where $\bar{\varepsilon}_m(i) = d_m(i) - \mathbf{w}^{\mathrm{H}}(k+1)\mathbf{x}_m(i)$. The first term on the right hand side of the equation is a penalization of the norm of the coefficient vector, which imparts shrinkage to the solution. This penalization slowly fades as k increases and more data is gathered, for a forgetting factor $0 < \lambda < 1$.

The gradient vector of $\xi_{\mathrm{IRLS}}(k)$ with respect to $\mathbf{w}(k+1)$ gives

$$
\begin{aligned}
\nabla_{\mathbf{w}(k+1)}\xi_{\mathrm{IRLS}}(k) =& 2M\lambda^k\mathbf{w}(k+1) - 2\sum_{m=1}^{M}\sum_{i=1}^{k}\lambda^{k-i}\bar{\varepsilon}_m^*(i)\mathbf{x}_m(i) \\
=& 2M\lambda^k\mathbf{w}(k+1) - 2\sum_{m=1}^{M}\sum_{i=1}^{k}\lambda^{k-i}d_m^*(i)\mathbf{x}_m(i) \\
&+ 2\sum_{m=1}^{M}\sum_{i=1}^{k}\lambda^{k-i}\mathbf{x}_m(i)\mathbf{x}_m^{\mathrm{H}}(i)\mathbf{w}(k+1) \\
=& 2\sum_{m=1}^{M}\left[\mathbf{R}_{\mathrm{D},m}(k)\mathbf{w}(k+1) - \mathbf{p}_{\mathrm{D},m}(k)\right],
\end{aligned} \tag{4.57}
$$

where the $\mathbf{R}_{\mathrm{D},m}(k)$ and $\mathbf{p}_{\mathrm{D},m}(k)$ are sample averages for estimating the local autocorrelation matrices of $\mathbf{x}_m(k)$ and cross-correlation vectors between $d_m(k)$ and $\mathbf{x}_m(k)$, taking in samples from $i = 1, \ldots, k$:

$$\mathbf{R}_{\mathrm{D},m}(k) = \lambda^k \mathbf{I} + \sum_{i=1}^{k} \lambda^{k-i} \mathbf{x}_m(i)\mathbf{x}_m^{\mathrm{H}}(i) \tag{4.58}$$

and

$$\mathbf{p}_{\mathrm{D},m}(k) = \sum_{i=1}^{k} \lambda^{k-i} d_m^*(i)\mathbf{x}_m(i). \tag{4.59}$$

If we make the gradient equal to 0, we obtain the normal equation for the distributed case, which can be rearranged as

$$\begin{aligned}
\mathbf{0} =& \lambda \sum_{m=1}^{M} \left[\lambda^{k-1}\mathbf{I} + \sum_{i=1}^{k-1} \lambda^{k-1-i}\mathbf{x}_m(i)\mathbf{x}^{\mathrm{H}}(i) \right] \mathbf{w}(k+1) \\
& - \lambda \sum_{m=1}^{M}\sum_{i=1}^{k-1} \lambda^{k-1-i} d_m^*(i)\mathbf{x}_m(i) \\
& - \sum_{m=1}^{M} \mathbf{x}_m(k)\left[d_m^*(k) - \mathbf{x}_m^{\mathrm{H}}(k)\mathbf{w}(k+1)\right] \\
=& \sum_{m=1}^{M} \left[\lambda \mathbf{R}_{\mathrm{D},m}(k-1) + \mathbf{x}_m(k)\mathbf{x}_m^{\mathrm{H}}(k)\right] \mathbf{w}(k+1) \\
& - \sum_{m=1}^{M} \left[\lambda \mathbf{p}_{\mathrm{D},m}(k-1) + d_m^*(k)\mathbf{x}_m(k)\right].
\end{aligned} \tag{4.60}$$

Although the local estimates $\mathbf{R}_{\mathrm{D},m}(k)$, where $m = 1, \ldots, M$, could benefit from the Woodbury matrix identity (or matrix inversion lemma) [20],[3] estimates of the centralized solution above requires solving $\sum_{m=1}^{M} \mathbf{R}_{\mathrm{D},m}(k)\mathbf{w}(k+1) = \sum_{m=1}^{M} \mathbf{p}_{\mathrm{D},m}(k)$, which is unmanageable locally because it implies access to data from all M nodes, from iteration $i = 1$ to k. However, through cooperation nodes can provide an incremental solution, yet exact after M cycles, to the distributed least squares problem.

Let us assume that node 1, arbitrarily chosen, has the knowledge of $\mathbf{w}(k)$, the solution generated at the previous iteration, and of the inverse of $\mathbf{R}_{\mathrm{D}}(k-1) = \sum_{m=1}^{M} \mathbf{R}_{\mathrm{D},m}(k-1)$, denoted here as $\mathbf{S}_{\mathrm{D}}(k-1)$. The latter requirement may impose a burden on the communication among adjacent nodes, to be relaxed later. Let us also assume that all nodes have access to their respective current data pairs, $\{\mathbf{x}_m(k),\ d_m(k)\}$. A local update to $\mathbf{w}(k)$ can be performed as follows. Let the implicit definition of the local update at node 1 be defined as

$$\left[\lambda \mathbf{R}_{\mathrm{D}}(k-1) + \mathbf{x}_1(k)\mathbf{x}_1^{\mathrm{H}}(k)\right] \mathbf{w}_1(k+1) = \left[\lambda \mathbf{p}_{\mathrm{D}}(k-1) + d_1^*(k)\mathbf{x}_1(k)\right]. \tag{4.61}$$

[3] $(\mathbf{A} + \mathbf{UCV})^{-1} = \mathbf{A}^{-1} - \mathbf{A}^{-1}\mathbf{U}(\mathbf{C}^{-1} + \mathbf{VA}^{-1}\mathbf{U})^{-1}\mathbf{VA}^{-1}$.

Using the Woodbury matrix identity for the matrix multiplying $\mathbf{w}_1(k+1)$ from the left, we have

$$\begin{aligned}\mathbf{w}_1(k+1) =& \frac{1}{\lambda}\left[\mathbf{S}_\text{D}(k-1)\right.\\ &\left.-\frac{\mathbf{S}_\text{D}(k-1)\mathbf{x}_1(k)\mathbf{x}_1^\text{H}(k)\mathbf{S}_\text{D}(k-1)}{\lambda+\mathbf{x}_1^\text{H}(k)\mathbf{S}_\text{D}(k-1)\mathbf{x}_1(k)}\right]\left[\lambda\mathbf{p}_\text{D}(k-1)+d_1^*(k)\mathbf{x}_1(k)\right]\\ =& \mathbf{S}_\text{D}(k-1)\mathbf{p}_\text{D}(k-1)-\frac{\mathbf{S}_\text{D}(k-1)\mathbf{x}_1(k)\mathbf{x}_1^\text{H}(k)\mathbf{S}_\text{D}(k-1)\mathbf{p}_\text{D}(k-1)}{\lambda+\mathbf{x}_1^\text{H}(k)\mathbf{S}_\text{D}(k-1)\mathbf{x}_1(k)}\\ &+\frac{1}{\lambda}d_1^*(k)\mathbf{S}_\text{D}(k-1)\mathbf{x}_1(k-1)\\ &-\frac{1}{\lambda}\frac{d_1^*(k)\mathbf{S}_\text{D}(k-1)\mathbf{x}_1(k)\mathbf{x}_1^\text{H}(k)\mathbf{S}_\text{D}(k-1)\mathbf{x}_1(k)}{\lambda+\mathbf{x}_1^\text{H}(k)\mathbf{S}_\text{D}(k-1)\mathbf{x}_1(k)}.\end{aligned} \tag{4.62}$$

The solution from the previous iteration $\mathbf{w}(k)$ can be identified in the equation above as $\mathbf{S}_\text{D}(k-1)\mathbf{p}_\text{D}(k-1)$, which can be rearranged as

$$\mathbf{w}_1(k+1) = \mathbf{w}(k) + \frac{e_1^*(k)\mathbf{S}_\text{D}(k-1)\mathbf{x}_1(k)}{\lambda+\mathbf{x}_1^\text{H}\mathbf{S}_\text{D}(k-1)\mathbf{x}_1(k)}, \tag{4.63}$$

where

$$e_1(k) = d_1(k) - \mathbf{w}^\text{H}(k)\mathbf{x}_1(k). \tag{4.64}$$

Now let us assume that node 1 passes on this local estimate to node 2, along with $\hat{\mathbf{S}}_{\text{D},1}(k)$, which is the inverse of $\hat{\mathbf{R}}_{\text{D},1}(k) = \lambda\mathbf{R}_\text{D}(k-1) + \mathbf{x}_1(k)\mathbf{x}_1^\text{H}(k)$ used in node 1 in the definition of $\mathbf{w}_1(k+1)$. Notice that $\hat{\mathbf{R}}_{\text{D},1}(k)$ is not the local estimate of the autocorrelation matrix; it is just one step toward the global estimate of the autocorrelation matrix incorporating the input data from node 1. Let $\mathbf{w}_2(k+1)$ be implicitly defined as

$$\left[\hat{\mathbf{R}}_{\text{D},1}(k)+\mathbf{x}_2(k)\mathbf{x}_2^\text{H}(k)\right]\mathbf{w}_2(k+1) = \left[\hat{\mathbf{p}}_1(k)+d_2^*(k)\mathbf{x}_2(k)\right]. \tag{4.65}$$

If we apply once again the Woodbury matrix identity to the inverse of $\hat{\mathbf{R}}_{\text{D},2}(k) = \hat{\mathbf{R}}_{\text{D},1}(k)+\mathbf{x}_2(k)\mathbf{x}_2^\text{H}(k)$ in the previous equation, we get

$$\begin{aligned}\mathbf{w}_2(k+1) =& \left[\hat{\mathbf{S}}_{\text{D},1}(k)\right.\\ &\left.-\frac{\hat{\mathbf{S}}_{\text{D},1}(k)\mathbf{x}_2(k)\mathbf{x}_2^\text{H}(k)\hat{\mathbf{S}}_{\text{D},1}(k)}{1+\mathbf{x}_2^\text{H}(k)\hat{\mathbf{S}}_{\text{D},1}(k)\mathbf{x}_2(k)}\right]\left[\hat{\mathbf{p}}_1(k)+d_2^*(k)\mathbf{x}_2(k)\right]\\ =& \mathbf{w}_1(k+1)+\frac{e_2^*(k)\hat{\mathbf{S}}_{\text{D},1}(k)\mathbf{x}_2(k)}{1+\mathbf{x}_2^\text{H}(k)\hat{\mathbf{S}}_{\text{D},1}(k)\mathbf{x}_2(k)},\end{aligned} \tag{4.66}$$

once we realize that $\mathbf{w}_1(k+1)$ can be identified in the equation above as $\hat{\mathbf{S}}_{\text{D},1}(k)\hat{\mathbf{p}}_{\text{D},1}(k)$, and

$$e_2(k) = d_2(k) - \mathbf{w}_1^\text{H}(k+1)\mathbf{x}_2(k). \tag{4.67}$$

The same procedure can be applied to all other nodes sequentially until the last one, node M, which will be able to produce $\mathbf{w}_M(k+1)$ as the solution of $\hat{\mathbf{R}}_{\mathrm{D},M}(k)\mathbf{w}_M(k+1) = \hat{\mathbf{p}}_{\mathrm{D},M}(k)$ with $\hat{\mathbf{S}}_{\mathrm{D},M}(k)$ the inverse of $\hat{\mathbf{R}}_{\mathrm{D},M-1}(k) + \mathbf{x}_M(k)\mathbf{x}_M^{\mathrm{H}}(k)$. Notice that

$$\begin{aligned}\hat{\mathbf{R}}_{\mathrm{D},M}(k) =& \hat{\mathbf{R}}_{\mathrm{D},M-1}(k) + \mathbf{x}_M(k)\mathbf{x}_M^{\mathrm{H}}(k)\\ =& \hat{\mathbf{R}}_{\mathrm{D},1}(k) + \mathbf{x}_2(k)\mathbf{x}_2^{\mathrm{H}}(k) + \cdots + \mathbf{x}_M(k)\mathbf{x}_M^{\mathrm{H}}(k)\\ =& \lambda\mathbf{R}_{\mathrm{D}}(k-1) + \mathbf{x}_1(k)\mathbf{x}_1^{\mathrm{H}}(k) + \mathbf{x}_2(k)\mathbf{x}_2^{\mathrm{H}}(k) + \cdots + \mathbf{x}_M(k)\mathbf{x}_M^{\mathrm{H}}(k)\\ =& \mathbf{R}_{\mathrm{D}}(k).\end{aligned} \tag{4.68}$$

Similarly, one can realize that $\hat{\mathbf{p}}_{\mathrm{D},M}(k) = \mathbf{p}_{\mathrm{D}}(k)$. Therefore, $\mathbf{w}_M(k+1) = \mathbf{w}(k+1)$, the centralized least squares solution. Algorithm 4.21 summarizes all the necessary steps.

In order to alleviate the bandwidth requirement of transmitting to the next adjacent node not only the local estimate of the coefficient vector but also the inverse of the partial estimate of the global autocorrelation matrix, Algorithm 4.21 can be slightly modified to employ $\hat{\mathbf{S}}_{\mathrm{D},m}(k-1)$ in place of $\hat{\mathbf{S}}_{\mathrm{D},m-1}(k)$ at node m in iteration k.

Algorithm 4.21 The incremental RLS algorithm

Initialization

For a network with M nodes arranged as a Hamiltonian path, set:

$0 < \lambda \leq 1$

$\mathbf{w}(1) = \mathbf{0}$

$\mathbf{S}_{\mathrm{D}}(0) - \mathbf{I}$

For $k > 0$ **do**

  $\hat{\mathbf{S}}_{\mathrm{D},0}(k) = \lambda^{-1}\mathbf{S}_{\mathrm{D}}(k-1)$

  $\mathbf{w}_0(k+1) = \mathbf{w}(k)$

  For $m = 1, \ldots, M$ **do**

    $e_m(k) = d_m(k) - \mathbf{w}_{m-1}^{\mathrm{H}}(k+1)\mathbf{x}_m(k)$

    $\mathbf{w}_m(k+1) = \mathbf{w}_{m-1}(k+1) + \frac{e_m^*(k)\hat{\mathbf{S}}_{\mathrm{D},m-1}(k)\mathbf{x}_m(k)}{1+\mathbf{x}_m^{\mathrm{H}}(k)\hat{\mathbf{S}}_{\mathrm{D},m-1}(k)\mathbf{x}_m(k)}$

    $\hat{\mathbf{S}}_{\mathrm{D},m}(k) = \hat{\mathbf{S}}_{\mathrm{D},m-1}(k) - \frac{\hat{\mathbf{S}}_{\mathrm{D},m-1}(k)\mathbf{x}_m(k)\mathbf{x}_m^{\mathrm{H}}(k)\hat{\mathbf{S}}_{\mathrm{D},m-1}(k)}{1+\mathbf{x}_m^{\mathrm{H}}(k)\hat{\mathbf{S}}_{\mathrm{D},m-1}(k)\mathbf{x}_m(k)}$

    If $m < M$

      Share $\{\mathbf{w}_m(k+1),\ \hat{\mathbf{S}}_{\mathrm{D},m}(k)\}$ with node $m+1$

    end

  end

  $\mathbf{S}_{\mathrm{D}}(k) = \hat{\mathbf{S}}_{\mathrm{D},M}(k)$

  $\mathbf{w}(k+1) = \mathbf{w}_M(k+1)$

end

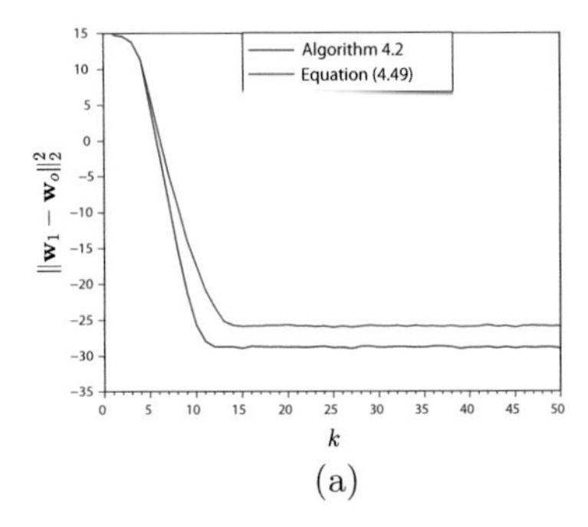

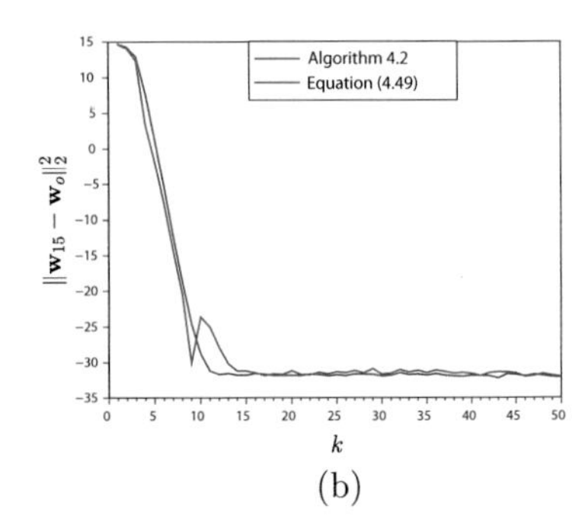

Figure 4.7 Squared norm of the coefficient-vector deviation for the two versions of the ILMS algorithm, the one in Algorithm 4.20 and the version implementing Equation (4.51): (a) node 1 and (b) node 15, all in dB.

Example 1

In order to test the incremental LMS algorithm and compare the performance of the version described in Algorithm 4.20 with that which uses $\bar{e}_m(k)$ as in Equation (4.51), we chose a network with $M = 15$ nodes connected as a Hamiltonian path. The reference was generated with a short linear model with order equal to 3, that is, $N = 4$. At each node the input signal was generated from Gaussian zero-mean and unitary variance independent random processes, and all nodes experienced equal signal-to-noise ratios, equal to 20 dB. The step size used was $\mu = 0.9$. Figures 4.7 (a) and (b) show the squared norm of the coefficient deviation, $\|\mathbf{w}_m(k) - \mathbf{w}_o\|^2$, averaged for 1000 runs in the ensemble.

Due to data reuse, performance of the estimators improves as the information is passed on from node to node within one iteration. However, there is a clear advantage in the approximation introduced in Algorithm 4.20 with the error calculated as in Equation (4.54), when compared to the counterpart that minimizes Equation (4.44) exactly and implements $\bar{e}_m(k)$ as in Equation (4.49).

Example 2

In this next example, Algorithms 4.20 and 4.21 were put to test in the same network with $M = 15$ nodes. However, in this experiment, each node experienced a different signal-to-noise ratio. The variances of the zero-mean independent additive white Gaussian noise contaminating the reference signals in the M nodes were randomly picked from a uniform distribution between 10^{-3} and 10^{-1}. The input signals were also independent zero-mean Gaussian random processes, but correlated by a filter with frequency response $H(z) = 1/(1 - 0.95z^{-1})$ and then normalized for unitary variance. The step size, μ, and the forgetting factor, λ, for the LMS and the RLS algorithms, respectively, were both chosen equal to 0.9.

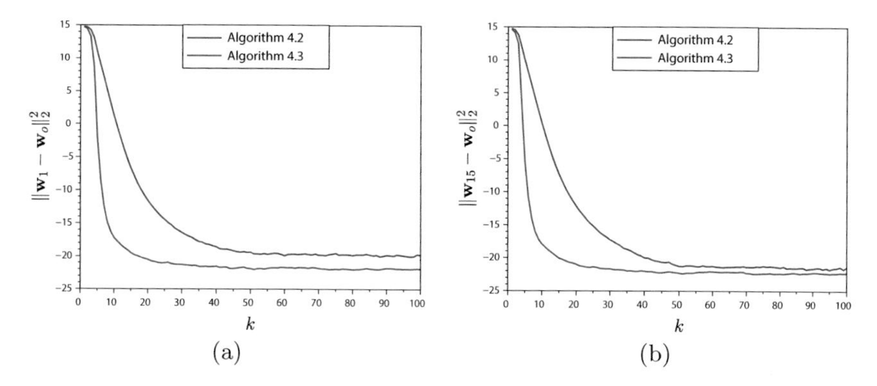

Figure 4.8 Squared norm of the coefficient-vector deviation for Algorithms 4.20 (ILMS) and 4.21 (IRLS) for correlated input signal: (a) node 1 and (b) node 15, all in dB.

Figures 4.8 (a) and (b) show the comparison of the incremental LMS and RLS algorithms with correlated input signals for nodes 1 and 15, respectively, also averaged for 1000 runs in the ensemble. Performance of the RLS algorithm does not change much as we move from node to node and, as expected, is superior to that of the LMS algorithm for correlated input signal.

4.3.2 Diffusion Update Strategies

The adaptive incremental strategies described in the previous section derived from natural developments of the minimization of global cost functions. The load of updating toward the solution becomes distributed: Each node makes a small contribution and passes the message onto the next node in line. The estimate is performed in two acts: (i) node m receives the estimate shared from node $m-1$ and (ii) node m updates the estimate using its own local data $\{\mathbf{x}_m(k),\ d_m(k)\}$.

Sequential message passing for incremental updating may be justifiable in several applications, but collaborating with a more generic neighborhood may pose advantages and be more akin to forming a better opinion pool, as described in Section 4.2.2. In this case, we may want to free the network from the token-ring constraint of the incremental strategy and allow the nodes to form cooperative neighborhoods such that data, as well as estimates, are exchanged among neighbors. As in DeGroot's algorithm, referred to in Section 4.2.3, we may diffuse information within the neighborhood for constructing a better estimate that benefits from the inherited spatial diversity.

Let node m have a neighborhood $\mathcal{N}_m$ with $|\mathcal{N}_m|$ nodes, including itself, as in Figure 4.9.

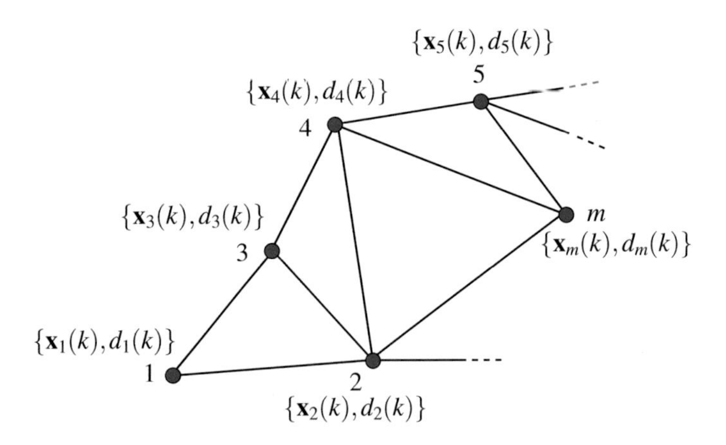

Figure 4.9 Network for diffusion adaptation.

A local cost function based on the minimum disturbance criterion for node m may be constructed as

$$\xi_m(k) = \sum_{l \in \mathcal{N}_m} p_{m,l} \|\boldsymbol{\phi}_m(k) - \mathbf{w}_l(k)\|^2, \qquad p_{m,l} \geq 0. \tag{4.69}$$

The optimal local solution for $\boldsymbol{\phi}_m(k)$ points toward an opinion pool formed as the convex combination of all estimates in the neighborhood:

$$\boldsymbol{\phi}_m(k) = \sum_{l \in \mathcal{N}_m} p_{m,l} \mathbf{w}_l(k), \tag{4.70}$$

where $\sum_{l \in \mathcal{N}_m} p_{m,l} = 1$ and $p_{m,l} \geq 0$. We may then rewrite Equation (4.15) in the distributed adaptive estimation context as

$$\begin{bmatrix} \boldsymbol{\phi}_1(k) \\ \boldsymbol{\phi}_2(k) \\ \vdots \\ \boldsymbol{\phi}_M(k) \end{bmatrix} = \begin{bmatrix} p_{1,1}\mathbf{I} & p_{1,2}\mathbf{I} & \cdots & p_{1,M}\mathbf{I} \\ p_{2,1}\mathbf{I} & p_{2,2}\mathbf{I} & \cdots & p_{2,M}\mathbf{I} \\ \vdots & & \ddots & \vdots \\ p_{M,1}\mathbf{I} & p_{M,2}\mathbf{I} & \cdots & p_{M,M}\mathbf{I} \end{bmatrix} \begin{bmatrix} \mathbf{w}_1(k) \\ \mathbf{w}_2(k) \\ \vdots \\ \mathbf{w}_M(k) \end{bmatrix}, \tag{4.71}$$

where $p_{m,l} > 0$ indicates that node l belongs to the neighborhood of node m and will contribute to its opinion pool. The key difference here, as compared to the strategy presented in Section 4.2.2, is that in the adaptive context new data is available at every iteration k, and the feedback–reassessment repeated routine shall incorporate this new information. There are several possibilities for the reassessment–feedback repeated routine in the adaptive context, but we shall examine here two of them in the form of adaptation algorithms: CtA and AtC [21]. Figure 4.10 outlines the general idea from a macro perspective and the major differences between the two approaches.

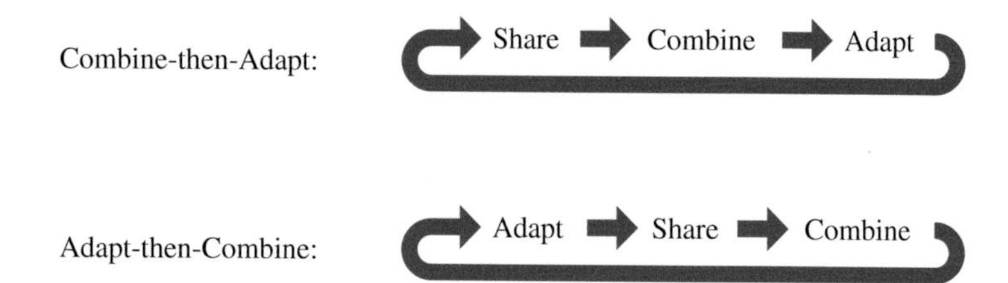

Figure 4.10 CtA and AtC diffusion strategies outlined.

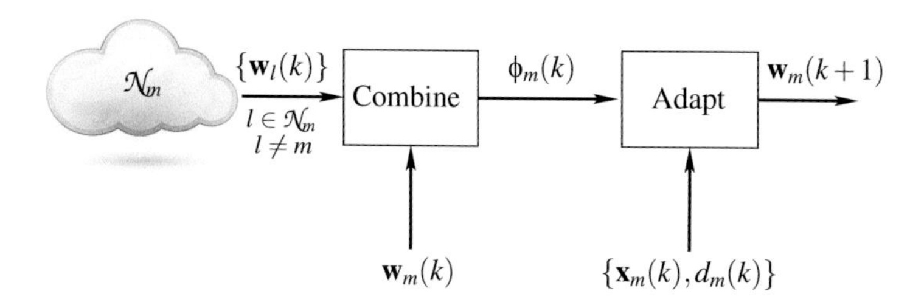

Figure 4.11 The Combine-then-Adapt diffusion strategy as seen from node m.

Adaptive Diffusion Strategies: The CtA LMS Algorithm

In this first strategy, node m forms an aggregate local estimate $\boldsymbol{\phi}_m(k)$ as a convex combination from the pool of opinions $\{\mathbf{w}_l(k)\}$ gathered in the neighborhood $\mathcal{N}_m$, as shown in Equation (4.71). In a second step, it incorporates local data $\{\mathbf{x}_m(k),\ d_m(k)\}$ in an adaptation step to construct $\mathbf{w}_m(k+1)$. Figure 4.11 shows the model with the steps and the corresponding intermediate results for the CtA diffusion strategy.

At iteration k, let $\boldsymbol{\phi}_m(k)$ be an available aggregate estimate at node m, built as a convex combination from the opinion pool formed in the neighborhood $\mathcal{N}_m$, as illustrated in Figure 4.9. The cost function to be minimized by the updated local estimate, $\mathbf{w}_m(k+1)$, can be constructed following the minimum disturbance approach:

$$\xi_{\mathrm{D1LMS},m}(k) = \|\mathbf{w}_m(k+1) - \boldsymbol{\phi}_m(k)\|^2_{\mathbf{P}_m(k)} + \mu|\varepsilon_m(k)|^2, \tag{4.72}$$

where

$$\varepsilon_m(k) = d_m(k) - \mathbf{w}_m^{\mathrm{H}}(k+1)\mathbf{x}_m(k) \tag{4.73}$$

and $\mathbf{P}_m(k)$ is given by Equation (4.47). Minimization of Equation (4.72) with respect to $\mathbf{w}_m(k+1)$ yields

$$\mathbf{w}_m(k+1) = \boldsymbol{\phi}_m(k) + \mu e^*_{\phi,m}(k)\mathbf{x}_m(k), \tag{4.74}$$

where

$$e_{\phi,m}(k) = d_m(k) - \boldsymbol{\phi}_m^{\mathrm{H}}(k)\mathbf{x}_m(k). \tag{4.75}$$

In Section 4.2.3, each expert had the prerogative of choosing the weights to be used in the assessment of its neighbors' opinions. However, in a distributed network, one may think of different strategies, some fixed for a given network topology, as in Equation (4.70), some based on sampling the environment. If no prior information is available, a simple strategy may be

$$p_{m,l} = \begin{cases} \frac{1}{|\mathcal{N}_m|}, & l \in \mathcal{N}_m, \\ 0, & \text{otherwise}, \end{cases} \tag{4.76}$$

which yields

$$\boldsymbol{\phi}_m(k) = \sum_{l \in \mathcal{N}_m} p_{m,l} \mathbf{w}_l(k), \tag{4.77}$$

as in Equation (4.71).

However fair, in a complex scenario, it may be advantageous to move away from this uniform weighting to a more informed matrix $\mathbf{P}$ in Equation (4.71). In this alternative strategy, let $\tilde{\mathbf{w}}_m(k)$ be the averaged estimate from the opinion pool of the neighborhood of node m, excluding itself,

$$\tilde{\mathbf{w}}_m(k) = \frac{1}{|\mathcal{N}_m| - 1} \sum_{\substack{l \in \mathcal{N}_m \\ l \neq m}} \mathbf{w}_l(k), \tag{4.78}$$

and the aggregate estimation be

$$\boldsymbol{\phi}_m(k) = \gamma_m(k)\mathbf{w}_m(k) + [1 - \gamma_m(k)]\tilde{\mathbf{w}}_m(k), \tag{4.79}$$

where $0 \leq \gamma_m(k) \leq 1$ is some monotonic, possibly smooth, function, for example, the sigmoid function:

$$\gamma_m(k) = \frac{1}{1 + e^{-\delta(k)}}. \tag{4.80}$$

Function $\gamma_m(k)$ offers a balance between node m's own assessment and that of its neighbors. The parameter controlling the sigmoid function, $\delta_m(k) \in \mathbb{R}$, can be made time-varying, chosen for pursuing the minimum squared error defined in Equation (4.75) [22],

$$\min_{\delta_m(k)} |e_{\phi,m}(k)|^2, \tag{4.81}$$

following a steepest descent path. In this case,

$$\delta_m(k+1) = \delta_m(k) - \mu_\delta \nabla_{\delta_m(k)} |e_{\phi,m}(k)|^2, \tag{4.82}$$

where

$$\begin{aligned} \nabla_{\delta_m(k)} |e_{\phi,m}(k)|^2 = &- \gamma_m(k)[1 - \gamma_m(k)] \left\{ e^*_{\phi,m}(k) \left[\mathbf{w}_m^{\mathrm{H}}(k) - \tilde{\mathbf{w}}_m^{\mathrm{H}}(k) \right] \mathbf{x}_m(k) \right. \\ &\left. + \, e_{\phi,m}(k) \mathbf{x}_m^{\mathrm{H}}(k) \left[\mathbf{w}_m(k) - \tilde{\mathbf{w}}_m(k) \right] \right\}. \end{aligned} \tag{4.83}$$

This strategy for updating $\delta_m(k)$ may converge too quickly to attracting stationary points that render $\nabla_{\delta_m(k)} |e_\phi(k)|^2 = 0$, for example, $\gamma_m(k) = 0$ or

Algorithm 4.22 The diffusion CtA LMS algorithm

Initialization

$\mu > 0$ and $\mu_\delta > 0$

$\delta_m(1) = 0, \ m = 1, \ \ldots, \ M$

For $k > 0$ **do**

  For $m = 1, \ \ldots, \ M$ **do**

    Share $\mathbf{w}_m(k)$ within $\mathcal{N}_m$

    Aggregate Step

    $\tilde{\mathbf{w}}_m(k) = \sum_{l \in \mathcal{N}_m, \ l \neq m} \mathbf{w}_l(k)/(|\mathcal{N}_m| - 1)$

    $\gamma_m(k) = 1/(1 + e^{-\delta_m(k)})$

    $\boldsymbol{\phi}_m(k) = \gamma_m(k)\mathbf{w}_m(k) + [1 - \gamma_m(k)]\tilde{\mathbf{w}}_m(k)$

    Adaptation Step

    $e_{\phi,m}(k) = d_m(k) - \boldsymbol{\phi}_m^{\mathrm{H}}(k)\mathbf{x}_m(k)$

    Calculate $\nabla_{\delta_m(k)}|e_{\phi,m}(k)|^2$ as in Equation (4.83)

    $\delta_m(k+1) = \delta_m(k) - \mu_\delta \nabla_{\delta_m(k)}|e_{\phi,m}(k)|^2$

    $\mathbf{w}_m(k+1) = \boldsymbol{\phi}_m(k) + \mu e^*_{\phi,m}(k)\mathbf{x}_m(k)$

  end

end

$\gamma_m(k) = 1$, which are not necessarily optimal solutions. In order to avoid this pitfall, it may be necessary to delay adaptation of $\delta_m(k)$ until all nodes have approached their solutions. Otherwise the steepest descent adaptation of Equation (4.82) may quickly put $\delta_m(k)$ on a nonreturning path that leads to either $\gamma_m(k) = 0$ or $\gamma_m(k) = 1$.

The CtA LMS algorithm can be summarized as in Algorithm 4.22.

Example 3

In order to test the diffusion CtA LMS algorithm and compare the performance of the version described in Algorithm 4.22 with that which uses a naive combination of the neighbors' estimates, as in Equation (4.76), we chose a network with $M = 3$ nodes and $N = 4$. At each node, the input signal was generated from Gaussian zero-mean random processes correlated by a filter with frequency response $H(z) = 1/(1 - 0.707z^{-1})$ and then normalized for unitary variance. The nodes experienced signal-to-noise ratios equal to 50 dB, 30 dB, and 10 dB, respectively. The step sizes were $\mu = 0.01$ and $\mu_\delta = 10$. Figures 4.12 (a) and (b) show $\gamma_m(k)$ and the coefficient-error norm, respectively, averaged for 1000 runs in the ensemble.

From Figure 4.12 (a), one can see that node $m = 1$ favors its own estimate in comparison to the average of the estimates of its neighbors, as $\gamma_1(k)$ raises to close to 0.9. The same happens to node $m = 2$, albeit to a lesser degree. However, node $m = 3$ faces a harsher environment with a 10 dB signal to noise ratio and favors the average of its neighbors' estimates in detriment of its own. The behavior is expected and the example illustrates the mechanism controlling

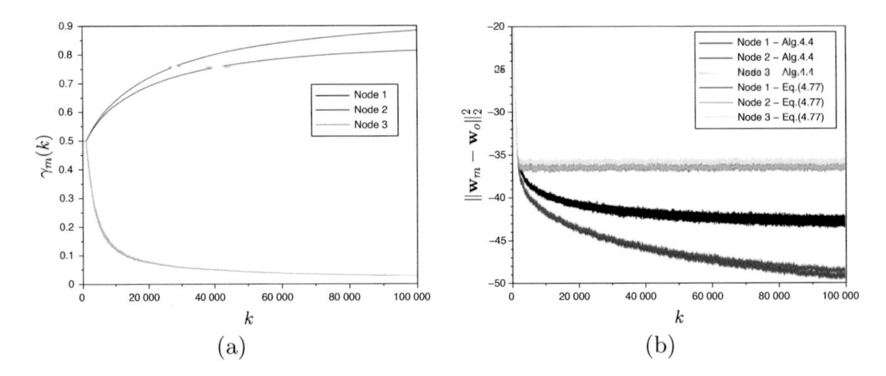

Figure 4.12 Coefficient error norm in dB for the diffusion CtA LMS algorithm employing fixed versus time-varying combination weights.

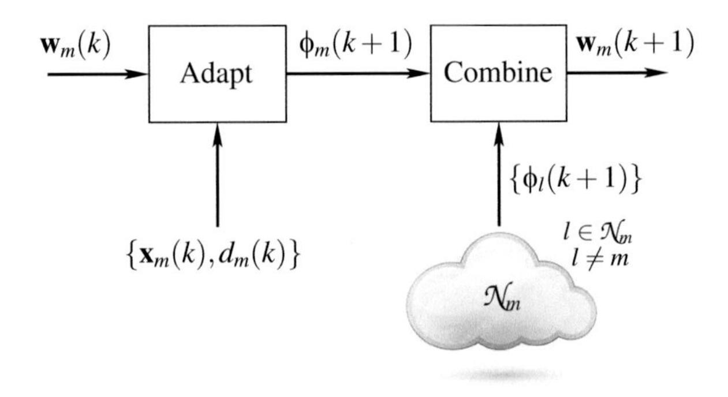

Figure 4.13 The Adapt-then-Combine diffusion strategy as seen from node m.

the weights employed by Algorithm 4.22 in the combination step. Figure 4.12 (b) shows the weight-error norm, where we clearly see the advantage of the mechanism that uses the time-varying parameter $\gamma_m(k)$. Each node automatically controls how much trust the opinion pool formed in the neighborhood deserves.

Adaptive Diffusion Strategies: The AtC LMS Algorithm

As an alternative to the CtA LMS algorithm, in the AtC LMS algorithm node m first incorporates local data $\{\mathbf{x}_m(k),\ d_m(k)\}$ in order to update the estimate from the previous iteration $\mathbf{w}_m(k)$ to form $\phi_m(k+1)$. In a second step, it constructs an aggregate, more informed, estimate $\mathbf{w}_m(k+1)$ as the combination of the opinions of the neighbors $\phi_l(k+1)$, which themselves are shared updated estimates incorporating their own data $\{\mathbf{x}_l(k), d_l(k)\}$, $l \in \mathcal{N}_m$, $l \neq m$. Figure 4.13 shows the model with the steps and the corresponding intermediate results for the AtC diffusion strategy.

The minimum disturbance cost function to be minimized in this case becomes

$$\xi_{\text{D2LMS},m}(k) = \|\boldsymbol{\phi}_m(k+1) - \mathbf{w}_m(k)\|^2_{\mathbf{P}_m(k)} + \mu|\varepsilon_{\boldsymbol{\phi},m}(k)|^2, \tag{4.84}$$

where

$$\varepsilon_{\boldsymbol{\phi},m}(k) = d_m(k) - \boldsymbol{\phi}_m^{\text{H}}(k+1)\mathbf{x}_m(k), \tag{4.85}$$

and $\mathbf{P}_m(k)$ is given by Equation (4.47). Minimization of Equation (4.84) with respect to $\boldsymbol{\phi}_m(k+1)$ yields

$$\boldsymbol{\phi}_m(k+1) = \mathbf{w}_m(k) + \mu e_m^*(k)\mathbf{x}_m(k), \tag{4.86}$$

where

$$e_m(k) = d_m(k) - \mathbf{w}_m^{\text{H}}(k)\mathbf{x}_m(k). \tag{4.87}$$

After adaptation, the aggregate estimate is constructed from the opinion pool in the neighborhood $\mathcal{N}_M$:

$$\mathbf{w}_m(k+1) = \sum_{l \in \mathcal{N}_m} p_{m,l}\boldsymbol{\phi}_l(k+1), \tag{4.88}$$

where $p_{m,l}$ can be chosen following the same guidelines used in the CtA LMS algorithm. For a fair, but uninformed, choice of weights, $p_{m,l}$ may be set as in Equation (4.76). Alternatively, the convex combination may use time-varying weights:

$$\mathbf{w}_m(k+1) = \gamma_m(k)\boldsymbol{\phi}_m(k+1) + [1-\gamma_m(k)]\tilde{\boldsymbol{\phi}}_m(k+1), \tag{4.89}$$

where $\tilde{\boldsymbol{\phi}}_m(k+1)$ is a simple averaged estimate from the opinion pool of the neighborhood of node m, excluding itself,

$$\tilde{\boldsymbol{\phi}}_m(k+1) = \frac{1}{|\mathcal{N}_m| - 1} \sum_{\substack{l \in \mathcal{N}_m \\ l \neq m}} \boldsymbol{\phi}_l(k+1). \tag{4.90}$$

Function $\gamma_m(k)$ assumes values within the interval $[0,\ 1]$, for example, the sigmoid function given by Equation (4.80), controlled by parameter $\delta_m(k)$, which is updated according to the steepest descent rule to minimize $|\varepsilon_m(k)|^2$, defined in Equation (4.73):

$$\delta_m(k+1) = \delta_m(k) - \mu_\delta \nabla_{\delta_m(k)}|\varepsilon_m(k)|^2, \tag{4.91}$$

where

$$\begin{aligned}\nabla_{\delta_m(k)}|\varepsilon_m(k)|^2 = &- \gamma_m(k)[1-\gamma_m(k)]\left\{\varepsilon_m^*(k)\left[\boldsymbol{\phi}_m^{\text{H}}(k+1) - \tilde{\boldsymbol{\phi}}_m^{\text{H}}(k+1)\right]\mathbf{x}_m(k)\right.\\ &\left.+ \varepsilon_m(k)\mathbf{x}_m^{\text{H}}\left[\boldsymbol{\phi}_m(k+1) - \tilde{\boldsymbol{\phi}}_m(k+1)\right]\right\}.\end{aligned} \tag{4.92}$$

The AtC LMS algorithm can be summarized as in Algorithm 4.23.

Algorithm 4.23 The diffusion AtC LMS algorithm

Initialization

$\mu > 0$ and $\mu_\delta > 0$

$\delta_m(1) = 0, \;\; m = 1, \;\ldots, \; M$

For $k > 0$ **do**

  For $m = 1, \;\ldots, \; M$ **do**

    Adaptation Step

    $e_m(k) = d_m(k) - \mathbf{w}_m^{\mathrm{H}}(k)\mathbf{x}_m(k)$

    $\boldsymbol{\phi}_m(k+1) = \mathbf{w}_m(k) + \mu e_m^*(k)\mathbf{x}_m(k)$

    Share $\boldsymbol{\phi}_m(k+1)$ within $\mathcal{N}_m$

    Aggregate Step

    $\tilde{\boldsymbol{\phi}}_m(k+1) = \sum_{l\in\mathcal{N}_m,\; l\neq m} \boldsymbol{\phi}_l(k+1)/(|\mathcal{N}_m| - 1)$

    $\gamma_m(k) = 1/(1 + e^{-\delta_m(k)})$

    $\mathbf{w}_m(k+1) = \gamma_m(k)\boldsymbol{\phi}_m(k+1) + [1 - \gamma_m(k)]\tilde{\boldsymbol{\phi}}_m(k+1)$

    Calculate $\nabla_{\delta_m(k)}|\varepsilon_m(k)|^2$ as in Equation (4.92)

    $\delta_m(k+1) = \delta_m(k) - \mu_\delta \nabla_{\delta_m(k)}|\varepsilon_m(k)|^2$

  end

end

The normalized version of the diffusion AtC LMS algorithm, herein referred to as the diffusion AtC normalized LMS algorithm, can be easily derived as

$$\boldsymbol{\phi}_m(k+1) = \mathbf{w}_m(k) + \frac{\mu e_m^*(k)}{\|\mathbf{x}_m(k)\|^2}\mathbf{x}_m(k). \tag{4.93}$$

Another variant of the diffusion AtC (N)LMS algorithm may include sharing data pairs within the neighborhood prior to the adaptation step. This procedure may improve convergence, but at the expense of increased local computation and traffic to and from neighbors. For the normalized version, for example, updating in Equation (4.93) becomes, for each node m,

$$\begin{aligned}
&\boldsymbol{\phi}_m(k+1) = \mathbf{w}_m(k)\\
&\text{For } l \in \mathcal{N}_m:\\
&\quad e_l(k) = d_l(k) - \boldsymbol{\phi}_m^{\mathrm{H}}(k+1)\mathbf{x}_l(k)\\
&\quad \boldsymbol{\phi}_m(k+1) \leftarrow \boldsymbol{\phi}_m(k+1) + \frac{\mu e_l^*(k)}{\|\mathbf{x}_l(k)\|^2}\mathbf{x}_l(k).
\end{aligned} \tag{4.94}$$

If this extra step is included, the total distributed adaptive filtering involves: (a) sharing local data with neighbors generating feedforward traffic, (b) updating the local estimate iteratively to use all available data, (c) diffusing the local estimate to neighbors generating feedback traffic, and (d) combining estimates from neighbors to approach consensus. Figure 4.10 redrawn to incorporate feedforward traffic becomes as in Figure 4.14.

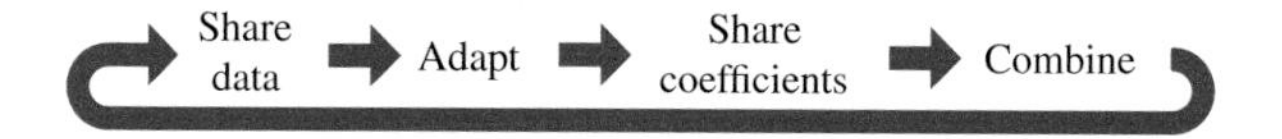

Figure 4.14 AtC diffusion strategy with feedforward and feedback traffic.

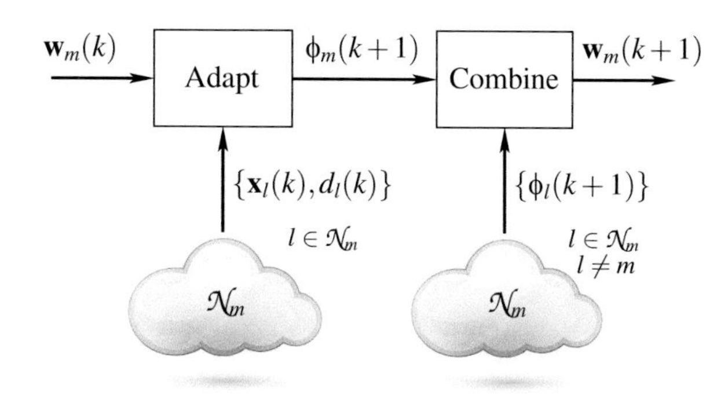

Figure 4.15 The Adapt-then-Combine diffusion strategy with feedforward and feedback traffic, as seen from node m.

From the point of view of node m, the diagram in Figure 4.13 becomes as seen in Figure 4.15.

The problem of increased computational load may be remedied with the set-membership techniques to be discussed in the next section.

4.4 Distributed Estimation with Selective Updating

Collaboration imparts diversity and possibly energy savings, for the nodes need to communicate only within the neighborhood. However, sharing data and aggregating estimates may impose unforeseen extra costs, which can be alleviated with selective cooperation. In addition, increased efficiency can be attained if nodes update only when needed, and cooperate only when advantageous. In fact, continuous updates of parameter estimates and unnecessary cooperation with neighbors may worsen quality of estimation. A remedy relies on selective data-dependent parameter estimation and cooperation. Cooperative SM adaptive filtering brings reduction in complexity at each node and also in data traffic among nodes.

In cooperative distributed adaptive filtering, as seen in the previous sections, all nodes share their information within the neighborhood. Before adaptation takes place, there may be feedforward traffic of data pairs, whereas before combination takes place, nodes exchange their estimates generating feedback traffic for consensus, as seen in Figure 4.14.

The concept of SM [23, 24] applied to time as well as to spatial updating yields substantial savings in computation per node and in traffic among nodes [25]. SM adaptation algorithms employ an upper bound on the output error to access innovation brought by new data. In [25], the SM strategy was extended for

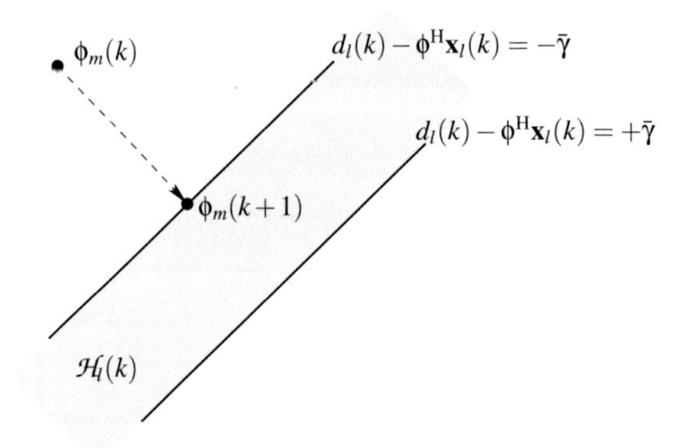

Figure 4.16 SM adaptation for the NLMS algorithm.

selective diffusion, whereby data to be sent to neighbors is first checked for innovation. If there is not enough innovation in the data, they do not need to be shared with neighbors, reducing feedforward traffic. In addition, local updates are not performed if deemed unnecessary and, consequently, need not be shared at every iteration. Therefore, the network experiences a drastic reduction in feedforward and feedback traffic, as well as in local computation.

There are different ways of checking for innovation and selectively deciding to cooperate with neighbors, or not, and to accept the contributions of neighbors, or not. According to the SM criterion, a data pair $\{\mathbf{x}_l(k),\ d_l(k)\}$ brings innovation to the estimate of parameter $\boldsymbol{\phi}_m(k)$ if it brings $\boldsymbol{\phi}_m(k)$ outside the constraint set $\mathcal{H}_l(k)$, defined as

$$\mathcal{H}_l(k) = \{\boldsymbol{\phi} \in \mathbb{C}^N : |d_l(k) - \boldsymbol{\phi}^{\mathrm{H}}\mathbf{x}_l(k)| \leq \bar{\gamma}\}. \tag{4.95}$$

If $|d_l(k) - \boldsymbol{\phi}_m^{\mathrm{H}}(k)\mathbf{x}_l(k)| \leq \bar{\gamma}$, then the data pair is deemed noninnovative and shall not be used to update $\boldsymbol{\phi}_m(k)$. On the other hand, if $|d_l(k) - \boldsymbol{\phi}_m^{\mathrm{H}}(k)\mathbf{x}_l(k)| > \bar{\gamma}$, then the principle of minimum disturbance suggests that adaptation should be just enough to bring $\boldsymbol{\phi}_m(k+1)$ to the nearest point within $\mathcal{H}_l(k)$, as illustrated in Figure 4.16.

For the diffusion AtC SM-NLMS algorithm in Equation (4.94), the modification should read as follows:

$$\begin{aligned}
&\boldsymbol{\phi}_m(k+1) = \mathbf{w}_m(k) \\
&\text{For } l \in \mathcal{N}_m : \\
&\quad e_l(k) = d_l(k) - \boldsymbol{\phi}_m^{\mathrm{H}}(k+1)\mathbf{x}_l(k) \\
&\quad \text{if } |e_l(k)| > \bar{\gamma}, \text{ then} \\
&\qquad\quad \boldsymbol{\phi}_m(k+1) \leftarrow \boldsymbol{\phi}_m(k+1) + \frac{\mu_l(k)e_l^*(k)}{\|\mathbf{x}_l(k)\|^2}\mathbf{x}_l(k),
\end{aligned} \tag{4.96}$$

where the step-size $\mu_l(k)$ is calculated according to the minimum-disturbance criterion as

$$\mu_l(k) = \begin{cases} 1 - \bar{\gamma}/|e_l(k)|, & \text{if} |e_l(k)| > \bar{\gamma}, \\ 0 & \text{otherwise}. \end{cases} \tag{4.97}$$

In the algorithm described above, the SM concept was applied to the local adaptation. As a consequence, the use of data pairs from the neighborhood is dependent upon the innovation they bring, controlled by the threshold $\bar{\gamma}$. If no updating is performed, sharing the coefficients is unnecessary, although sharing the data pair with the neighbors may be maintained. This strategy incorporates SM to the algorithm depicted in Figure 4.15.

If feedforward traffic is totally interrupted, the AtC SM-NLMS algorithm nonfeedforward traffic (AtC SM-NLMS-NFF) results. Still, data permeates throughout the network via the shared coefficients, and the SM concept is used for adaptation with local data. This strategy incorporates SM to the algorithm depicted in Figure 4.13. The AtC SM-NLMS algorithm is described in the listing Algorithm 4.24. For the NFF version, sharing is restricted to the coefficients, $\boldsymbol{\phi}_m(k+1)$, and updates use only the local data pair, $\{\mathbf{x}_m(k),\ d_m(k)\}$.

The two SM algorithms discussed above work in the extremes for the management of local data, either requiring complete feedforward traffic (AtC SM-NLMS algorithm) or no feedforward traffic at all (AtC SM-NLMS-NFF). A

Algorithm 4.24 The diffusion AtC SM-NLMS algorithm

Initialization

$\bar{\gamma} > 0$

For $k > 0$ **do**

  For $m = 1, \ldots, M$ **do**

    Share $\{\mathbf{x}_m(k),\ d_m(k)\}$ within $\mathcal{N}_m$

    Adaptation Step

    $\boldsymbol{\phi}_m(k+1) = \mathbf{w}_m(k)$

    For $l \in \mathcal{N}_m$ **do**

      $e_l(k) = d_l(k) - \boldsymbol{\phi}_m^{\mathrm{H}}(k+1)\mathbf{x}_l(k)$

      If $|e_l(k)| > \bar{\gamma}$

        $\mu_l(k) = 1 - \bar{\gamma}/|e_l(k)|$

        $\boldsymbol{\phi}_m(k+1) = \boldsymbol{\phi}_m(k+1) + \mu_l(k)e_l^*(k)\mathbf{x}_l(k)/\|\mathbf{x}_l(k)\|^2$

      end

    end

    If $\boldsymbol{\phi}_m(k+1) \neq \boldsymbol{\phi}_m(k)$

      Share $\boldsymbol{\phi}_m(k+1)$ within $\mathcal{N}_m$

    end

    Aggregate Step

    $\mathbf{w}_m(k+1) = f[\boldsymbol{\phi}_l(k+1)],\ l \in \mathcal{N}_m$

  end

end

compromise strategy may execute a preliminary innovation check in order to decide whether or not to communicate $\{\mathbf{x}_m(k),\ d_m(k)\}$ with the neighbors. As node m has knowledge of the latest estimate $\phi_l(k)$ from the neighborhood, $l \in \mathcal{N}_m$, a spatial innovation check could be performed onto each $\phi_l(k)$ using repeatedly the data pair $\{\mathbf{x}_m(k),\ d_m(k)\}$. This would provide insight if the data available at node m would be useful to node l. However, one may argue that this would add too much complexity and, in practical scenarios, feedforward traffic is likely to happen in broadcast within the neighborhood, not in unicast from m to l. Therefore the spatial innovation check as outlined above may not work as planned for reducing traffic. Instead, node m may perform a local spatial innovation check onto $\mathbf{w}_m(k)$ in order to decide whether or not to broadcast the data pair to all the neighbors. If equilibrium and consensus are to be expected, at least within $\mathcal{N}_m$, then testing locally the innovation brought by $\{\mathbf{x}_m(k),\ d_m(k)\}$ may be a good indication of its usefulness for the neighbors. Even during transient, when $\mathbf{w}_m(k)$ and $\phi_l(k)$ are expected to differ, the strategy may still work, for either vector should indicate the need for updating anyway.

The updating part of the AtC SM-NLMS algorithm with spatial innovation check (AtC SM-NLMS-SIC) is basically the same as in the AtC SM-NLMS algorithm of Equation (4.96), but the neighborhood to consider in the adaptation step is limited to the nodes that shared data pairs, $\mathcal{N}'_m$. The strategy applied above reduces feedforward traffic, for nodes evaluate their data before sharing it: "if it is not useful to me, chances are that it will not be useful to others." Feedback traffic, however, is not directly affected, for the coefficient estimates are shared with neighbors once they have been updated.

As an alternative to promote further reduction in traffic, the node may forgo feedback traffic unless the update happened with the local data pair, that is, with $\{\mathbf{x}_m(k),\ d_m(k)\}$. If only foreign data pairs contributed to updating the coefficients in the current iteration, the node may withhold transmission. Listing Algorithm 4.25 shows the algorithms with reduced feedback traffic (AtC SM-NLMS-SIC-RFB) and with spatial innovation check (AtC SM-NLMS-SIC).

Table 4.2 compares the algorithms in terms of feedforward and feedback traffic, as well as frequency of local updates.

4.4.1 Consensus in a Set-Membership Scenario

There are several possible ways to combine the estimates in the aggregate step of the algorithms, for example, as a convex combination with constant weights, or dynamically adjusting the weights given to the estimates in the formed opinion pool, or, more appropriately, coefficient-estimate pool. For the particular case of the SM-NLMS algorithm, one interesting approach to consensus is to combine estimates two at a time, sequentially, based on the idea of bounding spheroids [23], as explained below.

Algorithm 4.25 The diffusion AtC SM-NLMS-SIC and AtC SM-NLMS-SIC-RFB algorithms

Initialization

$\bar{\gamma} > 0$

For $k > 0$ **do**

  For $m = 1, \ldots, M$ **do**

    $e_m(k) = d_m(k) - \mathbf{w}_m^{\mathrm{H}}(k)\mathbf{x}_m(k)$

    If $|e_m(k)| > \bar{\gamma}$

      Share $\{\mathbf{x}_m(k),\ d_m(k)\}$ within $\mathcal{N}_m$

    end

    Adaptation Step

    $\boldsymbol{\phi}_m(k+1) = \mathbf{w}_m(k)$

    For $l \in \mathcal{N}'_m$ **do**

      $e_l(k) = d_l(k) - \boldsymbol{\phi}_m^{\mathrm{H}}(k+1)\mathbf{x}_l(k)$

      If $|e_l(k)| > \bar{\gamma}$

        $\mu_l(k) = 1 - \bar{\gamma}/|e_l(k)|$

        $\boldsymbol{\phi}_m(k+1) = \boldsymbol{\phi}_m(k+1) + \mu_l(k)e_l^*(k)\mathbf{x}_l(k)/\|\mathbf{x}_l(k)\|^2$

      end

    end

    If AtC SM-NLMS-SIC-RFB **and** $|e_m(k)| > \bar{\gamma}$

      Share $\boldsymbol{\phi}_m(k+1)$ within $\mathcal{N}_m$

    end

    Aggregate Step

    $\mathbf{w}_m(k+1) = f[\boldsymbol{\phi}_l(k+1)],\ l \in \mathcal{N}_m$

  end

end

Table 4.2 Updating and information sharing alternatives for diffusion AtC algorithms

Algorithm	Sharing data	Local adaptation	Sharing coefficients	Combination strategy
NLMS	Optional	Full	Full	Full
SM-NLMS	Full	If innovative	If updated	If shared
SM-NLMS-NFF	None	If innovative	If updated	If shared
SM-NLMS-SIC	If needed	If innovative	If updated	If shared
SM-NLMS-SIC-RFB	If needed	If innovative	If updated with local data	If shared

For all the SM-NLMS algorithm variations presented above, coefficient update, if deemed appropriate, follows the same rule, presented as a direct consequence of the minimum disturbance criterion. An alternative to this point-wise approach, which yields the same solution, is based on bounding spheroids.

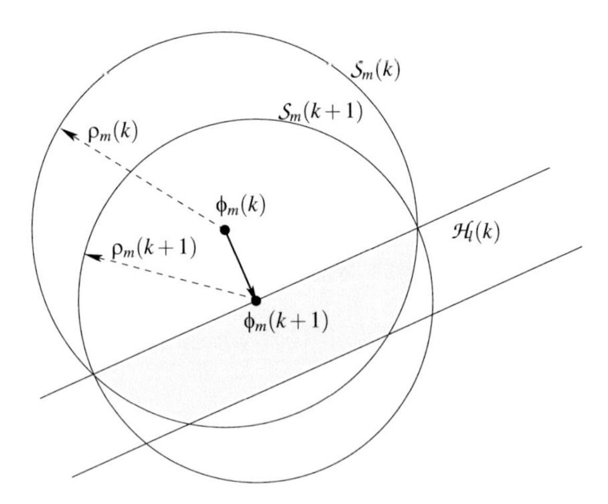

Figure 4.17 SM-NLMLS update based on bounding spheroids.

Let $\mathcal{S}_m(k)$ be the spheroid in the N-dimensional complex vector space $\mathbb{C}^N$, defined as

$$\mathcal{S}_m(k) = \{\boldsymbol{\phi} \in \mathbb{C}^N : \|\boldsymbol{\phi} - \boldsymbol{\phi}_m(k)\|^2 \leq \rho^2\}. \tag{4.98}$$

The updated local solution at node m shall be the center of a new spheroid, $\mathcal{S}_m(k+1)$, which is the smallest that contains the intersection between $\mathcal{S}_m(k)$ and the constraint set $\mathcal{H}_l(k)$, as defined in Equation (4.95). Figure 4.17 illustrates the idea. After update, the new center of the bounding spheroid moves to $\boldsymbol{\phi}_m(k+1)$ with radius $\rho_m(k+1)$ which is shrunk from $\rho_m(k)$ according to the expression

$$\rho_m^2(k+1) = \rho_m^2(k) - \frac{\mu_l^2(k)|e_l(k)|^2}{\|\mathbf{x}_l(k)\|^2}. \tag{4.99}$$

After the adaptation step, nodes m and l have local estimates $\boldsymbol{\phi}_m(k+1)$ and $\boldsymbol{\phi}_l(k+1)$, associated with respective bounding spheroids $\mathcal{S}_m(k+1)$ and $\mathcal{S}_l(k+1)$ of radii $\rho_m(k+1)$ and $\rho_l(k+1)$. For the aggregation step, the idea is to produce $\boldsymbol{\phi}_{ml}(k+1)$ as the center of the sphere that tightly bounds[4] the intersection of $\mathcal{S}_m(k+1)$ and $\mathcal{S}_l(k+1)$, as depicted in Figure 4.18.

Combination shall be carried out for all local estimates in the neighborhood, two at a time. As illustrated in Figure 4.18, at each step $\boldsymbol{\phi}_{ml}(k+1)$ should favor the local estimate associated with the smallest bounding spheroids.

The tightly bound spheroid $\mathcal{S}_{ml}(k+1)$ is obtained as [23]:

$$\begin{aligned}\mathcal{S}_{ml}(k+1) &= \{\boldsymbol{\phi} \in \mathbb{C}^N : (1-\lambda)\|\boldsymbol{\phi} - \boldsymbol{\phi}_m(k+1)\|^2 + \lambda\|\boldsymbol{\phi} - \boldsymbol{\phi}_l(k+1)\|^2 \\ &\qquad \leq (1-\lambda)\rho_m^2(k+1) + \lambda\rho_l^2(k+1)\} \\ &= \{\boldsymbol{\phi} \in \mathbb{C}^N : \|\boldsymbol{\phi} - \boldsymbol{\phi}_{ml}\|^2 \leq \rho_{ml}(k+1)\},\end{aligned} \tag{4.100}$$

[4] For a discussion of tightly bounded ellipsoids, see, for example, [26].

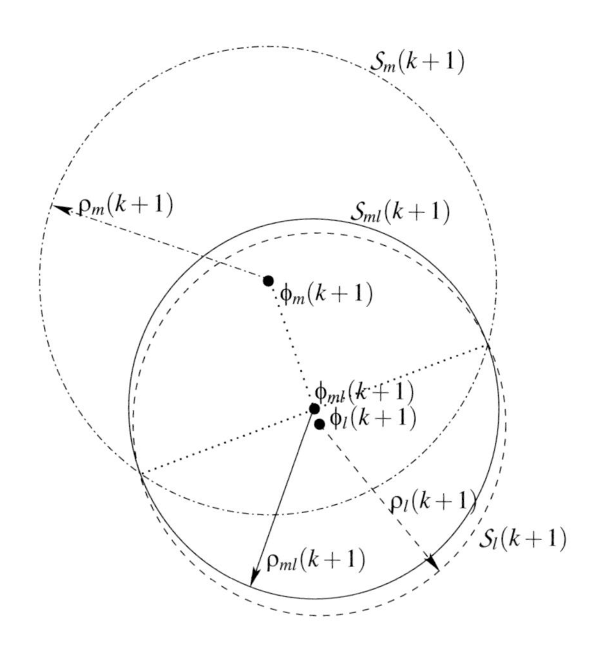

Figure 4.18 Pair-wise aggregation step for the SM-NLMS algorithms.

Algorithm 4.26 Pair-wise aggregation of local estimates at node m

---

For $l \in \mathcal{N}_m$ **do**

$$\lambda = \begin{cases} \frac{1}{2}\left[1 - \frac{\rho_m^2(k+1)-\rho_l^2(k+1)}{\|\phi_m(k+1)-\phi_l(k+1)\|^2}\right], & \text{if } \lambda \in (0,1), \\ 0 & \text{otherwise.} \end{cases}$$

$\phi_m(k+1) \leftarrow (1-\lambda)\phi_m(k+1) + \lambda\phi_l(k+1)$

$\rho_m^2(k+1) \leftarrow (1-\lambda)\rho_m^2(k+1) + \lambda\rho_l^2(k+1) - \lambda(1-\lambda)\|\phi_m(k+1) - \phi_l(k+1)\|^2$

end

$\mathbf{w}_m(k+1) = \phi_m(k+1)$

$\rho_m(k+1) = \rho_m(k+1)$

---

where λ is chosen for minimum $\phi_{ml}(k+1)$, as shown in Algorithm 4.26.

Example 4

In this example, different versions of the distributed NLMS algorithm were compared. The network had 10 nodes, $M = 10$, connected as shown in Figure 4.19(a), and in all of them the adaptive filter had four coefficients, $N = 4$. The input signals were generated from Gaussian zero-mean random processes correlated by a filter with frequency response $H(z) = 1/(1 - 0.707z^{-1})$, normalized for unitary variance. The nodes experienced additive independent zero-mean Gaussian noise whose variances are also indicated in Figure 4.19(a). The thresholds were set as $\sqrt{5\sigma_\eta^2}$ [25, 27], therefore chosen individually for each node, based on the signal-to-noise ratios.

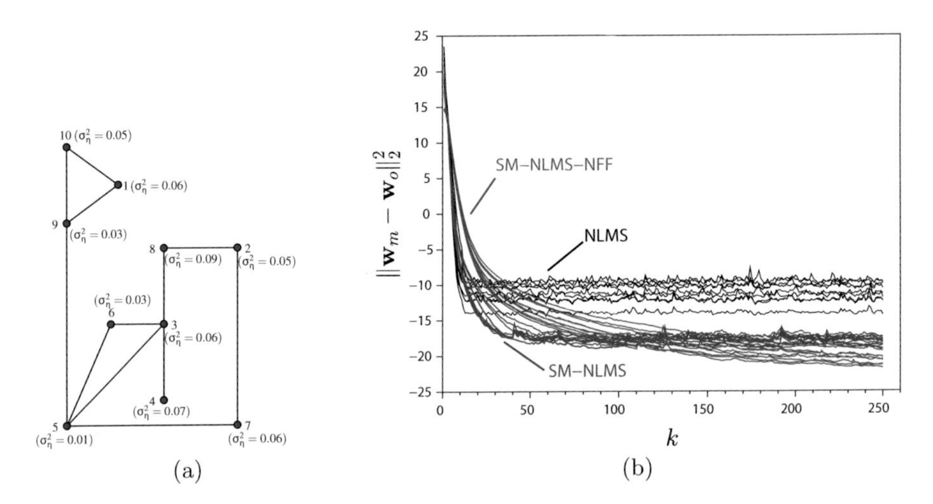

Figure 4.19 Network topology used in Example 4 (a), and squared norm of the coefficient-vector deviation in dB for algorithms NLMS, SM-NLMS, and SM-NLMS-NFF (b).

As seen in Figure 4.19(b), SM improves performance in steady state and saves resources otherwise wasted in unnecessary signal processing and communication. In the SM-NLMS-NFF algorithm, nodes do not share data pairs for adaptation, and communication is required for aggregation only. If, on the one hand, lack of diversity slows convergence when compared to the SM-NLMS algorithm, on the other hand, misadjustment in steady state is reduced due to not reusing data. Although not shown in Figure 4.19(b), performance of algorithms SM-NLMS-SIC and SM-NLMS-RFB resembled closely that of the SM-NLMS algorithm (c.f. Problem 4.9). Simple fixed convex combination was used for aggregation. Also not seen in the figure, but nevertheless worth mentioning is the fact that updating happened only on a fraction of time. For the SM-NLMS-NFF algorithm, updating was necessary in less than 5% of the time during the last 100 iterations.

4.5 Distributed Detection

Detection performance can also be significantly improved by cooperation [28]. Centralized detection based on information stemming from a large number of nodes may require a single fusion center to process a large amount of data, in addition to impose performance constraints on the nodes. The application, however, may require a different approach to the system, as compared to the problem of parameter estimation. This section discusses cooperative network for adaptive detection.

Let us consider a network of M nodes that sense the environment under two hypotheses: H_0, denoting absence of the event, or H_1 for the occurrence of the event. Let us also assume that nodes share their data within the neighborhood for a local assessment of the distributed information, generating feedforward traffic. At node m, this step allows a local binary decision u_m based on a linear combination of the sensed signals from the neighbors in $\mathcal{N}_m$. These binary decisions may now be fed back to the neighbors for aggregation in search for consensus.

In iteration k, the first step outlined above, which involves a soft linear combination of data, forms a test statistic $\mathrm{T}[\mathbf{x}_m(k)]$ defined as

$$\mathrm{T}[\mathbf{x}_m(k)] = \sum_{l\in\mathcal{N}_m} w_{ml}(k)x_{ml}(k) = \mathbf{w}_m^{\mathrm{T}}(k)\mathbf{x}_m(k) \underset{u_m(k)=1\ (H_1)}{\overset{u_m(k)=0\ (H_0)}{\lesseqgtr}} \tau_m, \tag{4.101}$$

where w_{ml} is the weight to be applied by node m to the data shared by the lth node, x_{ml}:

$$\mathbf{w}_m(k) = \left[w_1(k)\ w_2(k)\ \cdots\ w_{|\mathcal{N}_m|}(k)\right]^{\mathrm{T}} \tag{4.102}$$

and

$$\mathbf{x}_m(k) = \left[x_{m1}(k)\ x_{m2}(k)\ \cdots\ x_{m|\mathcal{N}_m|}(k)\right]^{\mathrm{T}}. \tag{4.103}$$

The test statistic is compared to a threshold τ_m to yield a local binary decision $u_m(k) = \{0,\ 1\}$. For the purposes of the development of the system herein, let us assume that all $x_{ml}(k)$, $m = 1, \ldots, M$, $l \in \mathcal{N}_m$, can be considered Gaussian random variables under each hypotheses, such that $\mathrm{T}[\mathbf{x}_m(k)]$ can also be considered Gaussian:

$$\mathrm{T}[\mathbf{x}_m(k)] \sim \begin{cases} \mathcal{N}[\mathbf{w}_m^{\mathrm{T}}(k)\boldsymbol{\mu}_{m,0},\ \mathbf{w}_m^{\mathrm{T}}(k)\boldsymbol{\Sigma}_{m,0}\mathbf{w}_m(k)], & \text{for } H_0 \\ \mathcal{N}[\mathbf{w}_m^{\mathrm{T}}(k)\boldsymbol{\mu}_{m,1},\ \mathbf{w}_m^{\mathrm{T}}(k)\boldsymbol{\Sigma}_{m,1}\mathbf{w}_m(k)], & \text{for } H_1 \end{cases}, \tag{4.104}$$

where $\mathcal{N}[\cdot,\ \cdot]$ above denotes a Gaussian random variable, and $\boldsymbol{\mu}_{m,0/1}$ and $\boldsymbol{\Sigma}_{m,0/1}$ are the mean vector and covariance matrix of vector $\mathbf{x}_m(k)$ under assumptions either H_0 or H_1, respectively.

After linear combination and under the Gaussian assumption, detection performance of node m can be evaluated as the probability of false alarm ($P_{f,m}(k)$), or false positive, and the probability of detection ($P_{d,m}(k)$), or true positive:

$$\begin{aligned} P_{f,m}(k) &= P\{\mathrm{T}[\mathbf{x}_m(k)] \geq \tau_m | H_0\} = Q\left[\frac{\tau_m - \mathbf{w}_m^{\mathrm{T}}(k)\boldsymbol{\mu}_{m,0}}{\sqrt{\mathbf{w}_m^{\mathrm{T}}(k)\boldsymbol{\Sigma}_{m,0}\mathbf{w}_m(k)}}\right], \\ P_{d,m}(k) &= P\{\mathrm{T}[\mathbf{x}_m(k)] \geq \tau_m | H_1\} = Q\left[\frac{\tau_m - \mathbf{w}_m^{\mathrm{T}}(k)\boldsymbol{\mu}_{m,1}}{\sqrt{\mathbf{w}_m^{\mathrm{T}}(k)\boldsymbol{\Sigma}_{m,1}\mathbf{w}_m(k)}}\right], \end{aligned} \tag{4.105}$$

where $Q(\cdot)$ is the complementary cumulative distribution function.

In a second step of the adaptive detection process, each node shares its binary decision, to be later hard-combined in order to render a local consensus, which can result from, for example, an OR-fusion rule, or from a voting decision.

4.5.1 Adaptive Weight Update

As seen in Equation (4.105), detection performance depends upon the coefficients, $\mathbf{w}_m(k)$, as well as the threshold, τ_m. In order to update the coefficient vector, we can minimize the mean squared error between the test statistic and a reference signal, as follows:

$$\mathbb{E}\{|e_m(k)|^2\} = \mathbb{E}\{|r_m(k) - \mathbf{w}_m^{\mathrm{T}}(k)\mathbf{x}_m(k)|^2\}, \tag{4.106}$$

where

$$r_m(k) = \begin{cases} \mathbf{1}^{\mathrm{T}}\boldsymbol{\mu}_{m,0}, & \text{for } H_0 \\ \mathbf{1}^{\mathrm{T}}\boldsymbol{\mu}_{m,1}, & \text{for } H_1, \end{cases} \tag{4.107}$$

and $\mathbf{1}$ is the vector with all elements equal to 1. The exact reference may not be available, and an estimate, $\hat{r}_m(k)$, should be used instead. According to [29–31], a local estimate for the reference, $\hat{d}_m(k)$, can be produced locally and shared within the neighborhood, together with $x_m(k)$. Based on the received data, node m can generate $\hat{r}_m(k)$ from an OR-fusion rule, or from a voting decision. The instantaneous threshold can also be estimated as

$$\hat{\tau}_m(k) = \mathbf{w}_m^{\mathrm{T}}(k)\hat{\boldsymbol{\mu}}_{m,0} + Q^{-1}(\epsilon)\sqrt{\mathbf{w}_m^{\mathrm{T}}(k)\hat{\boldsymbol{\Sigma}}_{m,0}\mathbf{w}_m(k)}, \tag{4.108}$$

which would guarantee a constant predefined probability of false alarm $P_{f,m}(k) = \epsilon$ if the exact values of $\boldsymbol{\mu}_{m,0}$ and $\boldsymbol{\Sigma}_{m,0}$ were available.

The LMS algorithm for the distributed detection is listed in Algorithm 4.27.

4.6 Conclusion

This chapter shows how to perform distributed learning with streaming data, starting with methodologies to combine the contributions of the experts to reach consensus. The fused and distributed combinations are exposed as alternative solutions.

The main adaptive filtering algorithms for distributed learning are discussed, in particular those which are based on the information-passing and information-diffusing approaches. The respective network topologies are very different and target different applications. In the former, information is passed on to the immediate neighbor in a ring, and the global solution is incrementally constructed. On the other hand, the latter allows establishment of neighborhoods where data and estimates are diffused. Here, the possibilities are many, for one can choose to combine estimates received from neighbors first and then adapt for a consensus estimate using local data, or one can adapt first and combine the estimates later.

Distributed learning brings diversity at the expense of more resource consumption in order to deal with the extra data. However, neighbor nodes usually exchange correlated data and coefficient estimates that are close to the network

Algorithm 4.27 The LMS algorithm for distributed detection

$\epsilon > 0$
For $k > 0$ **do**
  For $m = 1, \ldots, M$ **do**
    Share $\{x_m(k),\ \hat{d}_m(k)\}$ within $\mathcal{N}_m$
    Adaptation Step
    $\mathrm{T}[\mathbf{x}_m(k)] = \mathbf{w}_m^{\mathrm{T}}(k)\mathbf{x}_m(k)$
    If $\mathrm{T}[\mathbf{x}_m(k)] \geq \hat{\tau}_m(k)$
      $u_m(k) = 1$
    else
      $u_m(k) = 0$
    end
    Share $u_m(k)$ within $\mathcal{N}_m$
    $\hat{r}_m(k) = \mathrm{OR}\{\hat{d}_l(k)\}_{l \in \mathcal{N}_m}$
    $e_m(k) = \hat{r}_m(k) - \mathrm{T}[\mathbf{x}_m(k)]$
    $\mathbf{w}_m(k+1) = \mathbf{w}_m(k) + \mu e_m(k)\mathbf{x}_m(k)$
    Aggregate Step
    If $\sum_{l \in \mathcal{N}_m} u_l(k) > 0$ (for OR-fusion rule, or $> |\mathcal{N}_m|/2$ if voting rule)
      Accept H_1 hypothesis
    else
      Accept H_0 hypothesis
    end
  end
end

consensus. Therefore, little information may be available and buried under an abundance of noninnovative data. The SM approach offers an elegant and efficient strategy to save on local computation and inter-node communication. The chapter presents different alternatives for the SM normalized LMS algorithm, which offer excellent performance with different and controllable levels of resource expenditure.

The last section extends the concept of distributed adaptation from estimation to detection, where nodes build a test statistic to form a preliminary decision to be fed back to the neighborhood for improved performance. Here too, cooperation and adaptation run in tandem for improved performance.

Problems

4.1 For the pari-mutuel betting system, assume there are two bettors with budgets $b_1 = 0.2$ and $b_2 = 0.8$, and two horses. Let the matrix with the priors be

$$\mathbf{P} = \begin{bmatrix} 0.5 & 0.5 \\ 0.9 & 0.1 \end{bmatrix}.$$

Calculate the equilibrium points π_j, $j = 1,\ 2$, and the amounts bettors 1 and 2 shall place on each of the two horses using Equations (4.8) and (4.9). Discuss your results, from the point of view of the bettors. See how bettor 2 makes betting on horse 1 unattractive to bettor 1.

4.2 Still for the pari-mutuel betting system, assume there are two bettors with budgets $b_1 = 0.4$ and $b_2 = 0.6$, and three horses. Let the matrix with the priors be

$$\mathbf{P} = \begin{bmatrix} 0.3 & 0.5 & 0.2 \\ 0 & 0.4 & 0.6 \end{bmatrix}.$$

Calculate the equilibrium points π_j, where $j = 1,\ 2,\ 3$, and the amounts bettors 1 and 2 shall place on each of the three horses. You may use the Eisenberg and Gale algorithm.

4.3 Compare the performance of the two versions of the incremental LMS algorithm: with Equations (4.51) and (4.53). Consider a network with $M = 15$ nodes connected in a ring implementing an adaptive filter with $N = 10$ coefficients. The input signal vectors are delay lines $\mathbf{x}_m(k) = [x_m(k) \ \cdots \ x_m(k-N+1)]^{\mathrm{T}}$, with $x_m(k)$ drawn from independent and identically distributed (i.i.d.) samples of a random sequence with zero-mean and unitary variance Gaussian distribution, for all m and k. The observed reference signals obey the model $d_m(k) = \mathbf{w}_o^{\mathrm{T}}\mathbf{x}_m(k) + \eta_m(k)$, where $\mathbf{w}_o$ is a vector with equal entries and $\|\mathbf{w}_o\|^2 = 1$, and $\eta_m(k)$ are also zero-mean i.i.d. Gaussian samples with variance equal to 0.01. Examine the coefficient error norms $\|\mathbf{w}_o - \mathbf{w}_m(k)\|$ for all nodes and the squared output error $|e_M(k)|^2$ for the last node. Use ensemble averages of 100 simulations.

4.4 Redo Problem 4.3 for the input signal correlated by a moving average filter such that $x_m(k) = (x_m(k) + 2x_m(k-1) + x_m(k-2))/\sqrt{6}$, normalized for unitary variance.

4.5 Redo Problem 4.4 for the incremental RLS algorithm described in Algorithm 4.21.

4.6 For the network given in Figure 4.20(a), compare the performances of the diffusion CtA and AtC LMS algorithms. Assume the input signals are zero-mean Gaussian samples correlated by the same moving average filter given in Problem 4.4 and normalized for unitary variance. The model for the reference signal is a linear filter with $N = 4$ with all coefficients equal to 1. For both algorithms, use convex combination with equal weights, $p_{ml} = 1/|\mathcal{N}_m|$. Additive noise variances are all equal to 10^{-3}. Evaluate the coefficient-error norm for all nodes in ensemble averages of 100 simulations.

4.7 Use Algorithms 4.22 and 4.23 in a fully connected network with three nodes, each running adaptive filters of length $N = 4$ to identify a common system, also of length $N = 4$. Assume the input signals are zero-mean Gaussian samples correlated by the same moving average filter given in Problem 4.4, and normalized for unitary variance. The model for the reference signal is a linear filter with $N = 4$ with all coefficients equal to 1. Additive noise at the three nodes are i.i.d. samples of zero-mean Gaussian random processes with variances

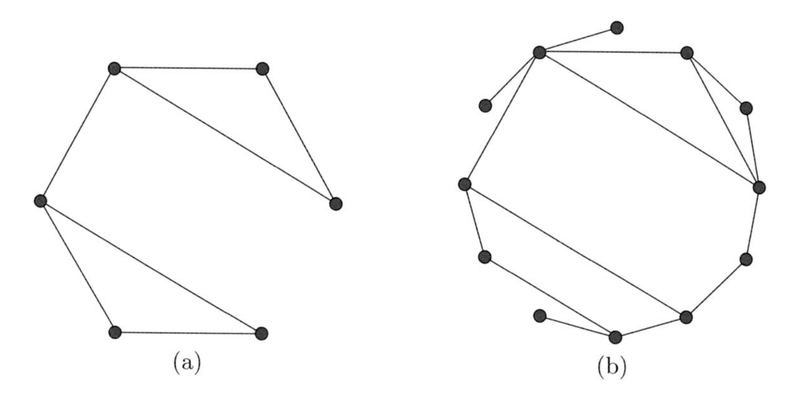

Figure 4.20 Network topologies for Problem 4.6 (a) and Problem 4.12 (b).

1, 1, and 10^{-3}, respectively. Evaluate the coefficient-error norm for all nodes in ensemble averages of 100 simulations. Explore different values of parameters μ and μ_δ, and comment of the final values of parameter $\gamma_m(k)$. It may be necessary to delay the adaptation of $\delta_m(k)$ for a meaningful assessment of the variable weights $\gamma_m(k)$.

4.8 Redo Problem 4.6 for the diffusion NLMS algorithms with full feedforward traffic and with no feedforward traffic.

4.9 For the network and the setup of Example 4, test and compare the SM-NLMS-SIC and the SM-NLMS-SIC-RFB algorithms. Compare also with the results shown in Figure 4.19(b).

4.10 Show the relation implicit in Equation (4.100):

$$\begin{aligned}\{\phi \in \mathbb{C}^N : (1-\lambda)\|\phi - \phi_m(k+1)\|^2 + \lambda\|\phi - \phi_l(k+1)\|^2 \\ \leq (1-\lambda)\rho_m^2(k+1) + \lambda\rho_l^2(k+1)\} \\ = \{\phi \in \mathbb{C}^N : \|\phi - \phi_{ml}\|^2 \leq \rho_{ml}(k+1)\}.\end{aligned}$$

4.11 Let a network with 10 nodes, all connected, be used to identify a 10-coefficient network. The input signals at all nodes are i.i.d. samples of a random sequence with zero-mean and unitary variance Gaussian distribution. Additive noise in all nodes are also samples from i.i.d. zero-mean Gaussian random variables, but with variances calculated such that the signal-to-noise ratios were all equal to 30 dB. Although all nodes are connected, feedforward and feedback traffic are subjected to capacity restrictions of the receiver. If only one neighbor is sharing data, or estimates, with node m, the probability of a correct reception is 100%. However, this probability decreases 10% with each additional neighbor attempting to get through node m, such that if all nine neighbors wish to communicate with node m, the probability of success is only 20%. Compare the performance of all the five algorithms listed in Table 4.2.

4.12 Assume that a 12-node network, as seen in Figure 4.20(b), is deployed to monitor a particular site for the occurrence of an event. Each node gathers 10^5 samples of energy estimates from i.i.d. χ^2 distributions, generated from zero-mean unitary variance Gaussian samples of a random sequence, in the case of H_0. For H_1, the mean value of the Gaussian random sequence is equal to 2, with the same unitary variance. Evaluate the performance in terms of probability of detection versus probability of false alarm of the distributed LMS algorithm that employs the soft and hard combination, as explained in Section 4.5 and compare with the adaptive version with a single node.

References

[1] E. Eisenberg and D. Gale, "Consensus of subjective probabilities: The pari-mutuel method," The Annals of Mathematical Statistics **30**, pp. 165–168 (1959).

[2] L. D. Brown and Y. Lin, "Racetrack betting and consensus of subjective probabilities," Statistics & Probability Letters **62**, pp. 175–187 (2003).

[3] T. Norvig, "Consensus of subjective probabilities: A convergence theorem," The Annals of Mathematical Statistics **38** pp. 221–225 (1967).

[4] M. Stone, "The opinion pool," The Annals of Mathematical Statistics **32**, pp. 1339–1342 (1961).

[5] R. L. Winkler, "The consensus of subjective probability distributions," Management Science **15**, pp. B-61–B-75 (1968).

[6] M. H. Degroot, "Reaching a consensus," Journal of the American Statistical Association **69**, pp. 118–121 (1974).

[7] O. C. Ibe, *Markov processes for stochastic modeling*, 2nd ed. (Elsevier, London, 2013).

[8] S. Kirkland, "On the sequence of powers of a stochastic matrix with large exponent," Linear Algebra and Its Applications **310**, pp. 109–122 (2000).

[9] E. Kani, N. J. Pullman, and N. M. Rice, "Powers of matrices," in Algebraic Methods. Kingston: Queen's University, 2019, ch. 7, pp. 311–370.

[10] M. Rabbat and R. Nowak, "Distributed optimization in sensor networks," in Third International Symposium on Information Processing in Sensor Networks, IPSN 2004, 2004, pp. 20–27.

[11] D. P. Spanos and R. M. Murray, "Distributed sensor fusion using dynamic consensus," in IFAC World Congress. Prague, Czech Republic: IFAC, 2005.

[12] J. B. Predd, S. R. Kulkarni, and H. V. Poor, "Distributed learning in wireless sensor networks," IEEE Signal Processing Magazine **23**, pp. 56–69 (2006).

[13] L. Xiao, S. Boyd, and S. Lall, "A scheme for robust distributed sensor fusion based on average consensus," in 4th International Symposium on Information Processing in Sensor Networks, IPSN 2005, vol. 2005, Los Angeles, CA, USA, 2005, pp. 63–70.

[14] L. Xiao, S. Boyd, and S. Lall, "A space-time diffusion scheme for peer-to-peer least-squares estimation," in Proceedings of the Fifth International Conference on Information Processing in Sensor Networks, IPSN '06, vol. 2006, pp. 168–176, 2006.

[15] F. S. Cattivelli, C. G. Lopes, and A. H. Sayed, "Diffusion recursive least-squares for distributed estimation over adaptive networks," IEEE Transactions on Signal Processing **56**, pp. 1865–1877 (2008).

[16] A. O. Hero, D. Cochran, and S. Member, "Sensor management: Past, present, and future," IEEE Sensors Journal **11**, pp. 3064–3075 (2011).

[17] A. Sayed, *Adaptation, learning, and optimization over networks* (Now Publishers, Delft, 2014).

[18] C. G. Lopes and A. H. Sayed, "Incremental adaptive strategies over distributed networks," IEEE Transactions on Signal Processing **55**, pp. 4064–4077 (2007).

[19] A. H. Sayed and C. G. Lopes, "Distributed recursive least-squares strategies over adaptive networks," in Fortieth Asilomar Conference on Signals, Systems and Computers, no. **1**, 2006, pp. 233–237.

[20] G. H. Golub and C. F. Van Loan, *Matrix computations*, 4th ed. (John Hopkins University Press, Baltimore, 2013).

[21] J. Chen and A. H. Sayed, "Diffusion adaptation strategies for distributed optimization and learning over networks," IEEE Transactions on Signal Processing **60**, pp. 4289–4305 (2012).

[22] C. G. Lopes and A. H. Sayed, "Diffusion least-mean squares over adaptive networks: formulation and performance analysis," IEEE Transactions on Signal Processing **56**, pp. 3122–3136 (2008).

[23] S. Gollamudi, S. Nagaraj, S. Kapoor, and Y. F. Huang, "Set-membership filtering and a set-membership normalized LMS algorithm with an adaptive step size," IEEE Signal Processing Letters **5**, pp. 111–114 (1998).

[24] S. Nagaraj, S. Gollamudi, S. Kapoor, and Y. F. Huang, "BEACON: An adaptive set-membership filtering technique with sparse updates," IEEE Transactions on Signal Processing **47**, pp. 2928–2941 (1999).

[25] S. Werner, Y.-F. Huang, M. L. R. de Campos, and V. Koivunen, "Distributed parameter estimation with selective cooperation," in 2009 IEEE International Conference on Acoustics, Speech and Signal Processing. Taipei: IEEE, 2009, pp. 2849–2852.

[26] W. Kahan, "Circumscribing an ellipsoid about the intersection of two ellipsoids," Canadian Mathematical Bulletin **11**, pp. 437–441 (1968).

[27] J. F. Galdino, J. A. Apolinário Jr., and M. L. R. de Campos, "A set-membership NLMS algorithm with time-varying error bound," in 2006 IEEE International Symposium on Circuits and Systems. Kos: IEEE, pp. 277–280.

[28] I. Sobron, W. A. Martins, M. L. R. de Campos, and M. Velez, "Incumbent and LSA licensee classification through distributed cognitive networks," IEEE Transactions on Communications **64** pp. 94–103 (2016).

[29] F. C. Ribeiro Jr., M. L. R. de Campos, and S. Werner, "Distributed cooperative spectrum sensing with adaptive combining," in 2012 IEEE International Conference on Acoustics, Speech and Signal Processing (ICASSP). Kyoto: IEEE, 2012, pp. 3557–3560.

[30] F. C. Ribeiro Jr., S. Werner, and M. L. R. de Campos, "Distributed cooperative spectrum sensing with selective updating," in European Signal Processing Conference (EUSIPCO), Bucharest, 2012, pp. 474–478.

[31] F. C. Ribeiro Jr., M. L. R. de Campos, and S. Werner, "Distributed cooperative spectrum sensing with double-topology," in 2013 IEEE International Conference on Acoustics, Speech and Signal Processing. Vancouver: IEEE, 2013, pp. 4489–4493.

5 Adaptive Beamforming

5.1 Introduction

This chapter deals with a class of adaptive filtering algorithms typically employed in an array of sensors in order to enhance the *signal of interest* (SOI) hitting the array from a given direction. We assume here that the mean is homogeneous such that, after a distance, the wavefront may be considered planar (far-field hypothesis). The array processor performing this job is named a *beamformer.* We leave the sensors (microphones, antennas, geophones, hydrophones, etc.) for the type of applications at hand and focus herein on the algorithm, basically shared by all types of sensors.

The motivation for this chapter is twofold: a brief introduction to the main concepts of array signal processing, especially those related to adaptive beamforming, and an overview of how to use linearly constrained adaptive algorithms for this application. The main applications of array signal processing are estimating the direction of arrival (DoA) of the incoming signal and enhancing the SOI from or to an assumed known direction (beamforming). Although both applications share many concepts, our main concern in this chapter is related to adaptive beamforming.

The following section contains a description of the basic principles related to narrowband (NB) beamforming. Then, we present several adaptive NB beamforming algorithms ranging from the classical constrained LMS to more advanced tools such as Householder-based and sparsity promoting adaptive beamformers. In the following, we address broadband adaptive beamforming and, finally, we provide valuable remarks before concluding the chapter.

5.2 Array Signal Processing

We start a review of array signal processing with the most usual beamforming approach, the NB beamformer. In NB beamforming, we assume the SOI has a center frequency in f_o and a frequency band Δf such that the ratio $\Delta f/f_\mathrm{o}$ is quite small. Assume we represent the transmitted signal by $x(t) = s(t)\mathrm{e}^{\mathrm{j}\Omega_\mathrm{o} t}$, with $\Omega_\mathrm{o} = 2\pi f_\mathrm{o}$, and $s(t)$, which contains the information transmitted, being a lowpass signal with maximum frequency component much lower than f_o. Then, a delayed version of such a signal, $x(t-\Delta t)$, may be obtained by multiplying $x(t)$ by a complex exponential $\mathrm{e}^{-\mathrm{j}\Omega_o \Delta t}$.

$$\begin{aligned} x(t)\mathrm{e}^{-\mathrm{j}\Omega_o \Delta t} &= s(t)\mathrm{e}^{\mathrm{j}\Omega_o t}\mathrm{e}^{-\mathrm{j}\Omega_o \Delta t} \\ &= s(t)\mathrm{e}^{\mathrm{j}\Omega_o (t-\Delta t)} \\ &\approx x(t-\Delta t), \end{aligned} \tag{5.1}$$

if $s(t) \approx s(t-\Delta t)$, which is valid for $\Delta f/f_o \ll 1$ such that $s(t)$ does not vary much within Δt.

This class of signals, those having energy only in the positive side of their spectrum, are known in the literature as *analytic* signals, their real and imaginary parts being related to each other by the Hilbert transform [1]. Equation (5.1) also implies that the multiplication for the complex exponential $\mathrm{e}^{-\mathrm{j}\Omega_o \Delta t}$ corresponds to a time delay Δt only for the case of $x(t)$ being an NB analytic signal. Indeed, NB analytic signals are easy to be delayed such that signals arriving from different sensors can be easily time aligned if we know their mutual delay or, equivalently, their directions of arrival (the angles formed by the wavefront and the imaginary line crossing the sensors), the velocity of propagation of the (assumed) plane wave, and the positions of the sensors.

Figure 5.1 illustrates a simplified diagram representing an NB beamformer implemented with a linear and equally spaced array of M sensors. After the analog signal $x_m(t)$ is converted to the digital domain, its analytic version is to be obtained. In Figure 5.1, the analytic discrete-time signal $x_m(k)$ is obtained after passing the output of the analog-to-digital (A/D) converter through an analytic bandpass filter centered in $\omega_o = \frac{2\pi f_o}{f_s}$, where f_s being the sampling frequency.

Before proceeding, an important remark is in order. The simplified diagram of the NB beamformer depicted in Figure 5.1, in some cases, might not be of much practical interest; an exception would be a class of NB signals centered in a relatively low frequency range. An RF front-end would undoubtedly have a more complicated diagram to deliver a discrete-time intermediate-frequency (IF) signal or a complex baseband (CBB) sequence. In the following, we detail a quadrature receiver from the antenna to the output CBB signal. An incoming signal within operational range is first tuned by an RF bandpass filter centered in $\Omega_o = 2\pi f_o$ (with a given bandwidth, B) and amplified by a low-noise amplifier (LNA). A mixer downconverts the tuned signal to an intermediate frequency. At this point, an A/D converter could be an option and the quadrature signal could be obtained in the discrete-time domain (TD). Yet, it could have a high sampling rate which would require a more efficient and usually expensive processor. Instead, as seen in Figure 5.2, we can use an IQ demodulator to obtain CBB components referred to as quadrature signal (the real part is known as in-phase and the imaginary part as quadrature). Assume that a received signal containing a delayed incoming SOI plus noise, $x(t) = s(t-\Delta t)\cos(\Omega_o(t-\Delta t)) + n(t)$, Δt representing the delay,

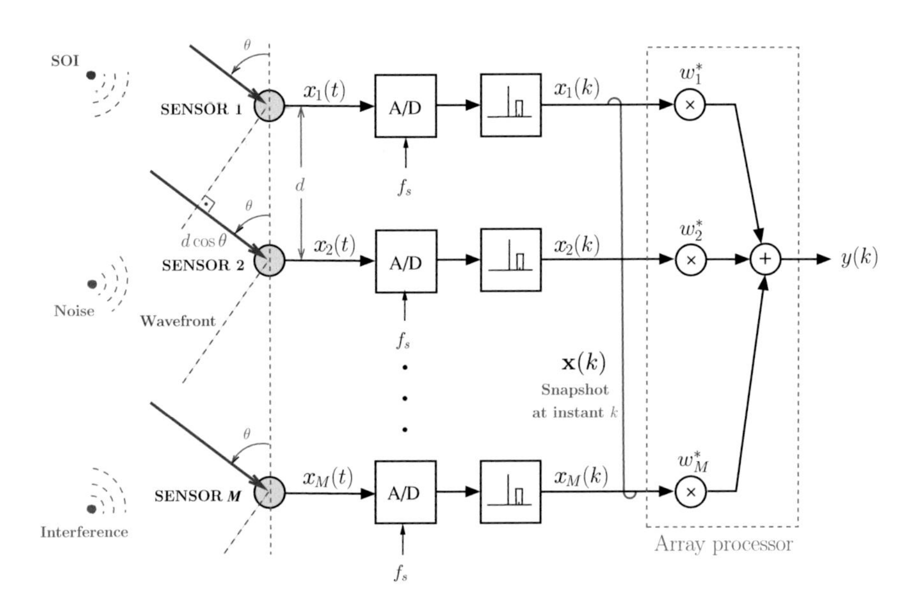

Figure 5.1 A simplified diagram of an NB beamformer implemented with a linear and equally spaced array of M sensors. The single sideband bandpass filter block guarantees that the signal from the m-th sensor, in the discrete-time domain, is narrowband and analytic. The angle θ corresponds to the direction of arrival of the signal of interest (SOI).

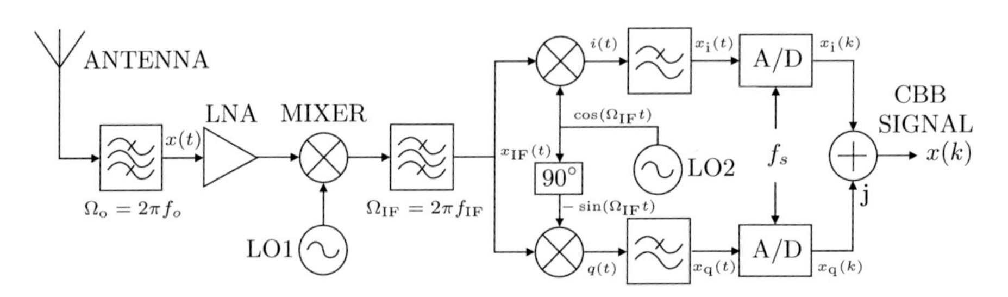

Figure 5.2 An example of a quadrature receiver architecture. LO1 is the local oscillator operating at frequency $f_o - f_{IF}$ while LO2 provides a fixed frequency f_{IF}.

is downconverted to the IF signal $x_{IF}(t) = s(t)\cos(\Omega_{IF}t - \Omega_o\Delta t) + n_{IF}(t)$; note here that $s(t) \approx s(t - \Delta t)$ and that we have dropped any possible multiplicative constant. The in-phase component comes from the multiplication of $x_{IF}(t)$ by cosinusoid $\cos(\Omega_{IF}t)$, followed by a low-pass filter, resulting in $x_i(t) = s(t)\cos(\Omega_o\Delta t) + n_i(t)$. The quadrature component corresponds to the multiplication of $x_{IF}(t)$ by the same cosinusoid shifted by 90°, which corresponds to $-\sin(\Omega_{IF}t)$, and a subsequent low-pass filter such that $x_q(t) = s(t)\sin(-\Omega_o\Delta t) + n_q(t)$. After analog to digital conversion, the discrete-time CBB signal is given as $x(k) = x_i(k) + jx_q(k)$, which corresponds to $x(k) = s(k)e^{-j\Omega_o\Delta t} + n(k)$, where $n(k) = n_i(k) + jn_q(k)$. This signal is centered in

$\omega = 0$ and has a bandwidth corresponding to $-\pi B/f_s < \omega < \pi B/f_s$ or, equivalently, $-B/2$ to $B/2$ (in Hz).

It is worth mentioning that the discrete-time IQ signal $x(k)$, in spite of being bandbase and having the same magnitude spectrum of the noisy SOI, maintains in its phase the information regarding the delay Δt. The receiver in Figure 5.2 can be used in an array such that, having $x_m(k) = s(k)\mathrm{e}^{-\mathrm{j}\Omega_o \Delta t_m} + n_m(k)$ for each channel, the snapshot vector for a single signal coming from a certain direction would be given as

$$\mathbf{x}(k) = \begin{bmatrix} x_1(k) \\ \vdots \\ x_M(k) \end{bmatrix} = s(k) \underbrace{\begin{bmatrix} \mathrm{e}^{-\mathrm{j}\Omega_o \Delta t_1} \\ \vdots \\ \mathrm{e}^{-\mathrm{j}\Omega_o \Delta t_m} \\ \vdots \\ \mathrm{e}^{-\mathrm{j}\Omega_o \Delta t_M} \end{bmatrix}}_{\mathbf{a}(\theta)} + \underbrace{\begin{bmatrix} n_1(k) \\ \vdots \\ n_m(k) \\ \vdots \\ n_M(k) \end{bmatrix}}_{\mathbf{n}(k)}, \tag{5.2}$$

where Δt_m corresponds to the delay (associated with the DoA θ) at the m-th antenna, $1 \leq m \leq M$, $n_m(k)$ is the corresponding noise filtered to have the maximum bandwidth allowed to the CBB signal, and vector $\mathbf{a}(\theta)$ contains all phase delays with respect to an arbitrary origin.

The array processor output is obtained from the *snapshot* at instant k, vector $\mathbf{x}(k) = [x_1(k)\, x_2(k) \cdots x_M(k)]^{\mathrm{T}}$, and the coefficient vector $\mathbf{w} = [w_1\, w_2 \cdots w_M]^{\mathrm{T}}$:

$$y(k) = \mathbf{w}^{\mathrm{H}}\mathbf{x}(k). \tag{5.3}$$

Before trying to obtain a suitable set of coefficients, the elements of vector $\mathbf{w}$, let us describe a simple model for the input vector of the array processor. We can observe that, defining c as the propagation speed, the delay between the signals of adjacent sensors acquired from the first wavefront hitting a uniform linear array (ULA) with an angle θ_1 is given by $\Delta t = \frac{d \cos \theta_1}{c}$. Assuming that G signals, with the first one considered the SOI and the remaining $G-1$ considered interferers, are arriving from G distinct directions, that different types of noise may be present, and also that the sensors are omnidirectional, we can write an expression for $x_m(t)$, $1 \leq m \leq M$, as

$$x_m(t) = \sum_{g=1}^{G} s_g(t) \cos\left(\Omega_o \left(t - (m-1)\frac{d\cos\theta_g}{c}\right) + \phi_g\right) + n_m(t), \tag{5.4}$$

where the g-th signal hitting the first sensor is expressed as $s_g(t)\cos(\Omega_o t + \phi_g)$ and ϕ_g is an arbitrary phase. We have used here the fact that $s(t) \approx s(t - m\Delta t)$, m from 1 to $M-1$. Also, note that $n_m(t)$ is the noise component of the m-th sensor output.

After the A/D converter and analytic bandpass filtering, the discrete-time snapshot can be represented as

$$\begin{aligned}
\mathbf{x}(k) &= \begin{bmatrix} x_1(k) \\ \vdots \\ x_M(k) \end{bmatrix} \\
&= s_1(k)\mathrm{e}^{\mathrm{j}(\omega_\mathrm{o}k+\phi_1)} \underbrace{\begin{bmatrix} 1 \\ \mathrm{e}^{-\mathrm{j}\frac{2\pi d\cos\theta_1}{\lambda}} \\ \vdots \\ \mathrm{e}^{-\mathrm{j}\frac{2\pi(M-1)d\cos\theta_1}{\lambda}} \end{bmatrix}}_{\mathbf{a}(\theta_1)} + \cdots \\
&\quad + s_G(k)\mathrm{e}^{\mathrm{j}(\omega_\mathrm{o}k+\phi_G)} \underbrace{\begin{bmatrix} 1 \\ \mathrm{e}^{-\mathrm{j}\frac{2\pi d\cos\theta_G}{\lambda}} \\ \vdots \\ \mathrm{e}^{-\mathrm{j}\frac{2\pi(M-1)d\cos\theta_G}{\lambda}} \end{bmatrix}}_{\mathbf{a}(\theta_G)} + \underbrace{\begin{bmatrix} n_1(k) \\ \vdots \\ n_M(k) \end{bmatrix}}_{\mathbf{n}(k)},
\end{aligned} \tag{5.5}$$

which, after defining matrix $\mathbf{A}$, becomes

$$\begin{aligned}
\mathbf{x}(k) &= \underbrace{\begin{bmatrix} \mathbf{a}(\theta_1) & \cdots & \mathbf{a}(\theta_G) \end{bmatrix}}_{\mathbf{A}} \underbrace{\begin{bmatrix} s_1(k)\mathrm{e}^{\mathrm{j}(\omega_\mathrm{o}k+\phi_1)} \\ \vdots \\ s_G(k)\mathrm{e}^{\mathrm{j}(\omega_\mathrm{o}k+\phi_G)} \end{bmatrix}}_{\mathbf{s}(k)} + \mathbf{n}(k) \\
&= \mathbf{A}\mathbf{s}(k) + \mathbf{n}(k),
\end{aligned} \tag{5.6}$$

where $\omega_\mathrm{o} = \frac{\Omega_\mathrm{o}}{f_s}$ is the central frequency in the discrete-TD (or digital operating frequency) and $\lambda = \frac{c}{f_\mathrm{o}}$ is the wavelength corresponding to the operating frequency f_o and the propagation speed c. Note that matrix $\mathbf{A}$ is $M \times G$ and vector $\mathbf{s}(k)$ is $G \times 1$. The columns of matrix $\mathbf{A}$ are usually known as *steering vectors*.

From this point, having the signal model given by Equation (5.6), we can start considering a proper coefficient vector to enhance the SOI. With this model, the output signal can be written as

$$y(k) = \sum_{g=1}^{G} \mathbf{w}^\mathrm{H}\mathbf{a}(\theta_g)s_g(k)\mathrm{e}^{\mathrm{j}(\omega_\mathrm{o}k+\phi_g)} + \mathbf{w}^\mathrm{H}\mathbf{n}(k). \tag{5.7}$$

A beamformer with fixed coefficients known as "Delay-and-Sum" (D&S) is obtained if we make $\mathbf{w}^\mathrm{H}\mathbf{a}(\theta_1) = 1$ such that the NB analytic input signals are properly delayed in order to align each copy from each sensor, resulting in an increase of the signal-to-noise and/or signal-to-interferer ratios, SNR and SIR,

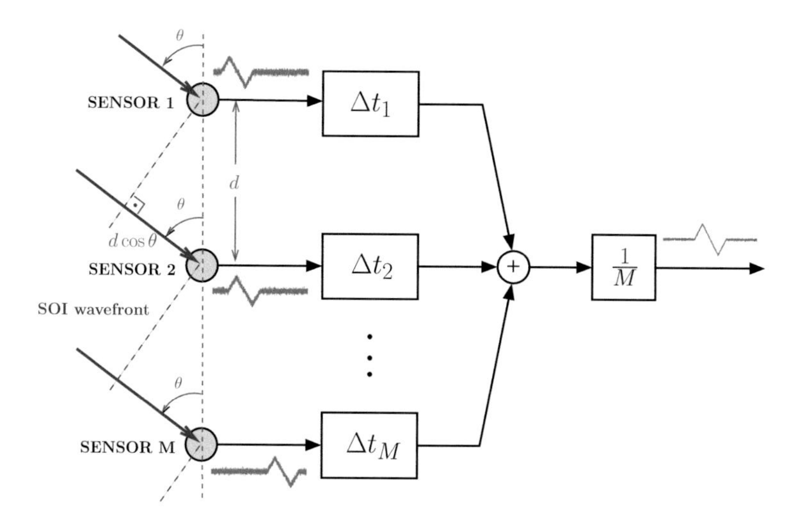

Figure 5.3 Delay and sum beamformer: using delays $\Delta t_m = (m-1)\frac{d\cos\theta}{c}$, $1 \leq m \leq M$, or $(m - \frac{M+1}{2})\frac{d\cos\theta}{c}$ if we are interested in symmetric coefficients, aligns the input signals and enhances the signal of interest (the one hitting the array with DOA equal to θ). For this figure, simulations using $M = 8$ sensors resulted in a SNR gain around 9 dB.

respectively. This D&S solution, whose expression for an incoming angle (DOA) equal to θ, is given by

$$\mathbf{w}_{\text{D\&S}} = \frac{1}{M}\mathbf{a}(\theta), \tag{5.8}$$

where $\mathbf{a}(\theta) = \begin{bmatrix} 1 & \mathrm{e}^{-\mathrm{j}\frac{2\pi d\cos\theta}{\lambda}} & \cdots & \mathrm{e}^{-\mathrm{j}\frac{2\pi(M-1)d\cos\theta}{\lambda}} \end{bmatrix}^{\mathrm{T}}$, which may be represented by the diagram depicted in Figure 5.3.

From the D&S beamforming, a plethora of options for the design of classical (or deterministic) beamformers can be found in the literature. Classical arrays are understood here as those which are not dependent on the statistics of the incoming data. Coefficient vectors can be synthesized by employing different techniques such as weighting, pattern sampling, manipulation of polynomial functions, least squares (LS) minimization of error pattern, and minimax-design, most of these techniques being exactly the same ones employed in FIR digital filter design.

Another class of beamformers is obtained if we assume we know statistical representations of the input signal. One of the most common approach is the minimum power distortionless response (MPDR) where, with respect to Figure 5.1, we want to minimize the output variance $\mathbb{E}\left[|y(k)|^2\right]$ subject to the *distortionless criterion* $\mathbf{w}^{\mathrm{H}}\mathbf{a}(\theta) = 1$.

Using the fact that $|y(k)|^2 = y(k)y^*(k) = \mathbf{w}^{\mathrm{H}}\mathbf{x}(k)\mathbf{x}^{\mathrm{H}}(k)\mathbf{w}$, we can form the cost function to be minimized

$$\mathbf{w}^{\mathrm{H}}\mathbf{R}_x\mathbf{w} \quad \text{subject to} \quad \mathbf{w}^{\mathrm{H}}\mathbf{a}(\theta) = 1, \tag{5.9}$$

where $\mathbf{R}_x = \mathbb{E}\left[\mathbf{x}(k)\mathbf{x}^{\mathrm{H}}(k)\right]$ and the term $\mathbf{w}^{\mathrm{H}}\mathbf{R}_x\mathbf{w}$, after minimization, corresponds to the *minimum output energy* (MOE).

At this point, instead of solving the problem in Equation (5.9), we recall the results of a more general case of minimizing $\mathbb{E}\left[|e(k)|^2\right]$ subject to $\mathbf{C}^{\mathrm{H}}\mathbf{w} = \mathbf{f}$, which is the solution for the linearly constrained adaptive filter (LCAF) problem [3]:

$$\mathbf{w}_{\mathrm{LCAF}} = \mathbf{R}_x^{-1}\mathbf{p} + \mathbf{R}_x^{-1}\mathbf{C}\left(\mathbf{C}^{\mathrm{H}}\mathbf{R}_x^{-1}\mathbf{C}\right)^{-1}\left(\mathbf{f} - \mathbf{C}^{\mathrm{H}}\mathbf{R}_x^{-1}\mathbf{p}\right), \tag{5.10}$$

with $\mathbf{p} = \mathbb{E}\left[d^*(k)\mathbf{x}(k)\right]$ and the dimension of the constrained vector $\mathbf{f}$ corresponding to the number of linear constraints.

For the case of an unconstrained optimization, the result corresponds to the Wiener solution $\mathbf{R}_x^{-1}\mathbf{p}$. On the other hand, if $d(k) = 0$ (null reference signal as in the typical case of beamforming) and $\mathbf{f} = 1$, the LCAF solution turns out to be the MPDR solution:

$$\mathbf{w}_{\mathrm{MPDR}} = \frac{\mathbf{R}_x^{-1}\mathbf{a}(\theta)}{\mathbf{a}^{\mathrm{H}}(\theta)\mathbf{R}_x^{-1}\mathbf{a}(\theta)}. \tag{5.11}$$

We note that $\mathbf{R}_x$, with the signal model defined in Equation (5.6) and assuming uncorrelated noise, can be written as

$$\mathbf{R}_x = \mathbf{A}\mathbf{R}_s\mathbf{A}^{\mathrm{H}} + \mathbf{R}_n, \tag{5.12}$$

where $\mathbf{R}_s = \mathbb{E}\left[\mathbf{s}(k)\mathbf{s}^{\mathrm{H}}(k)\right]$ and $\mathbf{R}_n = \mathbb{E}\left[\mathbf{n}(k)\mathbf{n}^{\mathrm{H}}(k)\right]$.

From the previous equation, we can express the variance (assuming a zero-mean stochastic process) of the array output as

$$\mathbb{E}\left[|y(k)|^2\right] = \mathbf{w}^{\mathrm{H}}\mathbf{A}\mathbf{R}_s\mathbf{A}^{\mathrm{H}}\mathbf{w} + \mathbf{w}^{\mathrm{H}}\mathbf{R}_n\mathbf{w}, \tag{5.13}$$

where we observe a portion of the output energy due to noise only.

The minimum variance distortionless response (MVRD) solution, also referred to as Capon beamformer, is defined as the coefficient vector that minimizes $\mathbf{w}^{\mathrm{H}}\mathbf{R}_n\mathbf{w}$ subject to $\mathbf{w}^{\mathrm{H}}\mathbf{a}(\theta) = 1$:

$$\mathbf{w}_{\mathrm{MVDR}} = \frac{\mathbf{R}_n^{-1}\mathbf{a}(\theta)}{\mathbf{a}^{\mathrm{H}}(\theta)\mathbf{R}_n^{-1}\mathbf{a}(\theta)}. \tag{5.14}$$

It is not uncommon to find in the literature the MPDR solution in Equation (5.11) addressed to as the MVDR solution.

In discrete-time signal processing, the concept of frequency response of a linear and time-invariant system comes from the output gain of the system for a single frequency input, that is, if $x(k) = \mathrm{e}^{\mathrm{j}\omega k}$ than the output will be $y(k) = H(\mathrm{e}^{\mathrm{j}\omega})\mathrm{e}^{\mathrm{j}\omega k}$, $H(\mathrm{e}^{\mathrm{j}\omega})$ being the *frequency response*. A similar concept is found in space-time filtering if a single plane wave with frequency f, digital frequency $\omega = \frac{2\pi f}{f_s}$ (f_s being the sampling frequency), hits the array from a given direction (θ). For the beamformer depicted in Figure 5.1, we can write the snapshot at time instant k as $\mathbf{x}(k) = \mathrm{e}^{\mathrm{j}\omega k}\mathbf{a}(\theta)$ and the output signal shall

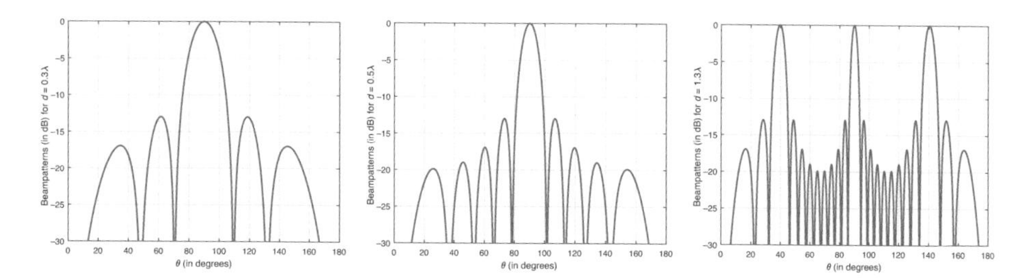

Figure 5.4 Beampatterns, in dB, for an M=10 sensor ULA beamformer steered at 90° for different distances between sensors: (left) $d = 0.3\lambda$, (center) $d = 0.5\lambda$, the most usual choice, and (right) $d = 1.3\lambda$ where we observe undesirable grating lobes.

be $y(k) = \mathbf{w}^{\mathrm{H}}\mathbf{a}(\theta)e^{j\omega k}$. The beampattern associated with an array coefficient vector ($\mathbf{w}$) may be then defined as the energy (usually expressed in dB) of this gain as a function of a DoA (angle θ for our ULA), or

$$\mathrm{BP}_{\mathrm{dB}}(\theta) = 10\log(|\mathbf{w}^{\mathrm{H}}\mathbf{a}(\theta)|^2). \tag{5.15}$$

A simple example is provided in Figure 5.4 where we observe the beampattern of a 10 sensor D&S beamformer steered to 90° (uniformly weighted linear array, i.e., $\mathbf{w} = \frac{1}{M}\mathbf{1}_{M\times 1}$) with different sensor spacing: 0.3λ, 0.5λ, and 1.3λ. Note that 0.5λ is the most effective since it presents the main lobe having a smaller beamwidth without the risk of *grating lobes*[1] entering the *visible region* ($0° \leq \theta \leq 180°$) as occurs for $d = 1.3\lambda$.

Another example, shown in Figure 5.5, corresponds to the beampatterns of the D&S, the MPDR, and the MVDR beamformers, for an $M = 8$ sensor ULA with sensor spacing 0.5λ. We consider $G = 2$ incoming signals modeled as in Equation (5.7) with $s_g(t)$ being sinusoids with frequencies $f_o/200$ (SOI) and $f_o/100$ (interferer) coming from 35° and 120°, respectively, and with SIR equal to 0 dB. The noise vector was assumed to have independent elements, with white Gaussian distribution, such that SNR=10 dB. We observe that: (1) the beampattern of the D&S beamformer does not have a null pointed at the interferer; (2) the beampattern of the MPDR beamformer places a null in the direction of the interferer; and (3) the MVDR beamformer, due to the fact that the zero-mean white Gaussian noise has an autocorrelation matrix corresponding to an identity matrix (multiplied by its variance), presents the same coefficient vector as the D&S beamformer and therefore shares the same beampattern.

These simple examples shed light on a few problems. In our simulations, the D&S solution carries no information on the direction of the interferer (certainly, if this direction θ_I is known, the reader will find in the literature ways

[1] A *grating lobe* is a lobe with the same height as the main lobe [37], usually outside the visible region, but which could move into this region when we increase the distance between sensors. This problem, spatial undersampling, is similar to that of aliasing or TD undersampling.

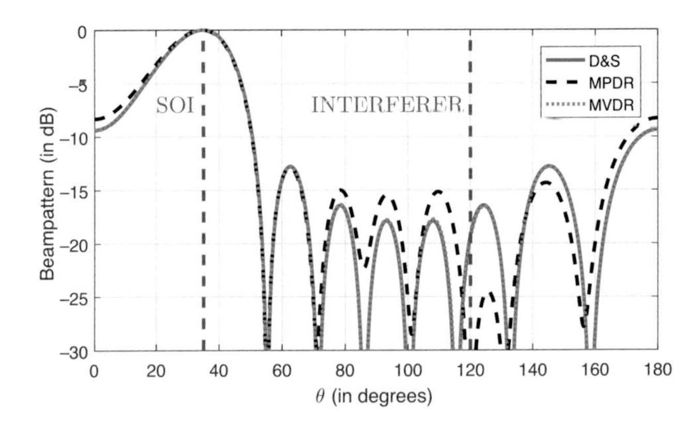

Figure 5.5 D&S and MPDR beampatterns, in dB, for an $M = 8$ sensor standard ($d = 0.5\lambda$) linear array steered to 35°. Note that the MPDR beamformer places a null at the direction of the interferer while both beamformers attain distortionless response (0 dB for the DoA of the SOI). In this case, the MVDR beamformer corresponds to the D&S solution.

to impose a null at θ_I), while the MPDR beamformer, from the information contained in the statistics of the incoming signals, manages not only to provide distortionless response but also to force a null in the direction of the interferer. Nevertheless, this feature comes at a price: We need to know $\mathbf{R}_x$ or estimate it from a set of previously buffered data or *sample support* (which introduces delay). Also, it is not guaranteed that, in every situation, the peak corresponds exactly to the direction of the SOI. Within this scenario, adaptive beamforming raises as an attractive option for it presents a distortionless solution, which approaches the optimal (MPDR, for instance) result as time goes by. In the next section, we present the linearly constrained adaptive algorithms employed in the implementation of an adaptive beamformer.

5.3 Narrowband Adaptive Beamforming

As pointed out in the former section, a LCAF minimizes the cost function $\mathbb{E}\left[|e(k)|^2\right]$ (MSE) subject to a set of p linear constraints given by

$$\mathbf{C}^{\mathrm{H}}\mathbf{w} = \mathbf{f}, \tag{5.16}$$

where $\mathbf{C}$ is an $M \times p$ matrix containing the p constraint vectors and $\mathbf{f}$ is a vector containing the values of the p constraints (a single constraint would lead to $\mathbf{f}$ being a scalar). Its general solution is given in Equation (5.10) while the MPDR solution in Equation (5.11) is obtained when we use $d(k) = 0$ such that we minimize $\mathbb{E}\left[|y(k)|^2\right]$ subject to $\mathbf{w}^{\mathrm{H}}\mathbf{a}(\theta)$=1, θ being the DoA of the SOI. The MPDR solution is able, even if an interferer is transmitting from another (unknown) direction, to place a null in that direction. Nevertheless, we can have more side information about the application and impose more than one constraint. A simple example corresponds to the case where we know the DoA of both, the SOI (θ_1) and the interferer (θ_2). In this hypothetical case, we make $\mathbf{C} = [\mathbf{a}(\theta_1)\ \ \mathbf{a}(\theta_2)]$ and $\mathbf{f} = [1\ \ 0]^{\mathrm{T}}$.

In NB beamforming, an adaptive filter can be easily employed to provide an output sample at every iteration k without the need of estimating matrix $\mathbf{R}_x$ and vector $\mathbf{p}$. A number of algorithms and structures that can perform this job shall be discussed in the following, all related to adaptive beamforming in a ULA of omnidirectional sensors.

5.3.1 LMS-Based Constrained Algorithms

The first linearly constrained adaptive algorithm, the constrained least-mean square (CLMS) [2], updates the coefficient vector at every iteration in order to receive a signal from a given direction (SOI) and attenuate possible interferences from other directions. Using Lagrange multipliers to include the constraints into the objective function, we may write

$$\xi(\mathbf{w}) = \mathbb{E}\left[|e(k)|^2\right] + \mathrm{Real}\left[\boldsymbol{\lambda}^{\mathrm{H}}(\mathbf{C}^{\mathrm{H}}\mathbf{w} - \mathbf{f})\right], \tag{5.17}$$

where Real $[x]$ means the real part of x.

This Lagrangian function can be minimized by applying a steepest descent method to update the coefficient vector at each iteration:

$$\mathbf{w}(k+1) = \mathbf{w}(k) - \frac{\mu}{2}\boldsymbol{\nabla}_{\mathbf{w}}\xi(\mathbf{w}), \tag{5.18}$$

$\boldsymbol{\nabla}_{\mathbf{w}}\xi(\mathbf{w})$ being the gradient vector, $\boldsymbol{\nabla}_{\mathbf{w}}\xi(\mathbf{w}) = -2\mathbb{E}\left[e^*(k)\mathbf{x}(k)\right] + \mathbf{C}\boldsymbol{\lambda}$, pointing toward the steepest ascent of the cost function.

Using the instantaneous estimate of the gradient vector, we find

$$\hat{\boldsymbol{\nabla}}_{\mathbf{w}}\xi(\mathbf{w}) = -2e^*(k)\mathbf{x}(k) + \mathbf{C}\boldsymbol{\lambda}, \tag{5.19}$$

which, using the constraint relation $\mathbf{C}^{\mathrm{H}}\mathbf{w}(k+1) = \mathbf{f}$, and Equation (5.19) for $\boldsymbol{\lambda}$, results in the closed-form expression for the CLMS algorithm,

$$\mathbf{w}(k+1) = \mathbf{P}\left[\mathbf{w}(k) + \mu e^*(k)\mathbf{x}(k)\right] + \mathbf{f}_{\mathrm{c}}, \tag{5.20}$$

where $e(k)$ is the *a priori* error, $\mathbf{P} = \mathbf{I}_{M\times M} - \mathbf{C}\left(\mathbf{C}^{\mathrm{H}}\mathbf{C}\right)^{-1}\mathbf{C}^{\mathrm{H}}$ is a projection matrix, and $\mathbf{f}_{\mathrm{c}} = \mathbf{C}\left(\mathbf{C}^{\mathrm{H}}\mathbf{C}\right)^{-1}\mathbf{f}$ is an $M \times 1$ vector. Note that the expression being multiplied by the projection matrix $\mathbf{P}$ corresponds to the updated expression of the (unconstrained) LMS algorithm.

The normalized version of the CLMS algorithm, the CNLMS algorithm, can be derived if we minimize, with respect to the step size (assumed now time-varying), the instantaneous *a posteriori* squared error at instant k [3]:

$$\frac{\partial\left(\varepsilon(k)\varepsilon^*(k)\right)}{\partial\mu_k^*} = 0, \tag{5.21}$$

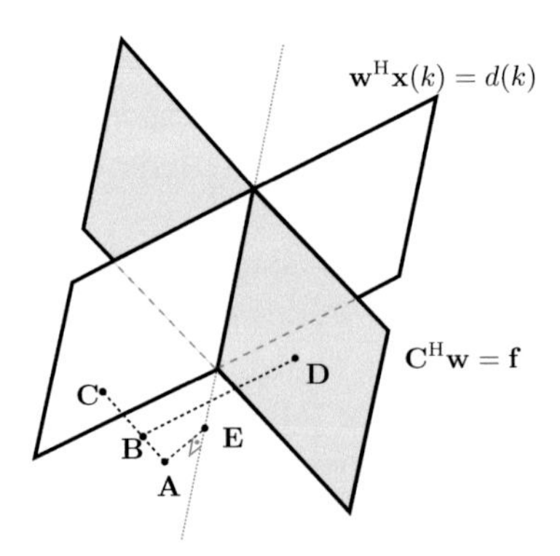

Figure 5.6 Letting point **A** be the coefficient vector before updating, $\mathbf{w}(k)$, we show its updated version, $\mathbf{w}(k+1)$, for: **B** the LMS algorithm; **C** the NLMS ($\mu = 1$) algorithm; **D** the CLMS algorithm; and **E** the CNLMS ($\mu = 1$) algorithm.

where the *a posteriori* error, as defined in Chapter 1, corresponds to $\varepsilon(k) = d(k) - \mathbf{w}^{\mathrm{H}}(k+1)\mathbf{x}(k)$.

Solving Equation (5.21) for μ_k, we obtain

$$\mu_k = \frac{1}{\mathbf{x}^{\mathrm{H}}(k)\mathbf{P}\mathbf{x}(k)}. \tag{5.22}$$

The closed-form updating expression of the CNLMS algorithm is given as [4]:

$$\mathbf{w}(k+1) = \mathbf{P}\left[\mathbf{w}(k) + \mu\frac{e^*(k)\mathbf{x}(k)}{\mathbf{x}^{\mathrm{H}}(k)\mathbf{P}\mathbf{x}(k) + \gamma}\right] + \mathbf{f}_{\mathrm{c}}, \tag{5.23}$$

where μ is a fixed convergence factor introduced to control misadjustment and γ is a parameter included to avoid overflow when the term in the denominator, $\mathbf{x}^{\mathrm{H}}(k)\mathbf{P}\mathbf{x}(k)$, becomes too small. Keep in mind that the step size μ used by the CNLM algorithm, although may range from 0 to 2 without divergence, is usually set in the range $0 < \mu \leq 1$ in order to obtain the best tradeoff between misadjustment and convergence speed. The coefficient vector $\mathbf{w}(k)$ is to be initialized with vector $\mathbf{f}_{\mathrm{c}}$.

Figure 5.6 shows a geometrical interpretation of the updating procedure for the LMS and the NLMS algorithms as well as for their constrained counterparts. We observe that the LMS algorithm updates in the direction of the input signal vector $\mathbf{x}(k)$, which is perpendicular to the hyperplane defined by the null *a posteriori* error, $\mathbf{w}^{\mathrm{H}}\mathbf{x}(k) - d(k) = 0$. Also, note that the CLMS coefficient vector corresponds to the projection of the LMS updated vector into the hyperplane defined by the linearly constraints, $\mathbf{C}^{\mathrm{H}}\mathbf{w} = \mathbf{f}$.

5.3.2 RLS-Based Constrained Algorithms

Similarly to the LCAF solution in Equation (5.10), if we now use a least squares (LS) cost function, the problem would be stated as minimizing $\sum_{i=0}^{k} \lambda^{k-i}|d(i) -$

$\mathbf{w}^{\mathrm{H}}\mathbf{x}(i)|^2$ subject to $\mathbf{C}^{\mathrm{H}}\mathbf{w} = \mathbf{f}$, where λ is the forgetting factor. The constrained least squares (CLS) solution, for time instant k, is then given by

$$\mathbf{w}(k+1) = \mathbf{R}_x^{-1}(k)\mathbf{p}(k) + \mathbf{R}_x^{-1}(k)\mathbf{C}[\mathbf{C}^{\mathrm{H}}\mathbf{R}_x^{-1}(k)\mathbf{C}]^{-1}[\mathbf{f} - \mathbf{C}^{\mathrm{H}}\mathbf{R}_x^{-1}(k)\mathbf{p}(k)], \tag{5.24}$$

where $\mathbf{R}_x(k)$ is the deterministic autocorrelation matrix, at instant k, of the input signal vector and $\mathbf{p}(k)$ is the cross-correlation vector, at instant k, between the input signal vector $\mathbf{x}(k)$ and the reference signal $d(k)$.

From the CLS optimal solution presented in Equation (5.24), we realize that it has two distinct terms: the first part corresponding to the unconstrained solution, $\mathbf{w}_{\mathrm{uc}}(k+1)$, plus an additional expression, $\mathbf{w}_{\mathrm{c}}(k+1)$, responsible for fulfilling the set of constraints imposed by $\mathbf{C}^{\mathrm{H}}\mathbf{w}(k+1) = \mathbf{f}$:

$$\mathbf{w}(k+1) = \mathbf{w}_{\mathrm{uc}}(k+1) + \mathbf{w}_{\mathrm{c}}(k+1), \tag{5.25}$$

where $\mathbf{w}_{\mathrm{uc}}(k+1) = \mathbf{R}_x^{-1}(k)\mathbf{p}(k)$ and the term in charge of ensuring the linear constraints is $\mathbf{w}_{\mathrm{c}}(k+1) = \mathbf{R}_x^{-1}(k)\mathbf{C}[\mathbf{C}^{\mathrm{H}}\mathbf{R}_x^{-1}(k)\mathbf{C}]^{-1}[\mathbf{f} - \mathbf{C}^{\mathrm{H}}\mathbf{R}_x^{-1}(k)\mathbf{p}(k)]$.

We present in the following a simple RLS-like form of the CLS optimal solution. We start by observing that $\mathbf{w}_{\mathrm{uc}}(k+1)$ already has a recursive expression corresponding to the unconstrained RLS algorithm:

$$\mathbf{w}_{\mathrm{uc}}(k+1) = \mathbf{w}_{\mathrm{uc}}(k) + e_{\mathrm{uc}}^*(k)\boldsymbol{\kappa}(k), \tag{5.26}$$

where $\boldsymbol{\kappa}(k) = \mathbf{R}_x^{-1}(k)\mathbf{x}(k)$, also expressed as $\mathbf{S}_{\mathrm{D}}(k)\mathbf{x}(k)$ or, equivalently, as $\frac{\mathbf{S}_{\mathrm{D}}(k-1)\mathbf{x}(k)}{\lambda+\mathbf{x}^{\mathrm{H}}(k)\mathbf{S}_{\mathrm{D}}(k-1)\mathbf{x}(k)}$, is the gain factor, $\mathbf{S}_{\mathrm{D}}(k) = \mathbf{R}_x^{-1}(k)$, and the *a priori* unconstrained error corresponds to $e_{\mathrm{uc}}(k) = d(k) - \mathbf{w}_{\mathrm{uc}}^{\mathrm{H}}(k)\mathbf{x}(k)$.

In order to simplify the notation, we define auxiliary matrices $\boldsymbol{\Gamma}(k)$ and $\boldsymbol{\Psi}(k)$:

$$\boldsymbol{\Gamma}(k) = \mathbf{R}_x^{-1}(k)\mathbf{C} = \mathbf{S}_{\mathrm{D}}(k)\mathbf{C} \tag{5.27}$$

$$\boldsymbol{\Psi}(k) = \mathbf{C}^{\mathrm{H}}\boldsymbol{\Gamma}(k) = \mathbf{C}^{\mathrm{H}}\mathbf{S}_{\mathrm{D}}(k)\mathbf{C} \tag{5.28}$$

such that $\mathbf{w}_{\mathrm{c}}(k+1) = \boldsymbol{\Gamma}(k)\boldsymbol{\Psi}^{-1}(k)[\mathbf{f} - \mathbf{C}^{\mathrm{H}}\mathbf{w}_{\mathrm{uc}}(k+1)]$.

The computation of $\boldsymbol{\Psi}^{-1}(k)$ is easily carried out if we have only one restriction ($p = 1$ such that $\boldsymbol{\Psi}(k)$ is scalar). With these definitions, Algorithm 5.28 depicts the constrained RLS (CRLS) algorithm. In this algorithm, vector $\boldsymbol{\psi}(k)$ is defined as $\mathbf{S}_{\mathrm{D}}(k-1)\mathbf{x}(k)$ in order to simplify the expressions and matrix $\mathbf{S}_{\mathrm{D}}(k)$ is updated according to the matrix inversion lemma[2] [3].

Note from Algorithm 5.28 that the updated coefficient vector satisfies the set of constraints for $\mathbf{C}^{\mathrm{H}}\mathbf{w}(k+1) = \mathbf{f}$, which holds regardless of the initialization. Also note that the updating equation of $\mathbf{w}(k+1)$ can be written as:

$$\mathbf{w}(k+1) = \underbrace{\left[\mathbf{I} - \boldsymbol{\Gamma}(k)\boldsymbol{\Psi}^{-1}(k)\mathbf{C}^{\mathrm{H}}\right]}_{\mathbf{P}(k)}\mathbf{w}_{\mathrm{uc}}(k+1) + \underbrace{\boldsymbol{\Gamma}(k)\boldsymbol{\Psi}^{-1}(k)\mathbf{f}}_{\mathbf{f}_{\mathrm{c}}(k)}, \tag{5.29}$$

which resembles the constrained LMS algorithm with time varying $\mathbf{P}$ and $\mathbf{f}_{\mathrm{c}}$.

[2] $[\mathbf{A} + \mathbf{BCD}]^{-1} = \mathbf{A}^{-1} - \mathbf{A}^{-1}\mathbf{B}\left[\mathbf{DA}^{-1}\mathbf{B} + \mathbf{C}^{-1}\right]^{-1}\mathbf{DA}^{-1}$, where $\mathbf{A}$, $\mathbf{B}$, $\mathbf{C}$ and, $\mathbf{D}$ have appropriate dimensions, and matrices $\mathbf{A}$ and $\mathbf{C}$ are nonsingular.

Algorithm 5.28 The constrained RLS algorithm (basic version)

Initialization

$\mathbf{w}_{\mathrm{uc}}(0)$, $\mathbf{S}_{\mathrm{D}}(-1) = \mathbf{R}_x^{-1}(-1)$, and $0 \ll \lambda < 1$

For $k \geq 0$ **do**

$\boldsymbol{\psi}(k) = \mathbf{S}_{\mathrm{D}}(k-1)\mathbf{x}(k)$

$\boldsymbol{\kappa}(k) = \frac{\boldsymbol{\psi}(k)}{\lambda + \mathbf{x}^{\mathrm{H}}(k)\boldsymbol{\psi}(k)}$

$\mathbf{S}_{\mathrm{D}}(k) = \frac{1}{\lambda}\left[\mathbf{S}_{\mathrm{D}}(k-1) - \frac{\boldsymbol{\psi}(k)\boldsymbol{\psi}^{\mathrm{H}}(k)}{\lambda + \mathbf{x}^{\mathrm{H}}(k)\boldsymbol{\psi}(k)}\right]$

$e_{\mathrm{uc}}(k) = d(k) - \mathbf{w}_{\mathrm{uc}}^{\mathrm{H}}(k)\mathbf{x}(k)$

$\mathbf{w}_{\mathrm{uc}}(k+1) = \mathbf{w}_{\mathrm{uc}}(k) + e_{\mathrm{uc}}^{*}(k)\boldsymbol{\kappa}(k)$

$\boldsymbol{\Gamma}(k) = \mathbf{S}_{\mathrm{D}}(k)\mathbf{C}$

$\boldsymbol{\Psi}(k) = \mathbf{C}^{\mathrm{H}}\boldsymbol{\Gamma}(k)$

$\mathbf{w}(k+1) = \mathbf{w}_{\mathrm{uc}}(k+1) + \boldsymbol{\Gamma}(k)\boldsymbol{\Psi}^{-1}(k)\left[\mathbf{f} - \mathbf{C}^{\mathrm{H}}\mathbf{w}_{\mathrm{uc}}(k+1)\right]$

end

Moreover, $\mathbf{P}(k)$ corresponds to a projection matrix in the subspace orthogonal to the subspace spanned by $\mathbf{R}_x^{-1}(k)\mathbf{C}$.

In order to disclose a more efficient version of the CRLS algorithm, we start by showing a recursive expression for $\boldsymbol{\Gamma}(k)$:

$$\boldsymbol{\Gamma}(k) = \frac{1}{\lambda}\left[\boldsymbol{\Gamma}(k-1) - \boldsymbol{\kappa}(k)\mathbf{x}^{\mathrm{H}}(k)\boldsymbol{\Gamma}(k-1)\right]. \tag{5.30}$$

Then, the updating of $\boldsymbol{\Psi}^{-1}(k)$ is carried out using the *Matrix Inversion Lemma* and the result is

$$\boldsymbol{\Psi}^{-1}(k) = \lambda\left[\boldsymbol{\Psi}^{-1}(k-1) + \frac{\boldsymbol{\Psi}^{-1}(k-1)\mathbf{C}^{\mathrm{H}}\boldsymbol{\kappa}(k)\mathbf{x}^{\mathrm{H}}(k)\boldsymbol{\Gamma}(k-1)\boldsymbol{\Psi}^{-1}(k-1)}{1 - \mathbf{x}^{\mathrm{H}}(k)\boldsymbol{\Gamma}(k-1)\boldsymbol{\Psi}^{-1}(k-1)\mathbf{C}^{\mathrm{H}}\boldsymbol{\kappa}(k)}\right]. \tag{5.31}$$

Aiming at simplifying the notation, we define the auxiliary vector

$$\boldsymbol{\ell}(k) = \frac{\boldsymbol{\Psi}^{-1}(k-1)\mathbf{C}^{\mathrm{H}}\boldsymbol{\kappa}(k)}{1 - \mathbf{x}^{\mathrm{H}}(k)\boldsymbol{\Gamma}(k-1)\boldsymbol{\Psi}^{-1}(k-1)\mathbf{C}^{\mathrm{H}}\boldsymbol{\kappa}(k)}, \tag{5.32}$$

such that

$$\boldsymbol{\Psi}^{-1}(k) = \lambda\left[\boldsymbol{\Psi}^{-1}(k-1) + \boldsymbol{\ell}(k)\mathbf{x}^{\mathrm{H}}(k)\boldsymbol{\Gamma}(k-1)\boldsymbol{\Psi}^{-1}(k-1)\right]. \tag{5.33}$$

From Equation (5.31), post-multiplying the whole equation by $\mathbf{C}^{\mathrm{H}}\boldsymbol{\kappa}(k)$ and dividing by λ, it is possible to prove that the result corresponds to

$$\boldsymbol{\ell}(k) = \frac{1}{\lambda}\boldsymbol{\Psi}^{-1}(k)\mathbf{C}^{\mathrm{H}}\boldsymbol{\kappa}(k). \tag{5.34}$$

Algorithm 5.29 The CRLS algorithm (version 1)

Initialization

$\mathbf{w}(0) \in \{\mathbf{w}|\mathbf{C}^{\mathrm{H}}\mathbf{w} = \mathbf{f}\}$, $\mathbf{S}_{\mathrm{D}}(-1)$, $\boldsymbol{\Gamma}(-1) = \mathbf{S}_{\mathrm{D}}(-1)\mathbf{C}$, $0 \ll \lambda < 1$, and $\boldsymbol{\Psi}(-1) = \mathbf{C}^{\mathrm{H}}\boldsymbol{\Gamma}(-1)$

For $k \geq 0$ **do**

Update $\boldsymbol{\kappa}(k)$ using a reduced complexity RLS algorithm

$\boldsymbol{\ell}(k) = \frac{\boldsymbol{\Psi}^{-1}(k-1)\mathbf{C}^{\mathrm{H}}\boldsymbol{\kappa}(k)}{1-\mathbf{x}^{\mathrm{H}}(k)\boldsymbol{\Gamma}(k-1)\boldsymbol{\Psi}^{-1}(k-1)\mathbf{C}^{\mathrm{H}}\boldsymbol{\kappa}(k)}$

$\boldsymbol{\Psi}^{-1}(k) = \lambda\left[\boldsymbol{\Psi}^{-1}(k-1) + \boldsymbol{\ell}(k)\mathbf{x}^{\mathrm{H}}(k)\boldsymbol{\Gamma}(k-1)\boldsymbol{\Psi}^{-1}(k-1)\right]$

$\boldsymbol{\Gamma}(k) = \frac{1}{\lambda}\left[\boldsymbol{\Gamma}(k-1) - \boldsymbol{\kappa}(k)\mathbf{x}^{\mathrm{H}}(k)\boldsymbol{\Gamma}(k-1)\right]$

$e(k) = d(k) - \mathbf{w}^{\mathrm{H}}(k)\mathbf{x}(k)$

$\mathbf{w}(k+1) = \mathbf{w}(k) + e^*(k)\boldsymbol{\kappa}(k) - \lambda e^*(k)\boldsymbol{\Gamma}(k)\boldsymbol{\ell}(k)$

end

All expressions obtained so far lead to a recursive expression for the CRLS coefficient vector updating process which, after algebraic manipulations, can be expressed as

$$\begin{aligned}\mathbf{w}(k+1) &= \mathbf{w}(k) + e^*(k)\boldsymbol{\kappa}(k) - \lambda e^*(k)\boldsymbol{\Gamma}(k)\boldsymbol{\ell}(k) \\ &= \mathbf{w}(k) + e^*(k)\mathbf{R}_x^{-1}(k)\mathbf{x}(k) \\ &\quad - e^*(k)\mathbf{R}_x^{-1}(k)\mathbf{C}[\mathbf{C}^{\mathrm{H}}\mathbf{R}_x^{-1}(k)\mathbf{C}]^{-1}\mathbf{C}^{\mathrm{H}}\mathbf{R}_x^{-1}(k)\mathbf{x}(k). \end{aligned} \tag{5.35}$$

Algorithm 5.29[3] shows this recursive version as seen in [5]. In Algorithm 5.29, it is worth noting that the initialization of $\mathbf{w}(k)$ belonging to the set all possible coefficient vectors satisfying the constraints, $\mathbf{C}^{\mathrm{H}}\mathbf{w} = \mathbf{f}$, is mandatory for the following reason. As we see, from the updating expression of the coefficient vector, vector $\mathbf{f}$ does not appear; it will be employed only in the initialization: $\mathbf{w}(0) = \mathbf{f}_{\mathrm{c}}$, for instance. This is a drawback of this algorithm for, in a finite precision environment, round-off errors will, after some iterations, probably cause the coefficients to drift from the constraint hyperplane. In order to avoid this drift, we can add a correction term as suggested in [5]. A second version of the CRLS algorithm, using a correction term in regular time intervals, is presented in Algorithm 5.30.

The CRLS algorithm is known for its fast convergence and widely used as a benchmark to compare the performance of other constrained algorithms. Yet, as a member of the RLS family of adaptive filters, it carries the burden of not having proven stability.

[3] The update of $\boldsymbol{\kappa}(k)$, as suggested in [5], could be carried out using a reduced complexity (order of M instead of M^2) approach whenever the input signal corresponds to a delay line (single input channel and not an input vector being formed from different sensors). If that is the case, the use of forward and backward linear prediction relations would lead to a reduced computational complexity as can be found in [3], Chapter 8 (Fast Transversal RLS Algorithms).

Algorithm 5.30 The CRLS algorithm (version 2)

Initialization

$\mathbf{w}(0) \in \{\mathbf{w}|\mathbf{C}^{\mathrm{H}}\mathbf{w} = \mathbf{f}\}$, $\mathbf{S}_{\mathrm{D}}(-1)$, $\mathbf{\Gamma}(-1) = \mathbf{S}_{\mathrm{D}}(-1)\mathbf{C}$, $0 \ll \lambda < 1$, and $\mathbf{\Psi}(-1) = \mathbf{C}^{\mathrm{H}}\mathbf{\Gamma}(-1)$

For $k \geq 0$ **do**

Update $\boldsymbol{\kappa}(k)$ using a reduced complexity RLS algorithm

$\boldsymbol{\ell}(k) = \frac{\mathbf{\Psi}^{-1}(k-1)\mathbf{C}^{\mathrm{H}}\boldsymbol{\kappa}(k)}{1-\mathbf{x}^{\mathrm{H}}(k)\mathbf{\Gamma}(k-1)\mathbf{\Psi}^{-1}(k-1)\mathbf{C}^{\mathrm{H}}\boldsymbol{\kappa}(k)}$

$\mathbf{\Psi}^{-1}(k) = \lambda\left[\mathbf{\Psi}^{-1}(k-1) + \boldsymbol{\ell}(k)\mathbf{x}^{\mathrm{H}}(k)\mathbf{\Gamma}(k-1)\mathbf{\Psi}^{-1}(k-1)\right]$

$\mathbf{\Gamma}(k) = \frac{1}{\lambda}\left[\mathbf{\Gamma}(k-1) - \boldsymbol{\kappa}(k)\mathbf{x}^{\mathrm{H}}(k)\mathbf{\Gamma}(k-1)\right]$

$e(k) = d(k) - \mathbf{w}^{\mathrm{H}}(k)\mathbf{x}(k)$

$\mathbf{w}(k+1) = \mathbf{P}\mathbf{w}(k) + \mathbf{f}_{\mathrm{c}} + e^*(k)\boldsymbol{\kappa}(k) - \lambda e^*(k)\mathbf{\Gamma}(k)\boldsymbol{\ell}(k)$

Only at regular time intervals:

$\mathbf{w}(k+1) = \mathbf{w}(k) - \mathbf{C}(\mathbf{C}^{\mathrm{H}}\mathbf{C})^{-1}e^*(k)\boldsymbol{\kappa}(k) + \lambda e^*(k)\mathbf{C}(\mathbf{C}^{\mathrm{H}}\mathbf{C})^{-1}\mathbf{\Gamma}(k)\boldsymbol{\ell}(k)$

end

5.3.3 Constrained Affine-Projection Algorithms

The constrained affine-projection (CAP) algorithm solves an optimization problem given as

$$\mathbf{w}(k+1) = \arg\min_{\mathbf{w}} \|\mathbf{w} - \mathbf{w}(k)\|^2 \text{ subject to } \begin{cases} \mathbf{C}^{\mathrm{H}}\mathbf{w} = \mathbf{f} \\ \mathbf{d}^*(k) - \mathbf{X}^{\mathrm{H}}(k)\mathbf{w} = \mathbf{0}, \end{cases} \tag{5.36}$$

where $\mathbf{d}(k) \in \mathbb{C}^{(L+1)\times 1}$ is the reference-signal vector and $\mathbf{X}(k) \in \mathbb{C}^{(N+1)\times(L+1)}$ is the input-signal matrix, defined as

$$\mathbf{d}(k) = [d(k)\, d(k-1)\, \cdots\, d(k-L)]^{\mathrm{T}} \text{ and} \tag{5.37}$$

$$\mathbf{X}(k) = [\mathbf{x}(k)\, \mathbf{x}(k-1)\, \cdots\, \mathbf{x}(k-L)], \tag{5.38}$$

$\mathbf{x}(k)$ being the input-signal vector as defined previously.

Lagrange multipliers can be used to solve this constrained minimization problem, leading to the CAP algorithm updating equation [6]

$$\mathbf{w}(k+1) = \mathbf{P}[\mathbf{w}(k) + \mathbf{X}(k)\, \mathbf{t}(k)] + \mathbf{f}_{\mathrm{c}}, \tag{5.39}$$

with

$$\mathbf{t}(k) = [\mathbf{X}^{\mathrm{H}}(k)\, \mathbf{P}\, \mathbf{X}(k)]^{-1}\mathbf{e}^*(k) \tag{5.40}$$

and

$$\mathbf{e}^*(k) = \mathbf{d}^*(k) - \mathbf{X}^{\mathrm{H}}(k)\mathbf{w}(k). \tag{5.41}$$

As previously defined, matrix $\mathbf{P}$ performs a projection onto the homogeneous hyperplane defined by $\mathbf{C}^{\mathrm{H}}\mathbf{w} = \mathbf{0}$ and vector $\mathbf{f}_{\mathrm{c}}$ moves the projected solution back to the constraint hyperplane. Hence, the coefficient vector $\mathbf{w}(k)$ can be written as $\mathbf{P}\mathbf{w}(k) + \mathbf{f}_{\mathrm{c}}$ which could be used to simplify (5.39). Nevertheless, due

Algorithm 5.31 The constrained affine-projection algorithm

Initialization

$\mathbf{w}(0) = \mathbf{f}_c$

For $k \geq 0$ **do**

$\mathbf{X}(k) = [\mathbf{x}(k)\, \mathbf{x}(k-1)\, \cdots\, \mathbf{x}(k-L)]$

$\mathbf{d}(k) = [d(k)\, d(k-1)\, \ldots\, d(k-L)]^{\mathrm{T}}$

$\mathbf{e}^*(k) = \mathbf{d}^*(k) - \mathbf{X}^{\mathrm{H}}(k)\mathbf{w}(k)$

$\mathbf{t}(k) = \left[\mathbf{X}^{\mathrm{H}}(k)\, \mathbf{P}\, \mathbf{X}(k) + \gamma \mathbf{I}\right]^{-1} \mathbf{e}^*(k)$

$\mathbf{w}(k+1) = \mathbf{P}\left[\mathbf{w}(k) + \mu \mathbf{X}(k)\, \mathbf{t}(k)\right] + \mathbf{f}_c$

end

to the accumulation of round-off errors, this simplification may cause the solution to drift away from the constraint hyperplane [2] and should be avoided not only here but also for the Constrained LMS algorithm. Note that, for the cases of $L = 0$ and $L = 1$, the recursions are equivalent to those of the constrained normalized LMS (CNLMS) and constrained binormalized LMS (CBNLMS) algorithms [4], respectively.

The equations of the CAP algorithm are summarized in Algorithm 5.31, where a step size μ is usually chosen between 0 and 1. In order to improve robustness, a diagonal matrix $\delta \mathbf{I}$ (δ being a small constant) is employed for regularization.

In order to obtain a data-selective version of the CAP algorithm, we recall from Chapter 1 the concept of a constraint set $\mathcal{H}(k)$, a set containing all vectors $\mathbf{w}$ for which the associated output error is upper bounded (in magnitude) by $\bar{\gamma}$, that is

$$\mathcal{H}(k) = \{\mathbf{w} \in \mathbb{C}^{(N+1)\times 1} : |d(k) - \mathbf{w}^{\mathrm{H}}\mathbf{x}(k)| \leq \bar{\gamma}\}. \tag{5.42}$$

From the previous equation, follows the exact membership set which corresponds to the intersection of the constraint sets over time:

$$\psi(k) = \bigcap_{i=0}^{k} \mathcal{H}(i). \tag{5.43}$$

A set-membership affine-projection (SM-AP) algorithm [7] uses the information provided by the current and the L past constraint sets which intersection is given by $\psi_{L+1}(k)$, such that $\psi(k) = \psi(k-L-1) \cap \psi_{L+1}(k)$, where

$$\psi_{L+1}(k) = \bigcap_{i=k-L}^{k} \mathcal{H}(i). \tag{5.44}$$

Next, we formulate the following optimization problem whenever $\mathbf{w}(k) \notin \psi_{L+1}(k)$:

$$\mathbf{w}(k+1) = \arg\min_{\mathbf{w}} \|\mathbf{w} - \mathbf{w}(k)\|^2 \text{ subject to}$$
$$\begin{cases} \mathbf{C}^{\mathrm{H}}\mathbf{w} = \mathbf{f} \\ \mathbf{d}^*(k) - \mathbf{X}^{\mathrm{H}}(k)\mathbf{w} = \bar{\gamma}^*(k), \end{cases} \tag{5.45}$$

where, as also seen in Chapter 1, vector $\bar{\gamma}(k) = [\bar{\gamma}_1(k)\ \bar{\gamma}_2(k)\ \ldots\ \bar{\gamma}_{L+1}(k)]^{\mathrm{T}}$ sets the point in $\psi_{L+1}(k)$ for the update. The elements of $\bar{\gamma}(k)$ are to be chosen such that $|\bar{\gamma}_i(k)| \leq \bar{\gamma}$ or $i \in \{1, \cdots, L+1\}$.

Using Lagrange multipliers, we can solve the previous optimization problem which solution is as

$$\mathbf{w}(k+1) = \begin{cases} \mathbf{P}\left\{\mathbf{w}(k) + \mathbf{X}(k)[\mathbf{X}^{\mathrm{H}}(k)\mathbf{P}\mathbf{X}(k)]^{-1}\left[\mathbf{e}(k) - \bar{\gamma}(k)\right]^*\right\} + \mathbf{f}_{\mathrm{c}} & \text{if } |e(k)| > \gamma \\ \mathbf{w}(k) & \text{otherwise,} \end{cases} \tag{5.46}$$

where

$$\mathbf{e}(k) = [e(k)\ \varepsilon(k-1)\ \ldots\ \varepsilon(k-L)]^{\mathrm{T}} \tag{5.47}$$

and $\varepsilon(k-i) = d(k-i) - \mathbf{w}^{\mathrm{H}}(k)\mathbf{x}(k-i)$ denotes the *a posteriori* error at iteration $k-i$.

The version of the SM-CAP algorithm in [8] chooses $\bar{\gamma}_i(k) = \varepsilon(k-i+1)$, for $i \neq 1$; with such a choice, all but the first element in the vector $\mathbf{e}(k) - \bar{\gamma}(k)$ of (5.46) are canceled, and $\bar{\gamma}_1(k) = \bar{\gamma}e(k)/|e(k)|$ such that the *a posteriori* error lies on the closest boundary of $\mathcal{H}(k)$. The resulting update recursion [8] is given as

$$\mathbf{w}(k+1) = \mathbf{P}\left[\mathbf{w}(k) + \mathbf{X}(k)\left[\mathbf{X}^{\mathrm{H}}(k)\mathbf{P}\mathbf{X}(k)\right]^{-1}\mu(k)e^*(k)\mathbf{u}_1\right] + \mathbf{f}_{\mathrm{c}}, \tag{5.48}$$

where $\mathbf{u}_1 = [1\,0\ \ldots\ 0]^{\mathrm{T}}$ is the pinning vector and

$$\mu(k) = \begin{cases} 1 - \bar{\gamma}/|e(k)| & \text{if } |e(k)| > \bar{\gamma} \\ 0 & \text{otherwise} \end{cases} \tag{5.49}$$

is a data-dependent step size.

The equations of the SM-CAP algorithm [8] are summarized in Algorithm 5.32. We note that, whenever an update is needed, part or all of the elements of matrix $\mathbf{X}^{\mathrm{H}}(k)\mathbf{P}\mathbf{X}(k)$ needs to be recalculated and this increases the computational complexity per update (comparing to the regular CAP algorithm); this is particularly true close to the steady-state solution, when updating is sparse in time. On the other hand, this may not be a serious problem since the reduced frequency of updating compensates for the increase in complexity introduced by $\mathbf{X}^{\mathrm{H}}(k)\mathbf{P}\mathbf{X}(k)$.

There are ways to alleviate the increased computational complexity by reusing all past calculations of the cross-correlations in $\mathbf{X}^{\mathrm{H}}(k)\mathbf{P}\mathbf{X}(k)$ even when updating is sparse in time; see [9] for further details. Also, the error magnitude upper bound $\bar{\gamma}$ is usually set to $\sqrt{5\sigma_n^2}$ [3, 10], σ_n^2 being the variance of the observation error in a classical adaptive filtering setup, while $L+1$, the number of *a priori*

Algorithm 5.32 The set-membership constrained affine-projection algorithm

Initialization

$\mathbf{w}(0) = \mathbf{f}_c$, γ, and L

For $k \geq 0$ **do**

$e(k) = d(k) - \mathbf{w}^{\mathrm{H}}(k)\mathbf{x}(k)$

If $|e(k)| > \bar{\gamma}$

$\mu(k) = 1 - \bar{\gamma}/|e(k)|$

$\mathbf{X}(k) = [\mathbf{x}(k)\, \mathbf{x}(k-1)\, \cdots\, \mathbf{x}(k-L+1)]$

$\mathbf{t}(k) = \left[\mathbf{X}^{\mathrm{H}}(k)\, \mathbf{P}\, \mathbf{X}(k) + \delta\mathbf{I}\right]^{-1} \mu(k)e^*(k)\mathbf{u}_1$

$\mathbf{w}(k+1) = \mathbf{P}\,[\mathbf{w}(k) + \mathbf{X}(k)\,\mathbf{t}(k)] + \mathbf{f}_c$

else

$\mathbf{w}(k+1) = \mathbf{w}(k)$

end

end

error constraint hyperplanes, is chosen according to the application, trading-off among computational complexity, misadjustment, and speed of convergence.

5.3.4 The Generalized Sidelobe Canceller

The term generalized sidelobe canceller (GSC) was coined in [11] for an adaptive filtering scheme in which the linear constraints are embedded in the structure. We start reviewing this concept by doing a transformation in the coefficient vector. The transformation is carried out with matrix $\mathbf{T} = [\mathbf{C}\ \ \mathbf{B}]$ such that

$$\mathbf{w}(k) = \mathbf{T}\,\overline{\mathbf{w}}(k) = [\mathbf{C}\ \ \mathbf{B}]\begin{bmatrix}\overline{\mathbf{w}}_{\mathrm{U}}(k)\\ -\overline{\mathbf{w}}_{\mathrm{L}}(k)\end{bmatrix} = \mathbf{C}\overline{\mathbf{w}}_{\mathrm{U}}(k) - \mathbf{B}\overline{\mathbf{w}}_{\mathrm{L}}(k). \tag{5.50}$$

Matrix $\mathbf{B}$ is usually called the *Blocking Matrix*, and we recall the fact that the constraints are satisfied if $\mathbf{C}^{\mathrm{H}}\mathbf{w}(k) = \mathbf{f}$ such that $\mathbf{C}^{\mathrm{H}}\mathbf{w}(k) = \mathbf{C}^{\mathrm{H}}\mathbf{C}\overline{\mathbf{w}}_{\mathrm{U}}(k) - \mathbf{C}^{\mathrm{H}}\mathbf{B}\overline{\mathbf{w}}_{\mathrm{L}}(k) = \mathbf{f}$. If we impose the condition $\mathbf{B}^{\mathrm{H}}\mathbf{C} = \mathbf{0}$, we have $\overline{\mathbf{w}}_{\mathrm{U}}(k) = (\mathbf{C}^{\mathrm{H}}\mathbf{C})^{-1}\mathbf{f}$, a constant vector. It means that once this portion of the transformed coefficient vector is initialized with this vector, it requires no further update and the adaptation process will be carried out only in the lower part of the structure, hereafter designated $\mathbf{w}_{\mathrm{GSC}}(k) = \overline{\mathbf{w}}_{\mathrm{L}}(k)$.

In Figure 5.7, we split the transformation matrix into two parts: a fixed branch and an adaptive branch. Note that the nonadaptive part corresponds to vector $\mathbf{f}_c = \mathbf{C}\overline{\mathbf{w}}_{\mathrm{U}}(k) = \mathbf{C}(\mathbf{C}^{\mathrm{H}}\mathbf{C})^{-1}\mathbf{f}$. Also, note all dimensions involved in this transformation as pointed out in Figure 5.7.

Note, given the basic assumption $\mathbf{C}^{\mathrm{H}}\mathbf{B} = \mathbf{0}$, that the overall coefficient vector will automatically satisfy the constraints. Due to this feature, the adaptive part of the transformed vector can be implemented with any suitable unconstrained adaptive algorithm. The overall equivalent coefficient vector is given by

$$\mathbf{w}(k) = \mathbf{f}_c - \mathbf{B}\mathbf{w}_{\mathrm{GSC}}(k). \tag{5.51}$$

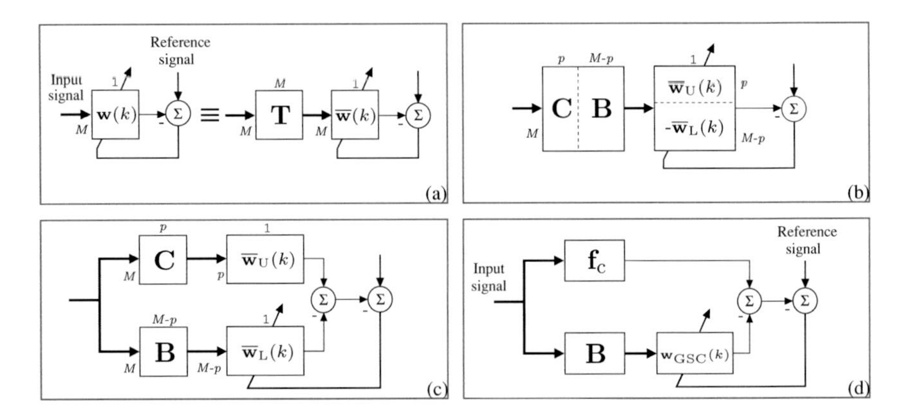

Figure 5.7 The transformation process: (a) $\mathbf{w}(k) = \mathbf{T}\overline{\mathbf{w}}(k)$; (b) $\mathbf{w}(k) = [\mathbf{C}\ \mathbf{B}]\left[\overline{\mathbf{w}}_{\mathrm{U}}^{\mathrm{T}}(k)\ \ -\overline{\mathbf{w}}_{\mathrm{L}}^{\mathrm{T}}(k)\right]^{\mathrm{T}}$; (c) $\mathbf{w}(k) = \mathbf{C}\overline{\mathbf{w}}_{\mathrm{U}}(k) - \mathbf{B}\overline{\mathbf{w}}_{\mathrm{L}}(k)$; and (d) $\mathbf{w}(k) = \mathbf{f}_{\mathrm{c}} - \mathbf{B}\mathbf{w}_{\mathrm{GSC}}(k)$.

From the previous equation and assuming we know the optimal solution for the equivalent (constrained) filter, we can express the GSC optimal coefficients as

$$\mathbf{w}_{\mathrm{OPT}} = \mathbf{f}_{\mathrm{c}} - \mathbf{B}\mathbf{w}_{\mathrm{GSC-OPT}}. \tag{5.52}$$

On the other hand, defining $\mathbf{R}_{\mathrm{GSC}} = \mathbb{E}\left[\mathbf{x}_{\mathrm{GSC}}(k)\mathbf{x}_{\mathrm{GSC}}^{\mathrm{H}}(k)\right] = \mathbf{B}^{\mathrm{H}}\mathbf{R}_x\mathbf{B}$ and $\mathbf{p}_{\mathrm{GSC}} = \mathbb{E}\left[d_{\mathrm{GSC}}^{*}(k)\mathbf{x}_{\mathrm{GSC}}(k)\right] = \mathbb{E}\left[[\mathbf{f}_{\mathrm{c}}^{\mathrm{H}}\mathbf{x}(k) - d(k)]^{*}\mathbf{B}^{\mathrm{H}}\mathbf{x}(k)\right]$, where $e_{\mathrm{GSC}}(k) = -e(k)$, it follows that

$$\mathbf{w}_{\mathrm{GSC-OPT}} = (\mathbf{B}^{\mathrm{H}}\mathbf{R}_x\mathbf{B})^{-1}(-\mathbf{B}^{\mathrm{H}}\mathbf{p} + \mathbf{B}^{\mathrm{H}}\mathbf{R}_x\mathbf{f}_{\mathrm{c}}). \tag{5.53}$$

In adaptive beamforming, the (most common) case where $d(k) = 0$ leads to the GSC structure depicted in Figure 5.8. For this case, the variables GSC input signal vector, $\mathbf{x}_{\mathrm{GSC}}(k)=\mathbf{B}^{\mathrm{H}}\mathbf{x}(k)$, GSC output signal $y_{\mathrm{GSC}}(k)=\mathbf{w}_{\mathrm{GSC}}^{\mathrm{H}}(k)\mathbf{x}_{\mathrm{GSC}}(k)$, and GSC reference signal $d_{\mathrm{GSC}}(k) = \mathbf{f}_{\mathrm{c}}^{\mathrm{H}}\mathbf{x}(k)$ are intuitive once compared to a standard adaptive filter setup. Comparing Figures 5.7 and 5.8, we note that, in order to have exactly the same setup used in other technical texts, we have dropped the negative sign that should exist if the reference signal is made equal to 0 in Figure 5.7. Although we defined $e_{\mathrm{GSC}}(k) = -e(k)$, the inversion of the sign in the error signal actually results in the same updating process because the error function to be minimized is always based on the absolute value.

The optimal overall coefficient error for this case where $d(k) = 0$, based on the unconstrained GSC filter and knowing that $\mathbf{p}_{\mathrm{GSC}} = \mathbf{B}^{\mathrm{H}}\mathbf{R}_x\mathbf{f}_{\mathrm{c}}$, is given as

$$\begin{aligned}
\mathbf{w}_{\mathrm{OPT}} &= \mathbf{f}_{\mathrm{c}} - \mathbf{B}\mathbf{w}_{\mathrm{GSC-OPT}} \\
&= \mathbf{f}_{\mathrm{c}} - \mathbf{B}(\underbrace{\mathbf{B}^{\mathrm{H}}\mathbf{R}_x\mathbf{B}}_{\mathbf{R}_{\mathrm{GSC}}})^{-1}\underbrace{\mathbf{B}^{\mathrm{H}}\mathbf{R}_x\mathbf{f}_{\mathrm{c}}}_{\mathbf{p}_{\mathrm{GSC}}} \\
&= [\mathbf{I} - \mathbf{B}(\mathbf{B}^{\mathrm{H}}\mathbf{R}_x\mathbf{B})^{-1}\mathbf{B}^{\mathrm{H}}\mathbf{R}_x]\mathbf{f}_{\mathrm{c}},
\end{aligned} \tag{5.54}$$

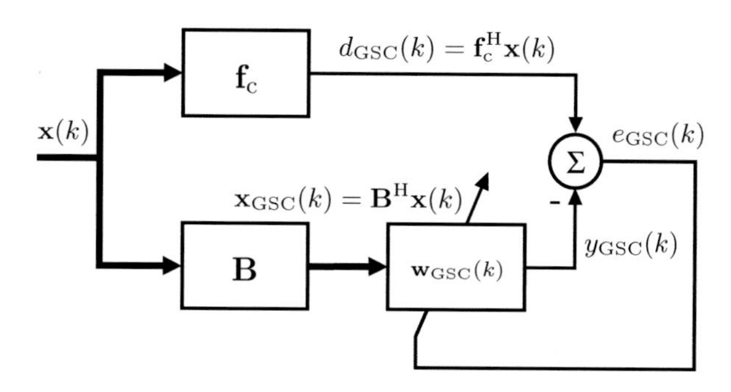

Figure 5.8 The most usual GSC structure.

which, assuming that $\mathbf{B}\left(\mathbf{B}^H\mathbf{R}_x\mathbf{B}\right)^{-1}\mathbf{B}^H\mathbf{R}_x+\mathbf{R}_x^{-1}\mathbf{C}\left(\mathbf{C}^H\mathbf{R}_x^{-1}\mathbf{C}\right)^{-1}\mathbf{C}^H=\mathbf{I}$ can be easily proved, leads to

$$\mathbf{w}_{\text{OPT}} = \mathbf{R}_x^{-1}\mathbf{C}(\mathbf{C}^H\mathbf{R}_x^{-1}\mathbf{C})^{-1}\mathbf{f}. \tag{5.55}$$

The last expression corresponds exactly to the constrained Wiener solution from Equation (5.10) when $\mathbf{p} = \mathbf{0}$. It is also worth mentioning that all main equations shown here for the MSE cost function are also valid for the LS case; we just have to replace $\mathbf{R}_x$ by $\mathbf{R}_x(k)$ and $\mathbf{p}$ by $\mathbf{p}(k)$, all previously defined for the CRLS algorithm.

We next address the relationship between the GSC using an unconstrained algorithm and its constrained counterpart. It can be somehow expected that both implementations lead to the same solution in steady state, approaching the Wiener solution whenever convergence is reached. In order to check the equivalence during convergence, we assume that both filters were initialized equivalently, that is, we guarantee that $\mathbf{w}(0) = \mathbf{f}_c - \mathbf{B}\mathbf{w}_{\text{GSC}}(0)$. We then evaluate the evolving of the coefficient vectors in time; this is carried out with the definition of the *difference vector* $\mathbf{\Delta w}(k) = \mathbf{w}(k+1) - \mathbf{w}(k)$.

For the CLMS algorithm, using the updating equation in (5.20) and considering that $\mathbf{P}\mathbf{w}(k) + \mathbf{f}_c = \mathbf{w}(k)$ in infinite precision environment, we express the difference vector as $\mathbf{\Delta w}(k) = \mu e^*(k)\mathbf{P}\mathbf{x}(k)$. For the GSC-LMS implementation, it is straightforward to see that the updating of the overall coefficient vector may be given by $\mathbf{w}(k+1) = \mathbf{f}_c - \mathbf{B}\mathbf{w}_{\text{GSC}}(k+1) = \mathbf{w}(k) + \mu e^*(k)\mathbf{B}\mathbf{B}^H\mathbf{x}(k)$ such that $\mathbf{\Delta w}(k) = \mu e^*(k)\mathbf{B}\mathbf{B}^H\mathbf{x}(k)$. Knowing that $\mathbf{P} = \mathbf{B}(\mathbf{B}^H\mathbf{B})^{-1}\mathbf{B}^H$, it can be easily seem that, in order to have identical (in infinite precision) update at every iteration, the GSC-LMS algorithm must use an orthonormal blocking matrix, that is, $\mathbf{B}^H\mathbf{B} = \mathbf{I}$.

On the other hand, it is seen in [12] that both CRLS and GSC-RLS algorithms are equivalent in infinite precision regardless of the orthogonality of the blocking matrix. Naturally, this equivalence requires equivalent initialization of all inner variables.

The GSC structure, although being an elegant approach due to the possibility of employing many unconstrained algorithms within a constrained framework, will or will not be used depending on practical implementation issues, mainly its computational load. Considering this issue, the structure of the blocking matrix $\mathbf{B}$ plays an important role since its choice determines the computational complexity and even the robustness against numerical instability [13] of the overall system. Given that the only requirement for the blocking matrix is having its columns forming a basis orthogonal to the constraints, $\mathbf{B}^{\mathrm{H}}\mathbf{C} = \mathbf{0}$, many options are possible. The computational cost of this choice is evaluated in the matrix-vector multiplication $\mathbf{Bx}(k)$ performed in every iteration.

We provide here a few examples of blocking matrices related to the simple case of a constraint matrix given by $\mathbf{C} = [1\,1\,1\,1]^{\mathrm{T}}$, which correspond to an incoming signal impinging a linear array of $M = 4$ sensors from broadside ($\theta = 90°$).

For this constraint matrix, we could choose simple blocking matrices such as

$$\mathbf{B}_1 = \begin{bmatrix} 1 & 0 & 0 \\ -1 & 1 & 0 \\ 0 & -1 & 1 \\ 0 & 0 & -1 \end{bmatrix} \text{ or } \mathbf{B}_2 = \begin{bmatrix} 1 & -1 & -1 \\ 1 & -1 & 0 \\ -1 & 1 & 0 \\ -1 & 1 & 1 \end{bmatrix}. \tag{5.56}$$

These two simple matrices with reduced complexity are viable blocking matrices, satisfying $\mathbf{B}^{\mathrm{H}}\mathbf{C} = \mathbf{0}$, but they are not orthonormal, $\mathbf{B}^{\mathrm{H}}\mathbf{B} \neq \mathbf{I}$. Although other possible blocking matrices are available in the literature [13, 14], we provide here two ways to obtain orthonormal blocking matrices using SVD and QR decomposition. Orthonormal blocking matrices can be generated with the following Matlab® commands (note that p is the number of constraints, $p = 1$ in this example):

```
[U,S,V]=svd(C);
B3=U(:,p+1:M);
[Q,R]=qr(C);
B4=Q(:,p+1:M);
```

And the results, for the constraint matrix of this example, are

$$\mathbf{B}_3 = \mathbf{B}_4 = \frac{1}{6}\begin{bmatrix} -3 & -3 & -3 \\ 5 & -1 & -1 \\ -1 & 5 & -1 \\ -1 & -1 & 5 \end{bmatrix}. \tag{5.57}$$

5.3.5 Constrained Conjugate Gradient Algorithms

We can find in the literature several adaptive algorithms based on the conjugate gradient (CG) method [15, 16]. These algorithms usually have fast convergence, close to algorithms such as the RLS and the LMS-Newton, but with lower computational complexity. The lower complexity comes from the fact that there is no need to perform matrix inversion, more specifically, the inverse of the autocorrelation matrix of the input signal. We start by considering the case of an exponentially decaying window that simultaneously attains fast convergence and low misadjustment. Furthermore, this class of algorithms provides stability and behaves, in a finite precision environment, similar to the steepest descent algorithm [16].

In [17], the equivalence between a constrained adaptive filter and its GSC counterpart is used to introduce a constrained conjugate gradient (CCG) algorithm from the modified conjugate gradient algorithm in [15]. We note that this algorithm uses a degenerated scheme of the CG algorithm in order to have only one iteration per coefficient-vector update.

The development of the CCG algorithm is found in [17] where the updating equation of the constrained LMS algorithm is first worked from the GSC equivalent filter. The algebraic manipulation used to relate both, now aiming at an updating expression for the CCG algorithm, is employed in the GSC CG updating equation which, starting from Equation (5.51), is given by:

$$\begin{aligned}\mathbf{w}(k+1) &= \mathbf{f}_c - \mathbf{B}\mathbf{w}_{\text{GSC}}(k+1)\\ &= \mathbf{f}_c - \mathbf{B}[\mathbf{w}_{\text{GSC}}(k) + \alpha_{\text{GSC}}(k)\mathbf{c}_{\text{GSC}}(k-1)]\\ &= \mathbf{P}\mathbf{w}(k) + \mathbf{f}_c - \underbrace{\alpha_{\text{GSC}}(k)}_{\alpha(k)}\underbrace{\mathbf{B}\mathbf{c}_{\text{GSC}}(k-1)}_{\mathbf{c}(k-1)},\end{aligned} \tag{5.58}$$

where $\alpha(k)$ is the step size that minimizes the CG cost function $\frac{1}{2}\mathbf{w}^{\text{H}}(k+1)$ $\mathbf{R}_x\mathbf{w}(k+1) - \text{Real}\left[\mathbf{p}^{\text{H}}\mathbf{w}(k+1)\right]$ [16] and $\mathbf{c}(k-1)$ is the direction vector that is updated as $\mathbf{c}(k) = \mathbf{g}(k) + \beta(k)\mathbf{c}(k-1)$, $\mathbf{g}(k)$ being the residual vector (it corresponds to the negative of the gradient of the CG cost function), while $\beta(k)$ provides the $\mathbf{R}_x$-orthogonality expressed as $\mathbf{c}^{\text{H}}(k-1)\mathbf{R}_x\mathbf{c}(i) = 0$, $i \neq k-1$. Algorithm 5.33 shows the resulting CCG algorithm where $\overline{\mathbf{R}}_x(k)$ corresponds to $\mathbf{P}\mathbf{R}_x(k)\mathbf{P}^{\text{H}} = \mathbf{P}\mathbf{R}_x(k)\mathbf{P}$ given that $\mathbf{P}$ is Hermitian, $\mathbf{R}_x(k)$ being the autocorrelation matrix estimated from an exponentially decaying data as in the case of the RLS algorithm. The performance in convergence rate of this algorithm is similar to that of the CRLS algorithm but, having no need to compute the inverse of matrix $\mathbf{R}_x(k)$, nor $\overline{\mathbf{R}}_x(k)$, it tends to be numerically stable.

We observe that the misadjustment of the CCG algorithm is expected to be equivalent to the misadjustment of the GSC-RLS algorithm for they minimize equivalent cost functions [15]. Its computational complexity can be expressed by $3(N+1)^2 + (10+4p)(N+1) + 1$ multiplications and two divisions [17], where p is the number of constraints. This computational load is expected to be higher than the CLMS algorithm, slightly smaller than the GSC-CG algorithm using orthogonal blocking matrix, and considerably smaller than the GSC-RLS algorithm.

More recently, in an attempt to reduce the peak computational complexity of the CCG algorithm, Wang [18] introduces a set membership conjugate gradient

Algorithm 5.33 The constrained conjugate-gradient algorithm

Initialization

λ, η with $(\lambda - 0.5) \leq \eta \leq \lambda$
δ small number
$\mathbf{w}(0) = \mathbf{f}_c$
$\overline{\mathbf{R}}_x(-1) = \mathbf{P}$
$\mathbf{g}(-1) = \mathbf{c}(-1) = [0 \ \cdots \ 0]^{\mathrm{T}}$

For $k \geq 0$ **do**

$\overline{\mathbf{x}}(k) = \mathbf{P}\mathbf{x}(k)$
$\overline{\mathbf{R}}_x(k) = \lambda \overline{\mathbf{R}}_x(k-1) + \overline{\mathbf{x}}(k)\overline{\mathbf{x}}^{\mathrm{H}}(k)$
$e(k) = d(k) - \mathbf{w}^{\mathrm{H}}(k)\mathbf{x}(k)$
$\alpha(k) = \eta \frac{\mathbf{c}^{\mathrm{H}}(k-1)\mathbf{g}(k-1)}{[\mathbf{c}^{\mathrm{H}}(k-1)\overline{\mathbf{R}}_x(k)\mathbf{c}(k-1)+\delta]}$
$\mathbf{g}(k) = \lambda \mathbf{g}(k-1) - \alpha(k)\overline{\mathbf{R}}_x(k)\mathbf{c}(k-1) - \overline{\mathbf{x}}(k)e^*(k)$
$\mathbf{w}(k+1) = \mathbf{P}\mathbf{w}(k) + \mathbf{f}_c - \alpha(k)\mathbf{c}(k-1)$
$\beta(k) = \frac{[\mathbf{g}(k)-\mathbf{g}(k-1)]^{\mathrm{H}}\mathbf{g}(k)}{[\mathbf{g}^{\mathrm{H}}(k-1)\mathbf{g}(k-1)+\delta]}$
$\mathbf{c}(k) = \mathbf{g}(k) + \beta(k)\mathbf{c}(k-1)$

end

algorithm to be used for adaptive beamforming. According to Wang [18], the number of multiplications is $N(2m+5)+\tau N(2m^2+9m+22)$, where m (here) is the number of sensors (the target application of this algorithm is beamforming), N (here) the number of snapshots, and τ $(0 \leq \tau \leq 1)$ the average update rate (related to the set-membership technique).

5.3.6 The Householder Constrained Algorithm

As previously seen, the CNLMS algorithm *a posteriori* output error is 0 for $\mu = 1$. The solution is at the intersection of the hyperplane defined by the constraints, $\mathcal{H}_o : \mathbf{C}^{\mathrm{H}}\mathbf{w} = \mathbf{f}$, with the hyperplane defined by the null *a posteriori* error $\mathcal{H}_1 : \mathbf{w}^{\mathrm{H}}\mathbf{x}(k) - d(k) = 0$. The updated coefficient vector $\mathbf{w}(k+1)$ is not merely a projection of the solution of the NLMS algorithm onto the hyperplane defined by the constraints as illustrated in Figure 5.9 (left). In a two-dimensional scenario, hyperplanes $\mathcal{H}_o$ and $\mathcal{H}_1$ become two lines and $\mathbf{w}(k)$ must belong to $\mathcal{H}_o$ and to the hyperplane defined by the null *a posteriori* error at instant $k-1$, that is, $\mathcal{H}_2 : \mathbf{w}^{\mathrm{H}}\mathbf{x}(k-1) - d(k-1) = 0$. This vector can be decomposed in two mutually orthogonal vectors, $\mathbf{f}_c$ and $\mathbf{P}\mathbf{w}(k)$.

Examining closely the decomposition of vectors $\mathbf{w}(k-1)$ and its updated versions (by constrained algorithms CLMS and CNLMS), we note that they are all located along the hyperplane $\mathcal{H}_o$. If the axes were rotated such that the rotated w_1 axis and $\mathbf{f}_c$ had the same direction, the component along this direction would not need to be updated [19]. This rotation can be carried out with an orthogonal rotation matrix $\mathbf{Q}$ used as the transformation matrix that generates a modified coefficient vector $\overline{\mathbf{w}}(k)$ as in

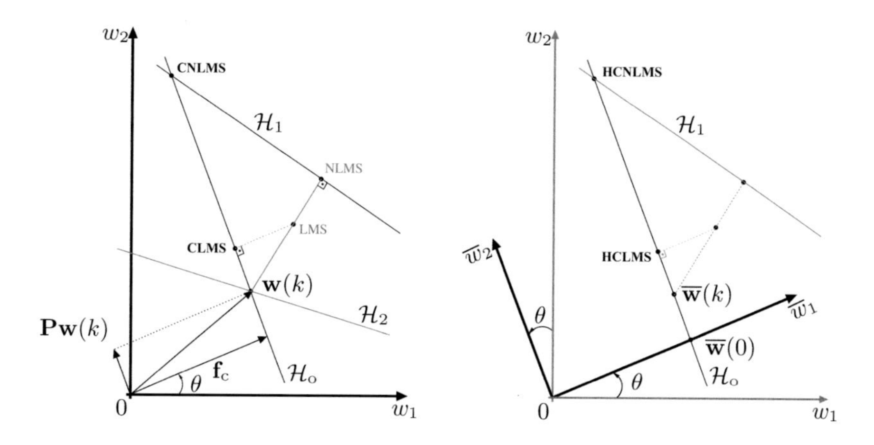

Figure 5.9 Coefficient vectors of constrained algorithms (left) and of the Householder constrained algorithms (right). Note that the hyperplanes are defined as: $\mathcal{H}_o : \mathbf{C}^H\mathbf{w} = \mathbf{f}$, $\mathcal{H}_1 : \mathbf{w}^H\mathbf{x}(k) = d(k)$, and $\mathcal{H}_2 : \mathbf{w}^H\mathbf{x}(k-1) = d(k-1)$.

$$\overline{\mathbf{w}}(k) = \mathbf{Q}\mathbf{w}(k). \tag{5.59}$$

We can visualize this in Figure 5.9 (right) if we imagine axis $\mathbf{w}_1$ and $\mathbf{w}_2$ rotated counterclockwise by angle θ. If we choose matrix $\mathbf{Q}$, $\mathbf{QQ}^H = \mathbf{Q}^H\mathbf{Q} = \mathbf{I}$, such that $\overline{\mathbf{C}} = \mathbf{QC}$ satisfies

$$\overline{\mathbf{C}}(\overline{\mathbf{C}}^H\overline{\mathbf{C}})^{-1}\overline{\mathbf{C}}^H = \begin{bmatrix} \mathbf{I}_p & \mathbf{0} \\ \mathbf{0} & \mathbf{0} \end{bmatrix}, \tag{5.60}$$

then $\overline{\mathbf{C}}^H\overline{\mathbf{w}}(k) = \mathbf{f}$ and the *transformed projection matrix* is given as

$$\begin{aligned} \overline{\mathbf{P}} &= \mathbf{QPQ}^H \\ &= \mathbf{I}_M - \bar{\mathbf{C}}(\bar{\mathbf{C}}^H\bar{\mathbf{C}})^{-1}\bar{\mathbf{C}}^H \\ &= \begin{bmatrix} \mathbf{0}_{p\times p} & \mathbf{0} \\ \mathbf{0} & \mathbf{I}_{M-p} \end{bmatrix}. \end{aligned} \tag{5.61}$$

If we initialize the transformed coefficient vector as $\overline{\mathbf{w}}(0) = \overline{\mathbf{C}}(\overline{\mathbf{C}}^H\overline{\mathbf{C}})^{-1}\mathbf{f} = \mathbf{Qf}_c$, its first p elements, $\overline{\mathbf{w}}_U(0)$, need not be updated.

The update equation of the Householder transform constrained LMS (HCLMS) algorithm [20] is obtained from

$$\begin{aligned} \overline{\mathbf{w}}(k+1) &= \mathbf{Qw}(k+1) = \mathbf{Q}\left\{\mathbf{P}\left[\mathbf{w}(k) + \mu e^*(k)\mathbf{x}(k)\right] + \mathbf{f}_c\right\} \\ &= \left[\mathbf{QPQ}^H\right]\left[\mathbf{Qw}(k)\right] + \mu e^*(k)\left[\mathbf{QPQ}^H\right]\left[\mathbf{Qx}(k)\right] + \mathbf{Qf}_c \\ &= \begin{bmatrix} \mathbf{0}_{p\times p} & \mathbf{0} \\ \mathbf{0} & \mathbf{I}_{M-p} \end{bmatrix}\overline{\mathbf{w}}(k) + \mu e^*(k)\begin{bmatrix} \mathbf{0}_{p\times p} & \mathbf{0} \\ \mathbf{0} & \mathbf{I}_{M-p} \end{bmatrix}\overline{\mathbf{x}}(k) + \begin{bmatrix} \overline{\mathbf{w}}_U(0) \\ \mathbf{0} \end{bmatrix} \\ &= \begin{bmatrix} \overline{\mathbf{w}}_U(0) \\ \overline{\mathbf{w}}_L(k+1) \end{bmatrix} = \begin{bmatrix} \overline{\mathbf{w}}_U(0) \\ \overline{\mathbf{w}}_L(k) \end{bmatrix} + \mu e^*(k)\begin{bmatrix} \mathbf{0} \\ \overline{\mathbf{x}}_L(k) \end{bmatrix}, \end{aligned} \tag{5.62}$$

where $\overline{\mathbf{w}}_{\mathrm{L}}(k+1)$ and $\overline{\mathbf{x}}_{\mathrm{L}}(k)$ denote the $M-p$ last elements of vectors $\overline{\mathbf{w}}(k+1)$ and $\overline{\mathbf{x}}(k)$, respectively. Note that vector $\overline{\mathbf{C}}(\overline{\mathbf{C}}^{\mathrm{H}}\overline{\mathbf{C}})^{-1}\mathbf{f}$ has only p nonzero elements.

Although the solution $\overline{\mathbf{w}}(k+1)$ corresponds to the original updated vector rotated by matrix $\mathbf{Q}$, the output signal and the output error are not modified by this orthogonal matrix. Thus, the HCLMS algorithm minimizes the same objective function minimized by the CLMS algorithm.

Matrix $\mathbf{Q}$ in (5.59) can be constructed with successive Householder transformations [21] applied onto each of the p columns of matrix $\mathbf{CL}$, $\mathbf{L}$ being the square-root factor[4] of matrix $(\mathbf{C}^{\mathrm{H}}\mathbf{C})^{-1}$, that is, $\mathbf{L}\mathbf{L}^{\mathrm{H}} = (\mathbf{C}^{\mathrm{H}}\mathbf{C})^{-1}$.

It was seen in [19] that if

$$\mathbf{Q} = \mathbf{Q}_p \cdots \mathbf{Q}_2\mathbf{Q}_1, \tag{5.63}$$

where

$$\mathbf{Q}_i = \begin{bmatrix} \mathbf{I}_{i-1\times i-1} & \mathbf{0}^{\mathrm{T}} \\ \mathbf{0} & \overline{\mathbf{Q}}_i \end{bmatrix}, \tag{5.64}$$

and $\overline{\mathbf{Q}}_i$ is an $(M-i+1)\times(M-i+1)$ Householder transformation matrix on the form $\overline{\mathbf{Q}}_i = \mathbf{I} - 2\overline{\mathbf{v}}_i\overline{\mathbf{v}}_i^{\mathrm{H}}$ [21], then (5.60) is satisfied.

From Equation (5.62), we see that the algorithm updates the coefficients in a subspace with reduced dimension. The entries of vector $\mathbf{w}(k)$ lying in the subspace of the constraints need not be updated. Due to the equivalence of Householder reflections and Givens rotations [22], a succession of Givens rotations could also be used. However, rotations are not as efficiently implemented as reflections and computational complexity might result in an algorithm with limited applicability.

A normalized version of the HCLMS algorithm, the HCNLMS algorithm [20], can be easily derived and its update equation is given by

$$\overline{\mathbf{w}}(k+1) = \begin{bmatrix} \overline{\mathbf{w}}_{\mathrm{U}}(0) \\ \overline{\mathbf{w}}_{\mathrm{L}}(k) \end{bmatrix} + \mu\frac{e^*(k)}{\overline{\mathbf{x}}_{\mathrm{L}}^{\mathrm{H}}(k)\overline{\mathbf{x}}_{\mathrm{L}}(k)}\begin{bmatrix} \mathbf{0} \\ \overline{\mathbf{x}}_{\mathrm{L}}(k) \end{bmatrix}. \tag{5.65}$$

Note that the Householder transformation allows normalization without the need of multiplication by a projection matrix, as required by the CNLMS in (5.23). The HC approach can be regarded as a GSC structure and, therefore, any unconstrained adaptive algorithm can be used to update $\mathbf{w}_{\mathrm{L}}(k)$. This can be observed in Algorithm 5.34, a pseudo-code of the HCLMS algorithm.

Aiming at an efficient implementation, the transformation of the input-signal vector at every iteration is carried out through p reflections given by

$$\overline{\mathbf{x}}(k) = \mathbf{Q}\mathbf{x}(k) = \mathbf{Q}_p \cdots \mathbf{Q}_2\mathbf{Q}_1\mathbf{x}(k), \tag{5.66}$$

where

$$\mathbf{Q}_i = \begin{bmatrix} \mathbf{I}_{i-1} & \mathbf{0}^{\mathrm{T}} \\ \mathbf{0} & \bar{\mathbf{Q}}_i \end{bmatrix} \tag{5.67}$$

[4] Since matrix $(\mathbf{C}^{\mathrm{H}}\mathbf{C})^{-1}$ is Hermitian, `L=(chol(inv(C'*C)))'` is an example that uses Cholesky decomposition of a Matlab® command for obtaining matrix $\mathbf{L}$.

Algorithm 5.34 The HCLMS algorithm

Initialization

$\mathbf{C}$, $\mathbf{f}$, $\mathbf{Q}$, and μ (step size)

$\overline{\mathbf{w}}(0) = \mathbf{Q}\mathbf{C}(\mathbf{C}^{\mathrm{H}}\mathbf{C})^{-1}\mathbf{f}$

For $k \geq 0$ **do**

$\overline{\mathbf{x}}(k) = \mathbf{Q}\mathbf{x}(k)$

$\overline{\mathbf{x}}_{\mathrm{L}}(k) = M - p$ last elements of $\overline{\mathbf{x}}(k)$

$\overline{\mathbf{w}}(k) = \begin{bmatrix} \overline{\mathbf{w}}_{\mathrm{U}}(0) \\ \overline{\mathbf{w}}_{\mathrm{L}}(k) \end{bmatrix}$

$e(k) = d(k) - \overline{\mathbf{w}}^{\mathrm{H}}(k)\overline{\mathbf{x}}(k)$

$\overline{\mathbf{w}}_{\mathrm{L}}(k+1) = \overline{\mathbf{w}}_{\mathrm{L}}(k) + \mu e^*(k)\overline{\mathbf{x}}_{\mathrm{L}}(k)$

end

Algorithm 5.35 Computation of $\overline{\mathbf{x}}_k = \mathbf{Q}\mathbf{x}_k$

$\overline{\mathbf{x}}_k = \mathbf{x}(k)$

For $i = 1 : p$ **do**

$\overline{\mathbf{x}}_k(i:M) = \overline{\mathbf{x}}_k(i:M) - 2\mathbf{V}(i:M,i)\left[\mathbf{V}^{\mathrm{H}}(i:M,i)\overline{\mathbf{x}}_k(i:M)\right]$

end

Return

$\overline{\mathbf{x}}(k) = \overline{\mathbf{x}}_k$

and matrix $\overline{\mathbf{Q}}_i = \mathbf{I} - 2\overline{\mathbf{v}}_i\overline{\mathbf{v}}_i^{\mathrm{H}}$ is an $(M-i+1) \times (M-i+1)$ Householder transformation matrix [21].

If we define the vector $\mathbf{v}_i^{\mathrm{H}} = [\mathbf{0}_{i-1}^{\mathrm{T}}\ \overline{\mathbf{v}}_i^{\mathrm{H}}]^{\mathrm{H}}$, where the $p \times 1$ vector $\mathbf{0}_{i-1}$ introduces $i-1$ zeros before $\overline{\mathbf{v}}_i$, we can construct matrix $\mathbf{V} = [\mathbf{v}_1\ \mathbf{v}_2\ \cdots\ \mathbf{v}_p]$, and the factored product in (5.66) could be implemented with the procedure described in Algorithm 5.35. Furthermore, the procedure for the calculation of the Householder vectors and the resulting $\mathbf{V}$ is described in Algorithm 5.36. In this table, $\mathbf{A}$ is the matrix to be triangularized which in our particular case of interest corresponds to $\mathbf{A} = \mathbf{CL}$, $\mathbf{L}$ being the square-root factor of the matrix $\left(\mathbf{C}^{\mathrm{H}}\mathbf{C}\right)^{-1}$ as proposed earlier in this section.

From Algorithm 5.35 we see that the computation of $\overline{\mathbf{x}}(k) = \mathbf{Q}\mathbf{x}_k$ using the product representation in (5.66) only involves $2Mp - p(p-1)$ multiplications. Table 5.1 shows the computational complexity for algorithms CLMS, CNLMS, HCLMS, HCNLMS as well as for the GSC implementation of algorithms CLMS and CNLMS. The computational complexity for the GSC implementation is given for two choices of the blocking matrix ($\mathbf{B}$). The first implementation uses a matrix $\mathbf{B}$ obtained by, for example, SVD leading to a less efficient implementation of the multiplication $\mathbf{B}\mathbf{x}(k)$. The second implementation, applicable only in certain problems, uses a $\mathbf{B}$ matrix constructed through a cascade of sparse matrices as presented in [23], rendering an implementation of the multiplication $\mathbf{B}\mathbf{x}(k)$ of lower computational complexity. This approach can be regarded as

Algorithm 5.36 Computing matrix $\mathbf{V}$ containing the Householder vectors

Available at start

$\mathbf{A}$ is an $M \times p$ matrix to be triangularized

Initialization

$\mathbf{V} = \mathbf{0}_{M \times p}$

For $i = 1 : p$ **do**

$\mathbf{x} = \mathbf{A}(i : M, i)$

$\mathbf{e}_1 = [1 \ \mathbf{0}_{1\times(M-i)}]^{\mathrm{T}}$

$\mathbf{v} = \mathrm{sign}\,(\mathbf{x}(1))\, \|\mathbf{x}\| \mathbf{e}_1 + \mathbf{x}$

$\mathbf{v} = \mathbf{v}/\|\mathbf{v}\|$

$\mathbf{A}(i : M, i : p) = \mathbf{A}(i : M, i : p) - 2\mathbf{v}\left(\mathbf{v}^{\mathrm{H}}\mathbf{A}(i : M, i : p)\right)$

$\mathbf{V}(i : M, i) = \mathbf{v}$

end

Return

$\mathbf{V}$

Table 5.1 Computational complexity of distinct constrained versions of algorithms LMS and NLMS

Algorithm	Multiplications	Divisions
CLMS	$(2p + 2)M + 1$	0
CNLMS	$(3p + 3)M + 1$	1
GSC-LMS	$M^2 - (p - 3)M - (2p - 1)$	0
GSC-NLMS	$M^2 - (p - 4)M - (3p - 1)$	1
GSC-LMS with $\mathbf{B}$ of [23]	$(2p + 3)M - p(p + 3) + 1$	0
GSC-NLMS with $\mathbf{B}$ of [23]	$(2p + 4)M - p(p + 4) + 1$	1
HCLMS	$(2p + 2)M - (p^2 - 1)$	0
HCNLMS	$(2p + 3)M - (p^2 + p - 1)$	1

a GSC structure and, therefore, any unconstrained adaptive algorithm can be used to update $\mathbf{w}_{\mathrm{L}}(k)$ [19].

Using the Householder transformation we are able to reduce the dimension of the subspace in which the (constrained) adaptive-filter coefficients are updated, thus obtaining a persistently exciting transformed input signal. Viewed under the perspective of a GSC structure, the transformation matrix can be factored into a matrix satisfying the constraints and a blocking matrix. Also, in terms of computational complexity, the HC structure is comparable with the most efficient implementations of blocking matrices found in the literature with the advantage of being unitary.

Example 5.1 At this point, it is interesting to give an example of the converging properties of the constrained adaptive filters when employed in adaptive beamforming. For this example, we consider a standard ULA with $M = 6$ equally spaced ($d = \lambda/2$) sensors and $G = 2$ users: a SOI (SOI) coming from

$\theta_1 = 60^\circ$ and a jamming signal coming from $\theta_2 = 110^\circ$. The received discrete-time signal vector is modeled as in Equation (5.6) with a signal-to-noise ratio (SNR) set to 20 dB and a SIR equal to 0 dB. We chose algorithms CLMS, CNLMS, and CRLS to be compared to the MPDR solution as given in Equation (5.11), with $\theta = \theta_1$. We assumed one constraint (that we knew the SOI direction, θ_1) and used the theoretic $\mathbf{R}_x$ given in Equation (5.12). Figure 5.10 shows the beampatterns of each class of algorithms. Only one curve for each algorithm is seen due to the fact that their different versions (constrained, GSC, and Householder constrained (HC)) are equivalent given equivalent initializations. We observe the CRLS algorithm presenting the best performance, its beampattern close to the theoretical optimum. Although not displayed in this figure, the CCG algorithm would present similar performance. Also seen in this figure, the CNLMS and the CLMS algorithms, due to the need of higher sampler support, presenting inferior performance, that is, not suppressing adequately the interferer coming from θ_2. It is worth noting that this experiment (a simulation) was carried out in infinite precision (Matlab®) environment, with a limited number of iterations (chosen in order to highlight the poor convergence rate of constrained LMS-like algorithms comparing to the CRLS algorithm) and one single run. Also, the parameters of all algorithms were optimized for the best possible performance.

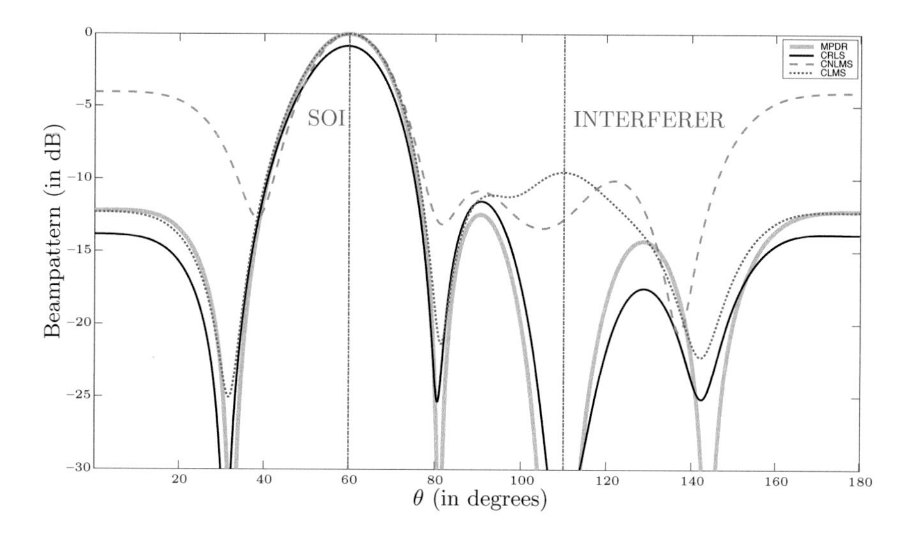

Figure 5.10 Beampatterns, in dB, of three adaptive beamformers: based on the constrained, GSC and Householder versions of the LMS algorithm, of the NLMS algorithm, and of the RLS algorithm. Note the fast converging capability of the CRLS algorithm, close to the (MPDR) solution even for the case of a limited sample support.

5.3.7 Sparsity-Promoting Adaptive Beamforming

Some adaptive beamforming applications, especially those with power supply limitations, favor the use of sparse (or thinned) arrays. In these applications, we are faced with a compromise between radiation pattern and number of sensors. Hence, algorithms able to turn off sensors without compromising their performances, might be a good choice. Examples of such beamforming applications are satellite communications and tactical mobile communication systems.

A number of algorithms, some of them discussed in Chapter 2, have been proposed for sparse system identification and are available in the technical literature [24–30]. Nevertheless, for beamformers, besides sparsity, the algorithm must be able to impose linear constraints to the solution. We start our discussion with an example of a sparsity-inducing adaptive beamforming algorithm, the ℓ_1-norm linearly constrained least mean square (ℓ_1-CLMS) algorithm [31, 32] which is an alternative version of the CLMS algorithm with an additional ℓ_1-norm constraint. Due to the limited performance of the ℓ_1-CLMS algorithm in ULA-based adaptive beamforming, we focus on its natural extension, the ℓ_1-norm linearly constrained normalized LMS (ℓ_1-CNLMS) algorithm [32, 33].

For the derivation of the ℓ_1-CLMS algorithm [31], a classical approach based on the deterministic minimization of the quadratic output error is employed with Lagrange multipliers to incorporate the constraints. An ℓ_1-norm penalty is added in order to favor coefficient shrinkage toward sparsity:

$$\min_{\mathbf{w}} \mathbb{E}\left[|e(k)|^2\right] \quad \text{subject to} \quad \begin{cases} \mathbf{C}^{\mathrm{H}}\mathbf{w} = \mathbf{f}, \\ \|\mathbf{w}\|_1 = t, \end{cases} \tag{5.68}$$

where t is the ℓ_1-norm constraint value that is to be satisfied by the coefficient vector.

A Lagrangian is then defined as

$$\xi(\mathbf{w}) = \mathbb{E}\left[|e(k)|^2\right] + \text{Real}\left[\boldsymbol{\lambda}_1^{\mathrm{H}}(\mathbf{C}^{\mathrm{H}}\mathbf{w} - \mathbf{f})\right] + \lambda_2(\|\mathbf{w}\|_1 - t), \tag{5.69}$$

from where the instantaneous estimate of the gradient may be expressed as

$$\hat{\boldsymbol{\nabla}}_{\mathbf{w}}\xi(\mathbf{w}) = -2e^*(k)\mathbf{x}(k) + \mathbf{C}\boldsymbol{\lambda}_1 + \lambda_2\text{sign}[\mathbf{w}], \tag{5.70}$$

sign[$\mathbf{w}$] being defined as

$$\text{sign}(w) \triangleq \begin{cases} \frac{w}{|w|} & \forall w \in \mathbb{C}^*, \\ 0 & \text{if } w = 0. \end{cases}$$

In order to simplify the notation, we define $\mathbf{s} = \text{sign}[\mathbf{w}]$, such that $\|\mathbf{w}(k)\|_1 = \mathbf{s}^{\mathrm{H}}(k)\mathbf{w}(k)$. Updating the coefficient vector according to the steepest descent approach [3], we obtain

$$\mathbf{w}(k+1) = \mathbf{w}(k) - \frac{\mu}{2}\left[-2e^*(k)\mathbf{x}(k) + \mathbf{C}\boldsymbol{\lambda}_1 + \lambda_2\mathbf{s}(k)\right]. \tag{5.71}$$

Solving for $\boldsymbol{\lambda}_1$, we premultiply (5.71) by $\mathbf{C}^{\mathrm{H}}$ and use the fact that $\mathbf{C}^{\mathrm{H}}\mathbf{w}(k+1) = \mathbf{C}^{\mathrm{H}}\mathbf{w}(k) = \mathbf{f}$, which leads to

$$\boldsymbol{\lambda}_1 = (\mathbf{C}^{\mathrm{H}}\mathbf{C})^{-1}\mathbf{C}^{\mathrm{H}}\left(2e^*(k)\mathbf{x}(k) - \lambda_2\mathbf{s}(k)\right). \tag{5.72}$$

In order to solve for λ_2, the sign of the updated coefficient vector, in Equation (5.71), would be required. An approximation is then carried out to avoid this unknown vector: given the constraints in Equation (5.68), $\mathbf{s}^{\mathrm{H}}(k)\mathbf{w}(k+1)$ is approximated to t which is especially likely to happen near convergence. Due to this approximation, the coefficient vector will not strictly obey the $\|\mathbf{w}(k+1)\|_1$ constraint and an ℓ_1-norm error shall occur, that is, $e_{\ell_1}(k) = t - t(k)$, where $t(k) = \mathbf{s}^{\mathrm{H}}(k)\mathbf{w}(k)$ is the actual ℓ_1-norm. Recalling that $\mathbf{s}^{\mathrm{H}}(k)\mathbf{s}(k) = M$, the number of sensors in our case, premultiplying (5.71) by $\mathbf{s}^{\mathrm{H}}(k)$, results in

$$t = t(k) - \frac{\mu}{2}\left\{-2e^*(k)\mathbf{s}^{\mathrm{H}}(k)\mathbf{x}(k) + \mathbf{s}^{\mathrm{H}}(k)\mathbf{C}\boldsymbol{\lambda}_1 + \lambda_2 M\right\}. \tag{5.73}$$

After isolating λ_2 from the previous equation, we obtain

$$\lambda_2 = \left(\frac{-2}{M\mu}\right)e_{\ell_1}(k) + \frac{2}{M}e^*(k)\mathbf{s}^{\mathrm{H}}(k)\mathbf{x}(k) - \frac{1}{M}\mathbf{s}^{\mathrm{H}}(k)\mathbf{C}\boldsymbol{\lambda}_1. \tag{5.74}$$

In order to have an update equation for the ℓ_1-CLMS algorithm, as proposed in [31, 32], we need to obtain the Lagrange multipliers $\boldsymbol{\lambda}_1$ and λ_2 by solving the system of equations given by (5.72) and (5.74). The resulting update expression is written as

$$\mathbf{w}(k+1) = \mathbf{P}\left[\mathbf{w}(k) + \mu e^*(k)\overline{\mathbf{P}}(k)\mathbf{x}(k)\right] + \mathbf{f}_{\mathrm{c}} + \mathbf{f}_{\ell_1}(k), \tag{5.75}$$

with

$$\begin{cases} \overline{\mathbf{P}}(k) = \left[\mathbf{I} - \left(\frac{\mathbf{Ps}(k)}{\|\mathbf{Ps}(k)\|^2}\right)\mathbf{s}^{\mathrm{H}}(k)\right]\mathbf{P}, \\ e_{\ell_1}(k) = t - \mathbf{s}^{\mathrm{H}}(k)\mathbf{w}(k), \text{ and} \\ \mathbf{f}_{\ell_1}(k) = e_{\ell_1}(k)\left(\frac{\mathbf{Ps}(k)}{\|\mathbf{Ps}(k)\|^2}\right). \end{cases}$$

A normalized version of the ℓ_1-CLMS algorithm can be readily obtained assuming a time-varying step size, μ_k, that makes the *a posteriori* error $\varepsilon(k)$ equal to 0. This error is defined as the one obtained from the updated coefficient vector, that is, $\varepsilon(k) = d(k) - \mathbf{w}^{\mathrm{H}}(k+1)\mathbf{x}(k)$ as opposed to the *a priori* error $e(k) = d(k) - \mathbf{w}^{\mathrm{H}}(k)\mathbf{x}(k)$. If we replace the updating equation of $\mathbf{w}(k)$ including μ_k in $\varepsilon(k) = d(k) - \mathbf{w}^{\mathrm{H}}(k+1)\mathbf{x}(k)$, we obtain

$$\varepsilon(k) = e(k)\left[1 - \mu_k\mathbf{x}^{\mathrm{H}}(k)\overline{\mathbf{P}}(k)\mathbf{x}(k)\right] - \mathbf{f}_{\ell_1}^{\mathrm{H}}(k)\mathbf{x}(k). \tag{5.76}$$

Equating this expression to 0 and solving for μ_k, we obtain

$$\mu_k = \mu_o\frac{\left[e^*(k) - \mathbf{x}^{\mathrm{H}}(k)\mathbf{f}_{\ell_1}(k)\right]}{e^*(k)\mathbf{x}^{\mathrm{H}}(k)\overline{\mathbf{P}}(k)\mathbf{x}(k) + \gamma}, \tag{5.77}$$

where a fixed convergence factor μ_o, usually between 0 and 1, was included in the expression for misadjustment control. Also, a regularization parameter γ is used in the denominator to avoid division by 0. If μ_o is set to 1, the *a posteriori* error shall be null at every iteration.

Replacing the fixed convergence factor μ in Equation (5.75) by the expression in Equation (5.77), we obtain the updating equation of the ℓ_1-CNLMS algorithm which is summarized in Algorithm 5.37.

From this algorithm, its natural extensions, also sparse-inducing and presenting faster convergence, can be found in the literature: the ℓ_1-norm linearly weighted constrained normalized LMS algorithm (ℓ_1-WCNLMS) [33] and the ℓ_1-norm constrained affine projections algorithm (ℓ_1-CAP) [34].

We observe in Algorithm 5.37 two regularization parameters, γ_1 and γ_2, as suggested by the authors of [33]. Moreover, one thing not shown in the algorithm is the shrinking procedure to be carried out whenever a coefficient is smaller than a given threshold Δ; when it happens, the order is decreased or, equivalently, sensors are turned off.

Example 5.2 In order to test this algorithm, we apply it to a standard ($d = \lambda/2$) ULA with $N = 30$ sensors with a threshold $\Delta = 5 \times 10^{-5}$. The step size for both ℓ_1-CNLMS and CNLMS algorithms was set to $\mu_o = 5 \times 10^{-3}$. The SOI

Algorithm 5.37 The ℓ_1-CNLMS algorithm

Initialization

μ_o (step size, usually between 0 and 1)

$\mathbf{C}$ and $\mathbf{f}$ (linear constraints)

$\mathbf{P} = \mathbf{I}_N - \mathbf{C}(\mathbf{C}^{\mathrm{H}}\mathbf{C})^{-1}\mathbf{C}^{\mathrm{H}}$

$\mathbf{f}_{\mathrm{c}} = \mathbf{C}(\mathbf{C}^{\mathrm{H}}\mathbf{C})^{-1}\mathbf{f}$

$\mathbf{w}(0) = \mathbf{f}_{\mathrm{c}}$

$t = \alpha\|\mathbf{w}_{\mathrm{opt}}\|_1$ (α is a parameter used to adjust the sparsity level)

$\gamma_1 = N$

$\gamma_2 = 10^{-9}$

For $k \geq 0$ **do**

$e(k) = d(k) - \mathbf{w}^{\mathrm{H}}(k)\mathbf{x}(k)$

$e_{\ell_1}(k) = t - \|\mathbf{w}(k)\|_1$

$\mathbf{s}(k) = \mathrm{sign}\left[\mathbf{w}(k)\right]$

$\overline{\mathbf{P}}(k) = \left[\mathbf{I} - \frac{\mathbf{P}\mathbf{s}(k)}{\|\mathbf{P}\mathbf{s}(k)\|^2+\gamma_1}\mathbf{s}^{\mathrm{H}}(k)\right]\mathbf{P}$

$\mathbf{f}_{\ell_1}(k) = e_{\ell_1}(k)\frac{\mathbf{P}\mathbf{s}(k)}{\|\mathbf{P}\mathbf{s}(k)\|^2+\gamma_1}$

$\mu_k = \mu_o\frac{\left[e^*(k)-\mathbf{x}^{\mathrm{H}}(k)\mathbf{f}_{\ell_1}(k)\right]}{e^*(k)\mathbf{x}^{\mathrm{H}}(k)\overline{\mathbf{P}}(k)\mathbf{x}(k)+\gamma_2}$

$\mathbf{w}(k+1) = \mathbf{P}\left[\mathbf{w}(k) + \mu_k e^*(k)\overline{\mathbf{P}}(k)\mathbf{x}(k)\right] + \mathbf{f}_{\mathrm{c}} + \mathbf{f}_{\ell_1}(k)$

end

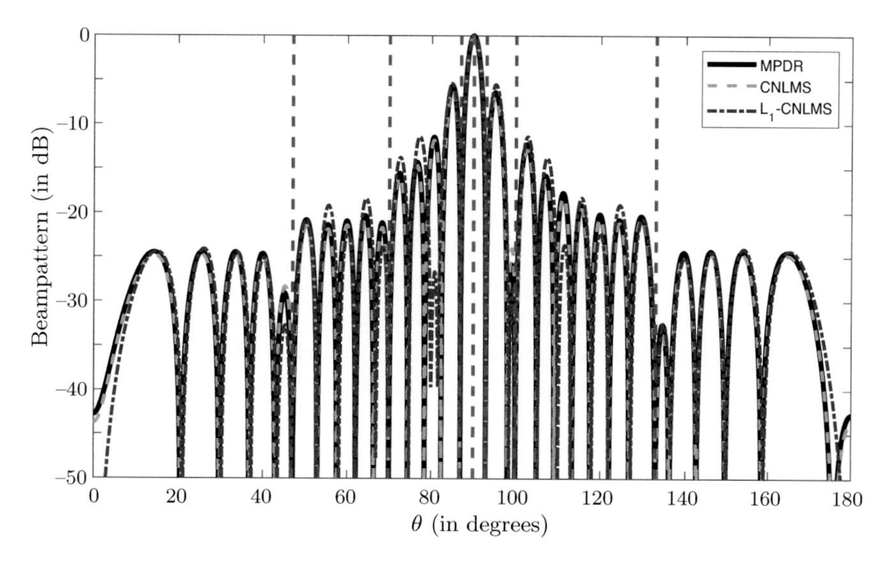

Figure 5.11 Beampattern, in dB, of ℓ_1-CNLMS algorithm compared to the beampatterns of the MDPR solution and of the CNLMS algorithm. Note the DoA of the signal of interest in 90° and the DoAs of the jammers in 47°, 70°, 87°, 93°, 100°, and 133°. For this case, the ℓ_1-CNLMS algorithm reduced its sensors from $N = 30$ to 24 (6 sensors were turned off).

DoA is 90° and interfering signals (jammers) are coming from 47°, 70°, 87°, 93°, 100°, and 133°. Given a unity noise variance, the energy of the incoming signals (modulated in QPSK) were set to 20 dB (SOI) and 40 dB each jammer. The constraint matrix and vector was such that only the DoA of the SOI was considered known.

The beampattern of the L1-CNLMS algorithm is depicted in Figure 5.11, which is compared to the beampatterns of the MPDR solution and of the CNLMS algorithm (using the same step size).

From this experiment, we can draw a number of conclusions. First, although having similar beampattern, the ℓ_1-CNLMS algorithm presents a slower convergence rate compared to the CNLMS algorithm. Fine tuning of parameters like α and μ_o might not be an easy task for a problem such as a ULA-based beamforming. The MPDR solution of such array is not sparse and therefore hard for a sparse-inducing algorithm to succeed in compressing the coefficient vector. We can expect better results for planar arrays, for instance. In this particular example, the choice of α led to six null coefficients (in average since it is not guaranteed to have the same number of null coefficients in every independent run). A final and important remark is the location of the sensors corresponding to null coefficients: here, as in the results found in [33], those in the middle of the array tend to be turned off by the ℓ_1-CNLMS algorithm. This is valid, of course, assuming that all sensors are working properly; otherwise, a defective sensor is expected to be detected and turned off regardless its position in the array.

5.4 Broadband Adaptive Beamforming

All adaptive beamforming algorithms seen so far in this chapter are specifically tailored for NB signals. Since the incoming NB signal is sampled in space by the sensors in different positions (usually uniformly spaced), we can perceive the NB beamformer as a spatial filter. For the case of wideband signals (those having the central frequency much smaller than the bandwidth), however, we have to resort to space-time filtering in order to steer the array beam toward the direction of interest for all frequencies of interest. This is vital since using a NB beamformer for a wideband signal shall degrade its performance considerably. We start this discussion by assuming that the SOI is wideband within a bandwidth from Ω_1 to Ω_2, or from $\omega_1 = \Omega_1/f_s$ to $\omega_2 = \Omega_2/f_s$ in the discrete-TD. One way to implement a broadband beamformer is employing a frequency transformation to decompose the input signals in NB components centered in the frequency bins within the input signal bandwidth. A structure that implements a frequency domain (FD) broadband beamformer is depicted in Figure 5.12.

Although another geometry is also conceivable, we assume that a linear array is employed. As observed in the diagram, (assumed real-valued) signals from each sensor are conveniently buffered and fed to the N-point IDFT whose outputs are processed by L linear combiners. Note that the (I)FFT, (inverse) fast Fourier transform, as seen in the figure, corresponds to an efficient implementation of the (I)DFT, the (inverse) discrete Fourier transform.

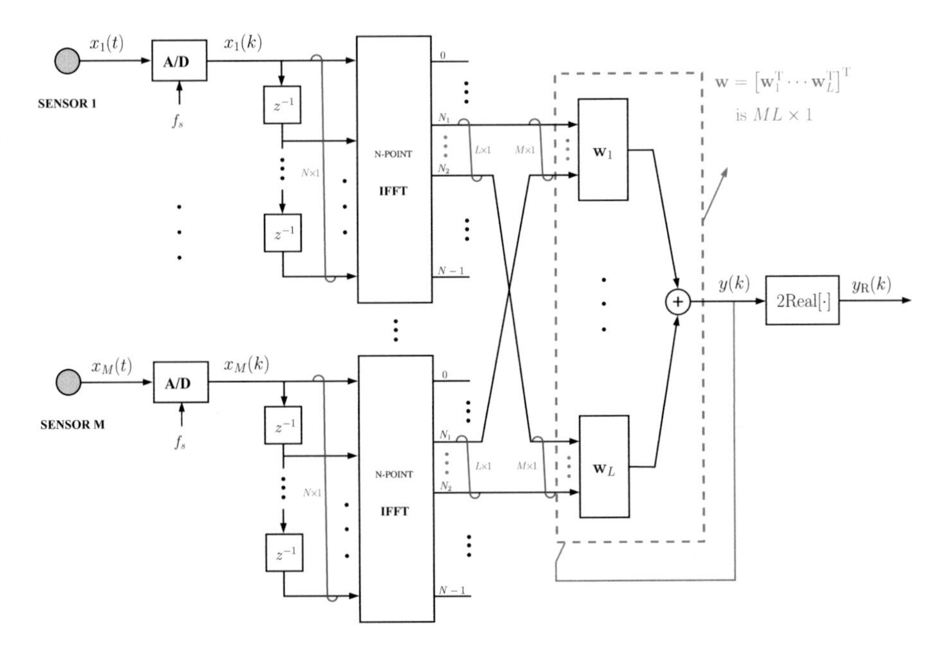

Figure 5.12 An example of an FFT-based broadband beamformer for real input signals.

Also note that the inverse DFT was employed instead of the (direct) DFT for a simple but important reason: the DFT, when sampling the discrete frequency, takes into account that input samples in a vector are ordered such that the most recent sample is the last element of the vector. If we have a sequence $x(k)$ with N samples given as a vector $\mathbf{x} = [x(0)\ x(1)\ \cdots\ x(N-1)]^{\mathrm{T}}$, its DFT may be represented as $\mathbf{W}_N\mathbf{x} = [X_0\ X_1 \cdots X_{N-1}]^{\mathrm{T}}$, where $\mathbf{W}_N$ is the N-point DFT matrix. The second bin, X_1, for example, corresponds to the digital frequency $\omega_1 = 2\pi/N$ or (in the continuous domain) $f_1 = f_s/N$ only if the last element of vector $\mathbf{x}$ corresponds to the most recent sample. In the present case, the vector assembled with the delay line has the most recent sample as its first element. Therefore, in order to have the n-th output representing an analytical NB signal with its central frequency equal to $\omega_n = 2\pi n/N$ (or, equivalently, $f_n = nf_s/N$, n from 0 to $N-1$), we use the IFFT (as seen in the figure). Each linear combiner corresponds to a NB beamformer, processing analytical NB signals from a given frequency bin of each IFFT output. The bins can be chosen according to the bandwidth of the incoming SOI, that is, $\omega_{N_1} = 2\pi N_1/N \geq \omega_1$ and $\omega_{N_2} = 2\pi N_2/N \leq \omega_2$.

Aiming at the proper constraints for the set of adaptive filters in Figure 5.12, we must find an expression for the output signal $y(k)$ when the input signal impinging the first sensor, in the discrete-TD, is given by $\mathrm{e}^{\mathrm{j}\omega_o k}$. By doing this, we shall be able to form the constraint matrix for a set of $L = N_2 - N_1 + 1$ frequencies.

For the ULA under consideration, assuming τ_θ as the delay (in samples and as a function of the DoA, θ) from each sensor to its closest neighbor, the input vector for the m-th N-point IFFT (m from 1 to M) would be given by $\mathrm{e}^{\mathrm{j}\omega_o k}\left[\mathrm{e}^{-\mathrm{j}\omega_o(m-1)\tau_\theta}\ \cdots\ \mathrm{e}^{-\mathrm{j}\omega_o[(m-1)\tau_\theta+(N-1)]}\right]^{\mathrm{T}}$. In the following, we express the IFFT output as the product of an $N \times N$ Inverse FFT matrix, $\mathbf{W}_N^{-1} = \frac{1}{N}\mathbf{W}_N^{\mathrm{H}}$, by the input vector. As seen in Figure 5.12, only a set of L bins of each IFFT output is used. The m-th vector $\mathbf{X}_m$ with the L selected bins may be expressed as the pre-multiplication of the IFFT output by an $L \times N$ selection matrix $\mathbf{S}$ (its elements are only ones and zeros) as in:

$$\mathbf{X}_m = \mathrm{e}^{\mathrm{j}\omega_o k}\mathbf{S}\mathbf{W}_N^{-1}\begin{bmatrix} \mathrm{e}^{-\mathrm{j}\omega_o(m-1)\tau_\theta} \\ \vdots \\ \mathrm{e}^{-\mathrm{j}\omega_o[(m-1)\tau_\theta+(N-1)]} \end{bmatrix}. \tag{5.78}$$

Samples of the output signal, in a block processing approach, could be obtained by applying an FFT to the output of all L linear combiners considering the null bins, the L bins from N_1 to N_2 and their complex conjugate (in order to respect the complex symmetry of the DFT for the case of real input signals). Taking into account that the first row of an N-point DFT matrix consists of ones only, the first element of the FFT output (in our case, the most recent sample of the output) could be obtained by adding all L outputs, taking their real part and multiplying the result by 2 (to account for the L complex conjugate terms). This is indeed what is represented in Figure 5.12 when $y_R(k)$ is obtained as 2Real $[y(k)]$. The output signal can be written as $y(k) = \mathbf{w}^{\mathrm{H}}\mathbf{X}(k)$, where $\mathbf{X}(k)$ is formed from all vectors $\mathbf{X}_m$ permuted by an $ML \times ML$ permutation matrix $\mathbf{P}_{ML}$ such that each l-th vector forming $\mathbf{w} = [\mathbf{w}_1 \ \cdots \ \mathbf{w}_L]^{\mathrm{T}}$ filters the signals from each subband as seen in Figure 5.12.

The output signal is then given as

$$y(k) = \mathbf{w}^{\mathrm{H}}\mathbf{x}_{ML}(k) = \mathbf{w}^{\mathrm{H}}\mathbf{P}_{ML}\begin{bmatrix}\mathbf{X}_1\\ \vdots \\ \mathbf{X}_M\end{bmatrix}, \tag{5.79}$$

which, given the assumed input signal, considering the steering vector $\mathbf{a}(\omega_{\mathrm{o}}, \theta) = \begin{bmatrix}1 & \mathrm{e}^{-\mathrm{j}\omega_{\mathrm{o}}\tau_\theta} & \cdots & \mathrm{e}^{-\mathrm{j}\omega_{\mathrm{o}}(M-1)\tau_\theta}\end{bmatrix}^{\mathrm{T}}$ and defining the *frequency domain delay-line* vector $\boldsymbol{\Delta}(\omega_{\mathrm{o}}) = \begin{bmatrix}1 & \mathrm{e}^{-\mathrm{j}\omega_{\mathrm{o}}} & \cdots & \mathrm{e}^{-\mathrm{j}(N-1)\omega_{\mathrm{o}}}\end{bmatrix}^{\mathrm{T}}$, can be rewritten as

$$y(k) = \mathrm{e}^{\mathrm{j}\omega_{\mathrm{o}}k}\mathbf{w}^{\mathrm{H}}\mathbf{P}_{ML}\mathbf{a}(\omega_{\mathrm{o}}, \theta) \otimes \mathbf{S}\mathbf{W}_N^{-1}\boldsymbol{\Delta}(\omega_{\mathrm{o}}). \tag{5.80}$$

Note, in the previous equation, that $\otimes$ means the *Kronecker* product and that we would have a distortionless condition (output being equal to input) whenever $\mathbf{w}^{\mathrm{H}}\mathbf{P}_{ML}\mathbf{a}(\omega_{\mathrm{o}}, \theta) \otimes \mathbf{S}\mathbf{W}_N^{-1}\boldsymbol{\Delta}(\omega_{\mathrm{o}}) = 1$. We therefore express the constraint vector $\mathbf{C}(\omega_{\mathrm{o}}) = \mathbf{P}_{ML}\mathbf{a}(\omega_{\mathrm{o}}, \theta) \otimes \mathbf{S}\mathbf{W}_N^{-1}\boldsymbol{\Delta}(\omega_{\mathrm{o}})$ and constrain the coefficient vector to $\mathbf{C}^{\mathrm{H}}(\omega_{\mathrm{o}})\mathbf{w} = 1$. From the constraint imposed to the coefficient vector for digital frequency ω_{o}, we extend this concept to all frequencies within the signal bandwidth which, in our implementation, correspond to the range $\omega_{N_1} = 2\pi N_1/N$ to $\omega_{N_2} = 2\pi N_2/N$, that is, we form the $ML \times L$ constraint matrix as follows.

$$\mathbf{C} = [\mathbf{C}(\omega_{N_1}) \ \ \mathbf{C}(\omega_{N_1+1}) \ \ \cdots \ \ \mathbf{C}(\omega_{N_2})], \tag{5.81}$$

where $\mathbf{C}(\omega_n) = \mathbf{P}_{ML}\mathbf{a}(\omega_n, \theta) \otimes \mathbf{S}\mathbf{W}_N^{-1}\boldsymbol{\Delta}(\omega_n)$, with $\omega_n = \frac{2\pi n}{N}$. The coefficient vector $\mathbf{w}$ shall comply with all constraints whenever subjected to $\mathbf{C}^{\mathrm{H}}\mathbf{w} = \mathbf{f}$, where the constraint vector, for L constraints, should be the $L \times 1$ vector $\mathbf{f} = [1 \ \cdots \ 1]^{\mathrm{T}}$.

With the constraint matrix and constraint vector well defined, a LCAF can be employed to act as a beamformer to an assumed known direction. The direction (θ) is passed to the structure by means of the delay τ_θ (in the steering vector) which, for a linear array, is given (in samples) by $\frac{d\cos\theta f_s}{v_p}$, where d is the distance between sensors, f_s is the sampling frequency, and v_p the propagation speed of the impinging wavefront.

Example 5.3 As a simple example of the expressions obtained so far, we present the broadband version of the delay and sum beamformer: starting from the expression of the optimum coefficient vector given in Equation (5.10) with L constraints and $\mathbf{p} = \mathbf{0}$,

$$\mathbf{w}_{\mathrm{MPDR}} = \mathbf{R}_{ML}^{-1}\mathbf{C}\left(\mathbf{C}^{\mathrm{H}}\mathbf{R}_{ML}^{-1}\mathbf{C}\right)^{-1}\mathbf{f}. \tag{5.82}$$

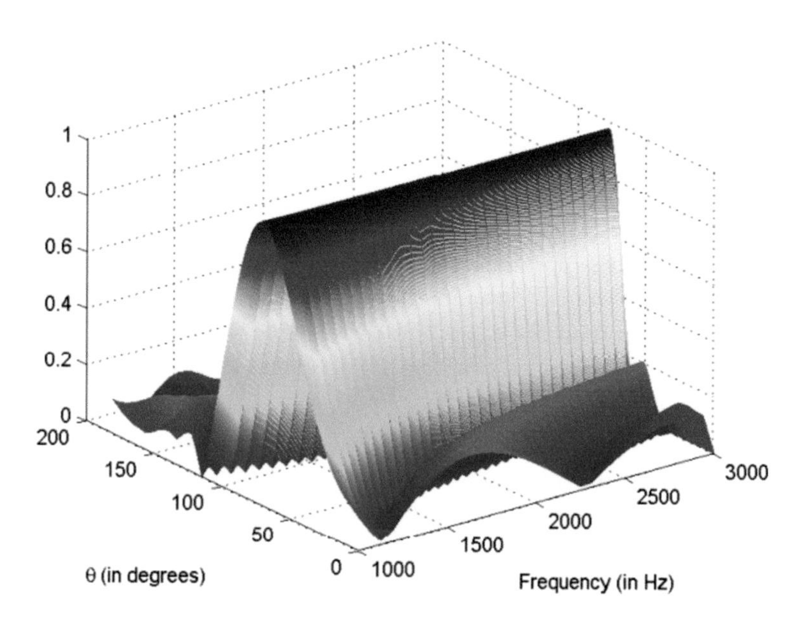

Figure 5.13 Beampattern of a broadband beamforming with fixed (Delay & Sum) coefficients.

Assuming that the signal hitting the first sensor in the discrete-TD is $e^{j\omega_o k}$, an overall autocorrelation matrix $\mathbf{R}_{ML} = \mathbb{E}\left[\mathbf{x}_{ML}(k)\mathbf{x}_{ML}^{\mathrm{H}}(k)\right]$, with $\mathbf{x}_{ML}(k)$ defined in Equation (5.79), would be an identity matrix $\mathbf{I}_{ML}$. The broadband D&S beamformer coefficient vector is then given by

$$\mathbf{w}_{\mathrm{D\&S}} = \mathbf{C}\left(\mathbf{C}^{\mathrm{H}}\mathbf{C} + \gamma\mathbf{I}_{ML}\right)^{-1}\mathbf{f}, \tag{5.83}$$

where a small regularization parameter, $\gamma = 10^{-12}$, was used to avoid ill-conditioned results. Setting the DoA to $\theta = 70°$ and using $d = \lambda/2$ for the highest frequency (in order to avoid spatial aliasing or, equivalently, grating lobes entering the visible region) ω_{N_2}, the beampattern for each frequency is shown in Figure 5.13. Note that $\mathbf{w}_{\mathrm{D\&S}}$ corresponds to vector $\mathbf{f}_{\mathrm{c}}$ used to initialize all constrained adaptive algorithms. For this experiment, we used: $M = 8$ sensors, $f_s = 8$ kHz, $N = 128$ (IFFT size) and assumed a broadband signal with frequencies ranging from $f_1 = 1$ kHz to $f_2 = 3$ kHz. Therefore, L corresponds to 33 equally spaced frequencies from f_1 to f_2.

We observe in Figure 5.13 that the beampatterns for higher frequencies present higher directivity; this can be observed from the trajectory of the first null (or of the first sidelobe), a curve opening from the highest to the lowest frequency. Another important observation is that the beampattern in this figure was evaluated only for the $L = 33$ frequencies of interest (from $N = 128$ points of the IFFT). This choice resulted in a well-behaved surface as seen in Figure 5.13. In case we evaluate the beampattern in more frequencies, with a much larger number of frequencies for instance, we would observe ripple or fluctuation in the beampatterns between those corresponding to the L frequencies shown in Figure 5.13.

Once we represented the input vector for the beamformer, we can obtain or estimate the optimum coefficient vector from Equation (5.82) and for this we need to obtain or estimate the autocorrelation matrix $\mathbf{R}_{ML}$. The approach used here is to treat the coefficient vector $\mathbf{w} = \begin{bmatrix}\mathbf{w}_1^{\mathrm{T}} & \cdots & \mathbf{w}_{\mathrm{L}}^{\mathrm{T}}\end{bmatrix}$ as a single vector and use any constrained adaptive filter seen in this chapter to approach the optimum solution after convergence.

In [35], it is shown the relationship between an FFT FD implementation of a broadband beamforming and its TD counterpart. In order to help the implementation in TD of a fixed FD coefficient vector, we show in the following, as in [36], the relationship between TD and FD coefficient vectors. Figure 5.14 illustrates a multichannel filter to be used as a TD implementation of an FD broadband beamformer.

The output of the TD implementation is written as $y(k) = \mathbf{w}_{TD}^{\mathrm{H}}\mathbf{x}(k)$, where, as seen in the diagram depicted in Figure 5.14, $\mathbf{x}(k) = \begin{bmatrix}\mathbf{x}_1^{\mathrm{T}}(k) & \cdots & \mathbf{x}_M^{\mathrm{T}}(k)\end{bmatrix}^{\mathrm{T}}$, with $\mathbf{x}_m(k) = [x_m(k) \;\cdots\; x_m(k-N+1)]^{\mathrm{T}}$. Conversely, from Figure 5.12, the

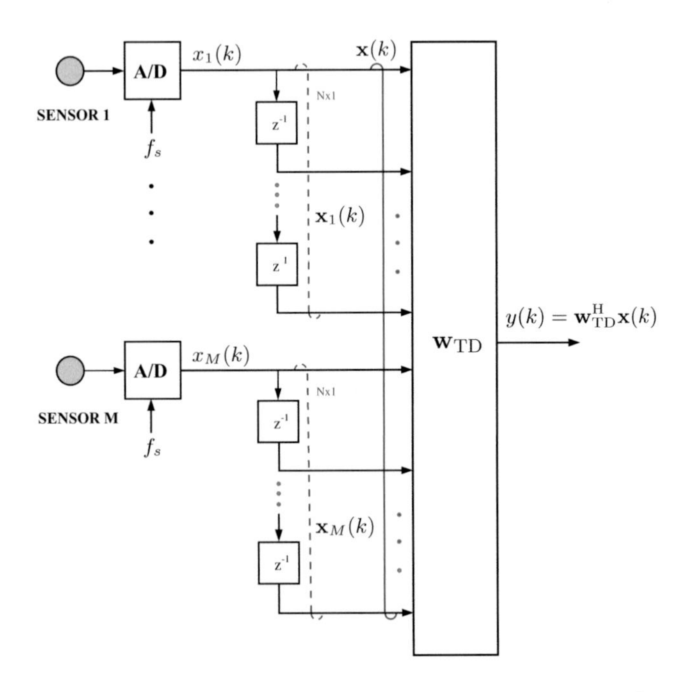

Figure 5.14 The FFT-based broadband beamformer of Figure 5.12 implemented in time domain as a multichannel digital filter.

output for M input signals $x_m(k)$, m from 1 to M, can be expressed as

$$y(k) = \mathbf{w}^{\mathrm{H}}\mathbf{P}_{ML}\begin{bmatrix}\mathbf{SW}_N^{-1}\mathbf{x}_1(k)\\ \vdots \\ \mathbf{SW}_N^{-1}\mathbf{x}_M(k)\end{bmatrix}$$

$$= \underbrace{\mathbf{w}^{\mathrm{H}}\,\underbrace{\mathbf{P}_{ML}\begin{bmatrix}\mathbf{SW}_N^{-1} & & \\ & \ddots & \\ & & \mathbf{SW}_N^{-1}\end{bmatrix}}_{\mathbf{T}^{\mathrm{H}}}}_{\mathbf{w}_{\mathrm{TD}}^{\mathrm{H}}}\mathbf{x}(k). \tag{5.84}$$

Therefore, we can write the equivalence of the coefficient vectors in both (time and frequency) domains as

$$\mathbf{w}_{\mathrm{TD}} = \mathbf{Tw}, \tag{5.85}$$

where $\mathbf{T} = \left[\mathbf{I}_{ML} \otimes \left(\mathbf{W}_N^{-\mathrm{H}}\mathbf{S}^{\mathrm{T}}\right)\right]\mathbf{P}_{ML}^{\mathrm{T}}$, with $\mathbf{W}_N^{-\mathrm{H}} = \left(\mathbf{W}_N^{-1}\right)^{\mathrm{H}}$.

From Equation (5.85), it is straightforward to obtain the constraints for the (discrete) TD: making $\mathbf{C}_{\mathrm{TD}}^{\mathrm{H}}\mathbf{w}_{\mathrm{TD}} = \mathbf{C}^{\mathrm{H}}\mathbf{w} = \mathbf{f}$ and replacing $\mathbf{w}_{\mathrm{TD}}$ as previously obtained, we find $\mathbf{C}_{\mathrm{TD}}^{\mathrm{H}}\mathbf{Tw} = \mathbf{C}^{\mathrm{H}}\mathbf{w}$ or $\mathbf{C} = \mathbf{T}^{\mathrm{H}}\mathbf{C}_{\mathrm{TD}}$. This last expression, due to the dimensions of the matrices, corresponds to an undetermined system with infinitely many possibilities for $\mathbf{C}_{\mathrm{TD}}$. One of them (minimum norm) is given as

$$\mathbf{C}_{\mathrm{TD}} = \mathbf{T}\left(\mathbf{T}^{\mathrm{H}}\mathbf{T}\right)^{-1}\mathbf{C}. \tag{5.86}$$

As an example of this equivalence, we show with the following Matlab® commands detailing how to compute the coefficient vectors ($\mathbf{w}_{\mathrm{D\&S}}$ and $\mathbf{w}_{\mathrm{TD}}$) and respective constraint matrices ($\mathbf{C}$ and $\mathbf{C}_{\mathrm{TD}}$) that produce the beampatterns in Figure 5.13.

```
% Assume we have: M, fs (sampling frequency), N, N1, N2,
% vp (propagation speed) and THETA (DoA in degrees)
IWN=inv(fft(eye(N)));    L=N2-N1+1;
S=[zeros(L,N1-1) eye(L) zeros(L,N-N2)]; % Bins N1 to N2
f1=N1*fs/N;    f2=N2*fs/N;    d=vp/(2*f2);
TAUtheta=d*cosd(THETA)*fs/vp; % delay between consecutive sensors
aux=zeros(1,L);    aux(1)=1;    PML=[ ];    C=[ ];
for cont=1:L
    PML=[PML
         kron(eye(M),aux)];
    aux=circshift(aux,[0 1]);
end
for n=N1:N2
    wn=2*pi*n/N;
    atheta=(exp(1i*(0:(M-1))*wn*TAUtheta))';
    vectDeltawn=(exp(1i*wn*(0:(N-1))))';
    cw=PML*kron(atheta,S*IWN*vectDeltawn);
    C=[C cw];
end
```

```
vectf=ones(L,1);
wDS=C*inv(C'*C+1e-12*eye(L))*vectf;
T=kron(eye(M),IWN'*S')*PML';
wTD=T*wDS;
CTD=T*inv(T'*T)*C;
```

Finally, having both FFT and TD implementations (and their respective constraint matrices, $\mathbf{C}$ and $\mathbf{C}_{\text{TD}}$), we can employ any constrained adaptive filter or a constrained structure (GSC or Householder) to run a broadband adaptive beamformer. The FFT was used here for its simplicity (and computational efficiency) but another filter bank could be assigned for this task. The TD implementation was presented here as way to understand in more details the main concepts behind an adaptive filter applied to an adaptive beamforming; but we remind the reader that it has a larger number of coefficients (MN instead of ML for the case of the IFFT implementation), which impacts its computational complexity.

5.5 Further Reading and Remarks

In the field of adaptive beamforming, there are still a number of topics not discussed here due basically to lack of space but which might deserve our attention.

First of all, in some applications, it might be interesting to impose additional constraints, others than the distortionless constraint. A few examples are quadratic constraints (which leads to a procedure named diagonal loading, useful for DoA mismatch and errors in the sensor positions), directional constraints (which helps to improve the robustness of a beamformer with respect to steering errors), and derivative constraints (for shaping the beampattern near the main lobe and possible nulls). Interested readers can find more detailed explanation in the classical literature [37].

Another brief and relevant comment is regarding the implementation of 2D (planar) and 3D arrays. We have, for the sake of simplicity, addressed the case of linear arrays but other geometry are mere extensions where the delay between sensors will be computed as a function of one or two angles, azimuth and elevation (or zenith as used in some references). A light reading on this topics and its importance to 5G wireless network is found in [38].

An array signal processor such as a beamformer is essentially a multidimensional (time, space, and frequency) system. In recent years, tensor algebra has been employed to take advantage of this inherent multidimensionality. A tensor-based beamformer is beyond the scope of this chapter but a few suggestions for this topic, as a starting point, would be [39–41].

A final observation is concerned to the numerical robustness of constrained adaptive beamforming: we may speculate that they tend to follow their unconstrained counterparts. Hence, constrained LMS-like algorithms, given a proper choice of step size, are assumed to keep their robustness in finite precision environment while keeping their main drawback, slow convergence for colored input signals. As for the faster constrained algorithms, Ramos [42] evaluated the numerical robustness of CRLS-like algorithms and the main results found therein are summarized in the following. For this evaluation, an experiment was carried out consisting of a seven sensor linear array with a look-direction set to broadside ($\theta = 90°$) and three jammers, while the SNR was set to 0 dB and each jammer-to-noise ratio (JNR) was made equal to 30 dB. The forgetting factor was set to 0.98. Regarding their stability, we comment on algorithms CRLS [43], CCG [17] and constrained *Quasi*-Newton (CQN) [44]. Each of these algorithms has a GSC and an HC counterparts. With the GSC and the HC structures, we used the unconstrained versions of the following algorithms: conventional RLS algorithm [3], conventional and inverse QRD-RLS (IQRDRLS) algorithm [45], CG algorithm [46], and quasi–Newton (QN) algorithm [47]. After a large number of snapshots, [42] concluded that, among this set of fast converging algorithms, only the CG and the QRD-RLS (both conventional and inverse versions) when used within the GSC or the Householder constrained structures, have not diverged after some time. It was observed that the only direct constrained algorithm with stable performance was the CCG algorithm; all other constrained versions (among those compared in this experiment) diverged. It was also observed that the CRLS algorithm took longer to diverge when the forgetting factor was increased toward 1.

5.6 Conclusion

This chapter dealt with adaptive filtering algorithms with linear constraints applied to adaptive beamforming. We carefully developed the introduction for a global understanding of the problem and it shall be helpful for a beginner as well as for a more experienced reader. The theory herein was devised for an array of generic sensors. One could apply the same approach with minor modifications to audio systems (microphone array) or RF systems (antenna array). The authors present the constrained algorithms in algorithmic formats and expect an interested reader to have no difficulties implementing them in various scenarios.

After addressing one example of a broadband beamformer, the chapter wraps up by discussing some practical issues related to algorithm implementation.

Problems

5.1 Carry out the details to derive Equation (5.20), the update equation of the constrained LMS algorithm.

5.2 Starting from the LS cost function $\xi_{\text{D}} = \sum_{i=0}^{k} \lambda^{k-i}|d(i) - \mathbf{w}^{\text{H}}\mathbf{x}(i)|^2$, and minimizing it subject to $\mathbf{C}^{\text{H}}\mathbf{w} = \mathbf{f}$, where λ is the forgetting factor, find the

CLS solution, for time instant k, as given in Equation (5.24). Before doing that, define $\mathbf{R}_x(k)$, the deterministic autocorrelation matrix of the input signal vector, and $\mathbf{p}(k)$, the cross-correlation vector between the input signal vector $\mathbf{x}(k)$ and the reference signal $d(k)$.

5.3 As for the CRLS algorithm, version 1, prove Equation (5.34) by post-multiplying Equation (5.33) by $\mathbf{C}^{\mathrm{H}}\boldsymbol{\kappa}(k)$ and dividing the result by λ.

5.4 Run the CAP algorithm and the set membership constrained affine projection (SM-CAP) algorithm in a system identification problem where the filter coefficients are constrained to preserve linear phase at every iteration. Reproduce the *Simulation 1* in [9] (reproduced below for convenience), using the same setup described therein (filling the gaps with reasonable choices), and plot the learning curves (MSE in dB obtained from an ensemble of 500 independent runs) for both algorithms, using: $L = 1$, $L = 2$, and $L = 4$ data reuses, variance of the observation error $\sigma_n^2 = 10^{-10}$, and an error bound $\gamma = \sqrt{6}\sigma_n$.

> *Simulation 1:*
> A simulation was carried out in a system identification problem where the filter coefficients were constrained to preserve linear phase at every iteration. For this example we chose $N = 11$ and, in order to fulfill the linear phase requirement, we made
>
> $$\mathbf{C} = \begin{bmatrix} \mathbf{I}_{(N-1)/2} \\ \mathbf{0}^{\mathrm{T}} \\ -\mathbf{J}_{(N-1)/2} \end{bmatrix}$$
>
> with $\mathbf{J}$ being a reversal matrix (an identity matrix with all rows in reversed order), and $\mathbf{f} = [0 \cdots 0]^{\mathrm{T}}$.
> This setup was employed to show the improvement of the convergence speed when L is increased. Due to the symmetry of $\mathbf{C}$ and the fact that $\mathbf{f}$ is a null vector, more efficient structures could be used. The input signal consists of colored noise with zero mean, unitary variance, and eigenvalue spread around 2068. The reference signal was obtained after filtering the input signal by a linear-phase FIR filter and adding observation noise with variance equal to $\sigma_n^2 = 10^{-10}$.
> The optimal coefficient vector used to compute the coefficient-error vector was obtained from $\mathbf{w}_{\mathrm{opt}} = \mathbf{R}^{-1}\mathbf{p} - \mathbf{R}^{-1}\mathbf{C}\left(\mathbf{C}^{\mathrm{T}}\mathbf{R}^{-1}\mathbf{C}\right)^{-1}\left(\mathbf{C}^{\mathrm{T}}\mathbf{R}^{-1}\mathbf{p} - \mathbf{f}\right)$ after replacing $\mathbf{R}^{-1}\mathbf{p}$ (the Wiener solution) by $\mathbf{w}_{us}$ (the FIR unknown system).
>
> The input signal was taken as colored noise generated by filtering white noise through a filter with a pole at $-\alpha = -0.99$. The autocorrelation matrix for this example is given by
>
> $$\mathbf{R} = \frac{\sigma_{\mathrm{WGN}}^2}{1-\alpha^2}\begin{bmatrix} 1 & -\alpha & (-\alpha)^2 & \cdots & (-\alpha)^{N-1} \\ -\alpha & 1 & -\alpha & \cdots & (-\alpha)^{N-2} \\ \vdots & \vdots & \ddots & & \vdots \\ (-\alpha)^{N-1} & (-\alpha)^{N-2} & \cdots & & 1 \end{bmatrix},$$
>
> where σ_{WGN}^2 is set such that $\frac{\sigma_{\mathrm{WGN}}^2}{1-\alpha^2}$ corresponds to the desired input signal variance σ_x^2, made equal to 1 in this experiment.

5.5 In order to better understand the transformation process of the GSC structure, prove Equation (5.53) in two different ways:

a) Using $\mathbf{w}_{\text{OPT}}$ of the LCAF and expliciting $\mathbf{w}_{\text{GSC-OPT}}$ in Equation (5.52);

b) Making $\mathbf{w}_{\text{GSC-OPT}} = \mathbf{R}_{\text{GSC}}^{-1}\mathbf{p}_{\text{GSC}}$ and using the definitions of $\mathbf{R}_{\text{GSC}}$ and $\mathbf{p}_{\text{GSC}}$ given before Equation (5.53).

5.6 In an infinite precision environment, prove that the CLMS and the GSC-LMS algorithms, given equivalent initializations, present identical updates at every iteration if the GSC structure uses an orthonormal blocking matrix, that is, $\mathbf{B}^{\text{H}}\mathbf{B} = \mathbf{I}$.

5.7 Repeat the simulation suggested for the CAP and SM-CAP algorithms (Problem 5.4) with the CCG algorithm given in Algorithm 5.33. This time, run an ensemble of 1000 independent runs and a more realistic variance of the observation noise ($\sigma_n^2 = 10^{-3}$). Also, make two figures: one with $\gamma = \sqrt{5\sigma_n^2}$ and another one with $\gamma = \sqrt{2\sigma_n^2}$; observe, when varying the threshold (error bound), the differences in the learning curves and in the number of updates.

5.8 Recalling that a HC algorithm employs an orthogonal rotation matrix $\mathbf{Q}$ that generates a modified coefficient vector, $\overline{\mathbf{w}}(k) = \mathbf{Q}\mathbf{w}(k)$ such that $\overline{\mathbf{C}} = \mathbf{Q}\mathbf{C}$ satisfies

$$\overline{\mathbf{C}}(\overline{\mathbf{C}}^{\text{H}}\overline{\mathbf{C}})^{-1}\overline{\mathbf{C}}^{\text{H}} = \begin{bmatrix} \mathbf{I}_p & \mathbf{0} \\ \mathbf{0} & \mathbf{0} \end{bmatrix},$$

and, consequently $\overline{\mathbf{C}}^{\text{H}}\overline{\mathbf{w}}(k) = \mathbf{f}$, show that the rotation matrix $\mathbf{Q} = \mathbf{Q}_p \cdots \mathbf{Q}_2\mathbf{Q}_1$, where $\mathbf{Q}_i = \begin{bmatrix} \mathbf{I}_{i-1 \times i-1} & \mathbf{0}^{\text{T}} \\ \mathbf{0} & \overline{\mathbf{Q}}_i \end{bmatrix}$, $\overline{\mathbf{Q}}_i$ being an $(M-i+1) \times (M-i+1)$ Householder transformation matrix on the form $\overline{\mathbf{Q}}_i = \mathbf{I} - 2\overline{\mathbf{v}}_i\overline{\mathbf{v}}_i^{\text{H}}$, is a viable solution.

5.9 Derive, from the equations given in this chapter, the updating expression of the householder constrained normalized LMS (HCNLMS) algorithm given by Equation (5.65) which is rewritten in the following. Also explain, in your words, the initialization suggested for the modified coefficient vector, that is, $\overline{\mathbf{w}}(0)$.

$$\overline{\mathbf{w}}(k+1) = \begin{bmatrix} \overline{\mathbf{w}}_{\text{U}}(0) \\ \overline{\mathbf{w}}_{\text{L}}(k) \end{bmatrix} + \mu \frac{e^*(k)}{\overline{\mathbf{x}}_{\text{L}}^{\text{H}}(k)\overline{\mathbf{x}}_{\text{L}}(k)} \begin{bmatrix} \mathbf{0} \\ \overline{\mathbf{x}}_{\text{L}}(k) \end{bmatrix}.$$

5.10 From Equations (5.72) and (5.74), derive the updating expression of the ℓ_1-CLMS algorithm given in Equation (5.75), rewritten in the following for convenience.

$$\mathbf{w}(k+1) = \mathbf{P}\left[\mathbf{w}(k) + \mu e^*(k)\overline{\mathbf{P}}(k)\mathbf{x}(k)\right] + \mathbf{f}_{\text{c}} + \mathbf{f}_{\ell_1}(k),$$

where

$$\begin{cases} \overline{\mathbf{P}}(k) = \left[\mathbf{I} - \left(\frac{\mathbf{Ps}(k)}{\|\mathbf{Ps}(k)\|^2}\right)\mathbf{s}^{\mathrm{H}}(k)\right]\mathbf{P}, \\ e_{\ell_1}(k) = t - \mathbf{s}^{\mathrm{H}}(k)\mathbf{w}(k), \text{ and} \\ \mathbf{f}_{\ell_1}(k) = e_{\ell_1}(k)\left(\frac{\mathbf{Ps}(k)}{\|\mathbf{Ps}(k)\|^2}\right). \end{cases}$$

5.11 Replicate the experiment that produced the beampatterns in Figure 5.13 using the information provided in the text. Instead of using $d = \lambda/2$ with respect to $f_2 = 3$ kHz, use $d = \frac{c}{2f_1}$ (with respect to f_1, the lowest frequency of interest). Describe the resulting beampatterns indicating the amount of grating lobes entering the visible region.

References

[1] A. V. Oppenheim and R. W. Schafer, *Discrete-Time Signal Processing*, 3rd ed. (Pearson, Upper Saddle River, 2009).

[2] O. L. Frost, III, An algorithm for linearly constrained adaptive array processing, Proceedings of the IEEE **60**, pp. 926–935 (1972).

[3] P. S. R. Diniz, *Adaptive Filtering: Algorithms and Practical Implementations*, 5th ed. (Springer, New York, 2020).

[4] J. A. Apolinário Jr., S. Werner, P. S. R. Diniz, and T. I. Laakso, Constrained normalized adaptive filters for CDMA mobile communications, 9th European Signal Process. Conf. (EUSIPCO 1998), Island of Rhodes, Greece, 1998, pp. 1–4.

[5] L. S. Resende, J. M. T. Romano, and M. G. Bellanger, A fast least-squares algorithm for linearly constrained adaptive filtering, IEEE Transactions on Signal Processing **44**, pp. 1168–1174 (1996).

[6] M. L. R. de Campos and J. A. Apolinário Jr., The constrained affine projection algorithm – development and convergence issues, First Balkan Conference on Signal Processing, Communications, Circuits and Systems, Istanbul, Turkey, 2000, pp. 1–4.

[7] S. Werner and P. S. R. Diniz, Set-membership affine projection algorithm, IEEE Signal Proc. Letters **8**, pp. 231–235 (2001).

[8] S. Werner, J. A. Apolinário Jr., and M. L. R. De Campos, The data-selective constrained affine-projection algorithm, IEEE International Conference on Acoustics, Speech, and Signal Processing (ICASSP 2001) **6**, Salt Lake City, USA, 2001, pp. 3745–3748.

[9] S. Werner, J. A. Apolinário Jr., M. L. R. de Campos, and P. S. R. Diniz, Low-complexity constrained affine-projection algorithms, IEEE Transactions on Signal Processing **53**, pp. 4545–4555 (2005).

[10] J. F. Galdino, J. A. Apolinário Jr., and M. L. R. de Campos, A set-membership NLMS algorithm with time-varying error bound, IEEE International Symposium on Circuits and Systems (ISCAS 2006), Island of Kos, Greece, 2006, pp. 277–280.

[11] L. J. Griffiths and C. W. Jim, An alternative approach to linearly constrained adaptive beamforming, IEEE Transactions on Antennas and Propagation **AP-30**, pp. 27–34 (1982).

[12] S. Werner, J. A. Apolinário Jr., and M. L. R. de Campos, On the equivalence of RLS implementations of LCMV and GSC processors, IEEE Signal Processing Letters **10**, pp. 356–359 (2003).

[13] C.-Y. Tseng and L. J. Griffiths, A systematic procedure for implementing the blocking matrix in decomposed form, Twenty-Second Asilomar Conference on Signals, Systems and Computers **2**, Pacific Grove, USA, 1988, pp. 808–812.

[14] Y. Chu, W.-H. Fang, and S.-H. Chang, A novel wavelet-based generalized sidelobe canceller, IEEE International Conference on Acoustics, Speech, and Signal Processing (ICASSP 1998) **4**, Seattle, USA, 1998, pp. 2497–2500.

[15] P. S. Chang and A. N. Willson, Adaptive filtering using modified conjugate gradient, 38th Midwest Symposium on Circuits and Systems, **1**, Rio de Janeiro, Brazil, 1995, pp. 243–246.

[16] P. S. Chang and A. N. Willson, Analysis of conjugate gradient algorithms for adaptive filtering, IEEE Transactions on Signal Processing **48**, pp. 409–418 (2000).

[17] J. A. Apolinário Jr., M. L. R. de Campos, and C. P. Bernal O., The constrained conjugate gradient algorithm, IEEE Signal Processing Letters **7**, pp. 351–354 (2000).

[18] L. Wang and R. C. de Lamare, Set-membership constrained conjugate gradient adaptive algorithm for beamforming, IET Signal Processing **6**, pp. 789–797 (2012).

[19] M. L. R. de Campos, S. Werner, and J. A. Apolinário Jr., Constrained adaptation algorithms employing Householder transformation, IEEE Transactions on Signal Processing **50**, pp. 2187–2195 (2002).

[20] M. L. R. de Campos, S. Werner, and J. A. Apolinário Jr., Householder-transform constrained LMS algorithms with reduced-rank updating, IEEE International Conference on Acoustics, Speech, and Signal Processing (ICASSP 1999) **4**, Phoenix, USA, 1999, pp. 1857–1860.

[21] G. H. Golub and C. F. Van Loan, *Matrix Computations*, 3rd ed. (The Johns Hopkins University Press, Baltimore, 1996).

[22] J. H. Wilkinson (ed.), *The Algebraic Eigenvalue Problem* (Oxford University Press, New York, 1988).

[23] C.-Y. Tseng and L. J. Griffiths, A systematic procedure for implementing the blocking matrix in decomposed form, Twenty-Second Asilomar Conference on Signals, Systems and Computers **2**, Pacific Grove, USA, 1988, pp. 808–812.

[24] D. Duttweiler, Proportionate normalized least-mean-squares adaptation in echo cancelers, IEEE Transactions on Speech and Audio Processing **8**, pp. 508–518 (2000).

[25] J. Benesty and S. L. Gay, An improved PNLMS algorithm, IEEE International Conference on Acoustics, Speech, and Signal Processing (ICASSP 2002) **2**, Orlando, USA, 2002, pp. 1881–1884.

[26] Y. Chen, Y. Gu, and A. O. Hero III, Sparse LMS for system identification, IEEE International Conference on Acoustics, Speech and Signal Processing (ICASSP 2009), Taipei, Taiwan, 2009, pp. 3125–3128.

[27] M. L. R. de Campos and J. A. Apolinário Jr., Shrinkage methods applied to adaptive filters, 2010 International Conference on Green Circuits and Systems, Shanghai, China, 2010, pp. 41–45.

[28] C. Paleologu, J. Benesty, and S. Ciochina, An improved proportionate NLMS algorithm based on the l0 norm, IEEE International Conference on Acoustics Speech and Signal Processing (ICASSP), Dallas, USA, 2010, pp. 309–312.

[29] Y. Kopsinis, K. Slavakis, and S. Theodoridis, Online sparse system identification and signal reconstruction using projections onto weighted ℓ_1 balls, IEEE Transactions on Signal Processing **59**, 936–952 (2011).

[30] M. V. S. Lima, T. N. Ferreira, W. A. Martins, and P. S. R. Diniz, Sparsity-aware data-selective adaptive filters, IEEE Transactions on Signal Processing **62**, pp. 4557–4572 (2014).

[31] J. F. de Andrade Jr., M. L. R. de Campos, and J. A. Apolinário Jr., An ℓ_1-norm linearly constrained LMS algorithm applied to adaptive beamforming, 7th IEEE Sensor Array and Multichannel Signal Processing Workshop (SAM 2012), Hoboken, USA, 2012, pp. 429–432.

[32] J. F. de Andrade Jr., M. L. R. de Campos, and J. A. Apolinário Jr., An ℓ_1-constrained normalized LMS algorithm and its application to thinned adaptive antenna arrays, IEEE International Conference on Acoustics, Speech, and Signal Processing (ICASSP 2013), Vancouver, Canada, 2013, pp. 3806–3810.

[33] J. F. Andrade Jr., M. L. R. de Campos, and J. A. Apolinário Jr., ℓ_1-constrained normalized LMS algorithms for adaptive beamforming, IEEE Transactions on Signal Processing **63**, pp. 6524–6539 (2015).

[34] J. F. Andrade Jr., M. L. R. Campos, and J. A. Apolinário Jr., An ℓ_1-norm linearly constrained affine projection algorithm, IEEE Sensor Array and Multichannel Signal Processing Workshop (SAM 2016), Rio de Janeiro, Brazil, 2016, pp. 1–5.

[35] L. C. Godara, Application of the fast Fourier transform to broadband beamforming, The Journal of the Acoustical Society of America **98**, pp. 230–240 (1995).

[36] J. A. Apolinário Jr. and M. L. R. de Campos, Sparse broadband acoustic adaptive beamformers for underwater applications, MTS/IEEE Oceans Conference, Aberdeen, Scotland, 2017, pp. 1–6.

[37] H. L. V. Trees, *Optimum Array Processing: Part IV of Detection, Estimation, and Modulation Theory* (John Wiley & Sons, Hoboken, 2002).

[38] S. M. Razavizadeh, M. Ahn, and I. Lee, Three-dimensional beamforming: A new enabling technology for 5G wireless networks, IEEE Signal Processing Magazine **31**, pp. 94–101 (2014).

[39] X. Gong and Q. Lin, Spatially constrained parallel factor analysis for semi-blind beamforming, Seventh International Conference on Natural Computation **1**, Shanghai, China, 2011, pp. 416–420.

[40] R. K. Miranda, J. P. C. L. da Costa, F. Roemer, A. L. F. de Almeida, and G. Del Galdo, Generalized sidelobe cancellers for multidimensional separable arrays, IEEE 6th International Workshop on Computational Advances in Multi-Sensor Adaptive Processing (CAMSAP 2015), Cancun, Mexico, 2015, pp. 193–196.

[41] L. N. Ribeiro, A. L. F. de Almeida, and J. C. M. Mota, Tensor beamforming for multilinear translation invariant arrays, IEEE International Conference on Acoustics, Speech and Signal Processing (ICASSP 2016), Shanghai, China, 2016, pp. 2966–2970.

[42] A. L. L. Ramos, J. A. Apolinário Jr., and M. L. R. de Campos, On numerical robustness of constrained RLS-like algorithms, Brazilian Telecommunication Symposium (SBrT 2004), Belém, Brazil, 2004.
[43] L. S. Resende, J. M. T. Romano, and M. G. Bellanger, A fast least-squares algorithm for linearly constrained adaptive filtering, IEEE Transactions on Signal Processing **44**, pp. 1168–1174 (1996).
[44] M. L. R. de Campos, S. Werner, J. A. Apolinário Jr., and T. I. Laakso, Constrained *quasi*-Newton algorithm for CDMA mobile communications, SBrT/IEEE International Telecommunications Symposium (ITS 1998) **1**, São Paulo, Brazil, 1998, pp. 371–376.
[45] J. A. Apolinário Jr. (ed.), *QRD-RLS Adaptive Filtering* (Springer, New York, 2009).
[46] Pi Sheng Chang and A. N. Willson, Analysis of conjugate gradient algorithms for adaptive filtering, IEEE Transactions on Signal Processing **48**, pp. 409–418 (2000).
[47] M. L. R. de Campos and A. Antoniou, A new *quasi*-Newton adaptive filtering algorithm, IEEE Transactions on Circuits and Systems II: Analog and Digital Signal Processing **44**, pp. 924–934 (1997).

6 Adaptive Filtering on Graphs

6.1 Introduction

This chapter presents a class of adaptive filtering algorithms that deal with multivariable signals whose snapshots are defined on irregular domains that admit a graph modeling. While Chapter 4 focused on decentralized data processing, here we describe centralized learning methods.

It is an undisputed fact that traditional digital signal processing (DSP) thrives when applied to uniform, Euclidean domains. The technology we use in a daily basis, ranging from audio codecs and digital TVs to high-end biomedical equipment and smartphones, overwhelmingly supports this statement. The uniformity of intervals in-between adjacent samples of a series, mostly taken for granted, plays in fact a key role for rendering Fourier representations and related tools simple, yet powerful. Unfortunately, not all practical domains have such a desirable property. For instance, when positioning sensors along a large area, geography and topography may hinder us from drawing a perfect, uniform rectangular grid out of the sensors. Likewise, when studying social networks, one has to account for the fact that not everyone has the same number of friends. Energy and transportation networks are also a case in point in which the nodes, distribution centers or road connections, are spatially structured in a nonuniform fashion.

Graph signal processing (GSP) is a set of techniques for manipulating signals defined on nonuniformly structured domains [1]. Graphs, being highly flexible abstract constructions, can be the stage to a wide range of phenomena, including the Internet of things (IoT), social networks, array processing [2], and big data analytics [3, 4].

This chapter is divided into two main parts. The first one briefly introduces the fundamental aspects of GSP that are required to fully grasp the key ideas of online learning and filtering of graph signals. It introduces in an intuitive manner the abstract concepts of graphs, graph signals, graph Fourier transform (GFT), graph filters, and sampling/reconstruction of graph signals. The second part describes the main adaptive graph filtering algorithms, which comprise two different families of adaptive graph filters, namely tapped memory line (TML) filters and interpolation filters. More specifically, the second part shows how

LMS-, NLMS-, and RLS-based algorithms can be derived for each family of adaptive graph filters.

6.2 Fundamentals of Graph Signal Processing

In this section, we present the fundamental ideas concerning the GSP framework. We start by describing the mathematical graph structure and defining graph signals. Then, we explain the extension of traditional DSP ideas, such as Fourier representations and linear filtering, to the graph domain. We finish this section by addressing compact representations of bandlimited graph signals and obtaining proper methods for sampling and reconstruction of these signals.

6.2.1 Graphs and Graph Signals

Graph is a mathematical structure made up of two classes of elements: *vertices* (or *nodes*) and *edges* (or *links*). These elements arise naturally in many applications and allow useful abstractions for solving practical problems. For instance, the seven-bridges-of-Konigsberg problem, tackled by Leonard Euler in 1736, served as the groundwork for the further development of the mathematical branch known as *graph theory*. Since the city of Konigsberg — now Kaliningrad, Russia — was formed by four landmasses connected by seven bridges in Euler's time, a practical question arose: Is it possible for a person to walk through the town so that each bridge would be crossed exactly once?

Euler conceived a convenient representation to solve the problem, wherein the landmasses become vertices and the bridges are seen as edges connecting these vertices, as illustrated in Figure 6.1. This is an example of a graph representation through which Euler proved that it was impossible to cross each bridge of Konigsberg once, and only once, because this graph has nodes of odd *degree*, that is, with an odd number of edges connecting them.

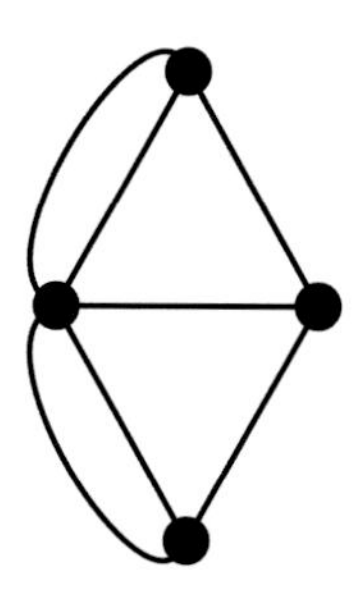

Figure 6.1 Graph representation of Konigsberg city in Euler's time.

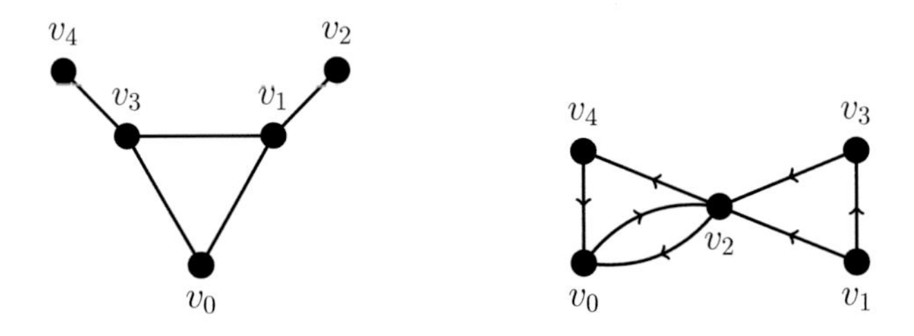

Figure 6.2 Examples of five-node graphs: (left) undirected $\mathcal{G}_a$ and (right) *digraph* $\mathcal{G}_b$.

Graphs

A *graph* is an ordered pair $\mathcal{G} = (\mathcal{V}, \mathcal{E})$, with $\mathcal{V}$ denoting the *set of vertices* and $\mathcal{E} \subset \mathcal{V} \times \mathcal{V}$, the *set of edges*. The set $\mathcal{V}$, also called the *graph domain*, is a finite list, which we will write as $\{v_0, v_1, \ldots, v_{N-1}\}$. The actual order of the node labeling does not affect the results of the main theorems of graph theory.[1] Although a general graph might present any number of connections between two vertices $v_n, v_m \in \mathcal{V}$, we assume that the relation between two nodes is represented by at most two edges, denoted as (v_n, v_m) or (v_m, v_n). Thus, the N-node structure $\mathcal{G} = (\mathcal{V}, \mathcal{E})$ has at most N^2 edges. Note that the edge definition accounts for a directionality in the pairwise relation between nodes: The edge (v_n, v_m) indicates a connection starting from v_n to v_m. Graphs that make explicit use of this directionality are called *directed graphs* or *digraphs*, and the edges in those graphs are usually depicted with an arrow. When this directionality is not important, as in *undirected graphs*, we do not make difference between the pairs (v_n, v_m) and (v_m, v_n) and depict a single edge with a simple line between those connected nodes. Figure 6.2 illustrates some of these concepts. These two graphs $\mathcal{G}_a = (\mathcal{V}_a, \mathcal{E}_a)$ and $\mathcal{G}_b = (\mathcal{V}_b, \mathcal{E}_b)$ have the same set of vertices $\mathcal{V}_a = \mathcal{V}_b = \{v_0, v_1, v_2, v_3, v_4\}$, but different sets of edges and different directionality properties.

As suggested by Euler's approach to solve the seven-bridges-of-Konigsberg problem, illustrated in Figure 6.1, a graph can be used to represent relations between real-world objects in practical applications. More specifically, we are interested in *quantifiable* relations for which an intensity value can be assigned for each pair of real-world objects. As the relation between a pair of objects usually depends on the specific objects, the underlying intensity value of the connection is not necessarily the same throughout the whole graph. Therefore, it is useful defining an auxiliary function to indicate stronger or weaker node links in a graph. We consider that each edge $(v_n, v_m) \in \mathcal{E}$ is associated with a

[1] Indeed, any result that does depend on the order of the labeling is actually a result about the labeling, rather than about the graph.

non-null weight $a_{nm} \in \mathbb{R}_*$, which can be understood as a *proximity* or *similarity metric*.[2]

By taking the weight a_{nm} as an element of the n-th row and m-th column, wherein $a_{nm} = 0$ whenever $(v_n, v_m) \notin \mathcal{E}$, we can build an $N \times N$ matrix called the *adjacency matrix* $\mathbf{A}$. Additionally, an alternative graph representation is given by the *Laplacian matrix*

$$\mathbf{L} = \mathbf{K} - \mathbf{A}\,, \tag{6.1}$$

where $\mathbf{K}$ is a diagonal matrix with the nodes' degrees as diagonal entries:

$$k_{nn} = \sum_{m \in \mathcal{N}} a_{nm}\,, \quad \text{with } \mathcal{N} = \{0, 1, \ldots, N-1\}\,. \tag{6.2}$$

Example 6.1 Assign link weights to the *bull graph* $\mathcal{G}_a$ in Figure 6.2 and determine the corresponding adjacency and Laplacian matrices.

Solution

The simplest way of assigning edge weights is to consider that $a_{nm} = 1$ if $(v_n, v_m) \in \mathcal{E}_a$ and $a_{nm} = 0$ if $(v_n, v_m) \notin \mathcal{E}_a$, which results in the following adjacency and Laplacian matrices:

$$\mathbf{A}_a = \begin{bmatrix} 0 & 1 & 0 & 1 & 0 \\ 1 & 0 & 1 & 1 & 0 \\ 0 & 1 & 0 & 0 & 0 \\ 1 & 1 & 0 & 0 & 1 \\ 0 & 0 & 0 & 1 & 0 \end{bmatrix}, \quad \mathbf{L}_a = \begin{bmatrix} 2 & -1 & 0 & -1 & 0 \\ -1 & 3 & -1 & -1 & 0 \\ 0 & -1 & 1 & 0 & 0 \\ -1 & -1 & 0 & 3 & -1 \\ 0 & 0 & 0 & -1 & 1 \end{bmatrix}. \tag{6.3}$$

If we wish to model some connections between nodes as stronger than others, we could use, for instance, the weights $\{a_{01}, a_{03}, a_{12}, a_{13}, a_{34}\} = \{2, 2, 1, 2, 1\}$, thus obtaining the adjacency and Laplacian matrices

$$\mathbf{A}'_a = \begin{bmatrix} 0 & 2 & 0 & 2 & 0 \\ 2 & 0 & 1 & 2 & 0 \\ 0 & 1 & 0 & 0 & 0 \\ 2 & 2 & 0 & 0 & 1 \\ 0 & 0 & 0 & 1 & 0 \end{bmatrix}, \quad \mathbf{L}'_a = \begin{bmatrix} 4 & -2 & 0 & -2 & 0 \\ -2 & 5 & -1 & -2 & 0 \\ 0 & -1 & 1 & 0 & 0 \\ -2 & -2 & 0 & 5 & -1 \\ 0 & 0 & 0 & -1 & 1 \end{bmatrix}. \tag{6.4}$$

Note that all matrices are symmetric, since the graph is undirected.

In this chapter, we will consider only undirected graph structures, where $a_{nm} = a_{mn}$ — both $\mathbf{A}$ and $\mathbf{L}$ are symmetric. A last restriction is in order. In a graph, an edge from a node to itself is called a *self-loop*, which happens whenever $a_{nn} \neq 0$ for some vertex v_n. Since GSP applications are concerned

[2] In general, we could consider $a_{nm} \in \mathbb{C}_*$, but we will restrict ourselves to the real case for the sake of simplicity.

with relations between elements, and self-loops are not useful to represent those relations, it is usual to disregard self-loops.

Graph Signals

A *graph signal* (GS) is a mapping from the graph nodes to scalar values:

$$x : \mathcal{V} \to \mathbb{C} \qquad \text{or} \qquad \mathbf{x} \in \mathbb{C}^{\mathcal{V}}$$

in which x is a signal on the graph whose vertices are $\mathcal{V}$. The difference in typeface from x to $\mathbf{x}$ is to emphasize that the signal can be seen as a function from the domain of vertices to the field of complex numbers; or as a vector with complex-valued entries. In the first case, the value of the signal on the n-th vertex is denoted as $x(v_n)$, and in the second, the n-th entry of the vector is denoted as x_n. In this case, one may loosely refer to node v_n as node n.

It is sometimes convenient to write

$$x : \mathcal{G} \to \mathbb{C} \qquad \text{or} \qquad \mathbf{x} \in \mathbb{C}^{\mathcal{G}}$$

using $\mathcal{G}$ in the place of $\mathcal{V}$, when we want to emphasize that the domain of the signal has the structure of a graph, instead of being just a list of elements with no connectivity information.

Visualizing real signals on one-dimensional domains is usually done via a Cartesian plot: We plot the independent variable (e.g., time) in the axis of abscissae, and the signal values in the axis of ordinates. Graph domains, on the other hand, usually represent, at best, two-dimensional spaces, which makes the Cartesian plots more difficult to depict on two-dimensional media and to interpret them – visualizing 3D plots is always hard. Consider Example 6.2 below.

Example 6.2 Represent the data in Table 6.1 as a graph signal.

Solution

If we consider that there is a link between cities only when their Euclidean distance $D_{\mathrm{E}}(v_n, v_m) = \sqrt{(\Delta x)^2 + (\Delta y)^2}$ is smaller than 3, we then conclude that the bull graph in Figure 6.2 can be used for the abstraction of the current

Table 6.1 Raw data to build a GS for Example 6.2

City (node)	Coordinates (x, y)	Temperature (°C)
v_0	$(0, 0)$	10
v_1	$(1, \sqrt{3})$	15
v_2	$(2, 1+\sqrt{3})$	20
v_3	$(-1, \sqrt{3})$	15
v_4	$(-2, 1+\sqrt{3})$	20

scenario. Although the city distances might be more accurately reported by assigning different values of edge weights for different distances, the temperature values are still not properly represented by the use of the graph structure *alone*, thus requiring a GS representation. In this case, $\mathbf{x} = \begin{bmatrix}10 & 15 & 20 & 15 & 20\end{bmatrix}^{\mathrm{T}}$. This *real* GS can be represented as in Figure 6.3.

Graph Structure Inference

One of the key steps for applying GSP tools to solve actual practical problems is constructing the graph itself. Although some practical applications exhibit an explicit graph structure representation, such as the Konigsberg problem in Figure 6.1, there are cases in which one has a signal dataset obtained from individual objects but does not know how these objects relate to each other, that is, the graph structure remains implicit. This scenario was illustrated in Example 6.2, where we did not have the explicit graph structure for precisely describing the relations among cities, but we followed a set of heuristics and tried to estimate the actual structure by assuming that nearby regions ($D_{\mathrm{E}}(v_n, v_m) \leq 3$) behave in a more similar way than far-away regions.

Depending on which data is available and what kind of restrictions are imposed by the problem at hand, the system designer might be responsible for setting up any number of the following graph features: nodes, links between nodes, and link weights. There is no definite answer on how each design choice influences the overall performance of the resulting GSP techniques.

It is often useful to assign edges only to pairs of vertices whose similarities do not exceed some threshold, for having less edges makes the adjacency matrix sparse, thus reducing the computational burden of the algorithms that will process these matrices, besides making the visualizations cleaner. For example, by considering an N-node sensor network wherein each sensor is taken as a graph vertex v_n, we can adopt the Euclidean distance $D_{\mathrm{E}}(v_n, v_m)$ between vertices v_n and v_m to evaluate their relatedness, since we expect that closer sensors provide a more similar measurement. Alternatively, when each vertex

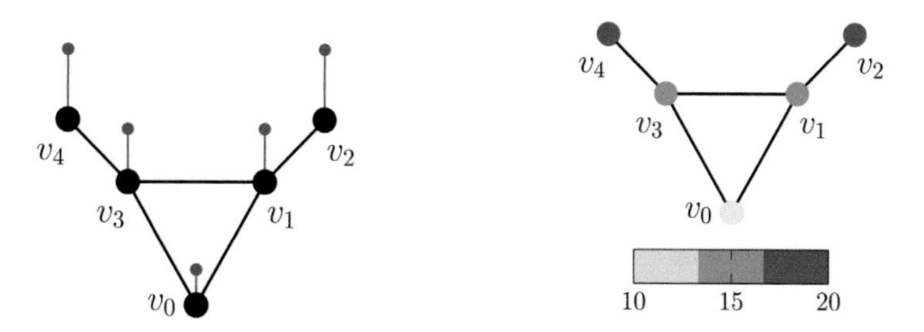

Figure 6.3 Graph signal representations of the data in Table 6.1: (left) bar-coded and (right) color-coded.

is represented by its geographical coordinates in a wide area, it might be useful to consider the *Haversine formula* to evaluate the great-circle distance between points using their latitude and longitude coordinates. For instance, if vertices v_n and v_m have latitudes φ_n^{lat} and φ_m^{lat}, and longitudes ψ_n^{lon} and ψ_m^{lon}, the Haversine distance $D_{\text{H}}(v_n, v_m)$ between these two nodes is given by

$$D_{\text{H}}(v_n, v_m) = 2r_{\text{E}} \arcsin\left\{\sqrt{\sin^2\left(\frac{\varphi_n^{\text{lat}} - \varphi_m^{\text{lat}}}{2}\right) + \cos(\varphi_n^{\text{lat}})\cos(\varphi_m^{\text{lat}})\sin^2\left(\frac{\psi_n^{\text{lon}} - \psi_m^{\text{lon}}}{2}\right)}\right\}, \tag{6.5}$$

where $r_{\text{E}} \approx 6360$ km represents the approximate Earth radius. Since we wish to obtain higher weights for closer nodes, a suitable mapping for evaluating the edge weight a_{nm} is to use the Gaussian kernel weighting function

$$a_{nm} = \text{e}^{-\frac{D^2(v_n, v_m)}{2\theta^2}}, \tag{6.6}$$

where $\theta > 0$ is a parameter that depends on the application. Moreover, in a similar way to the simpler inference method, we only evaluate Equation (6.6) if the distance $D(v_n, v_m)$ – either Euclidean or Haversine – is smaller than a predefined threshold.

Another way of forcing the sparsity of $\mathbf{A}$ is by connecting each node to its K-nearest (most similar) nodes. Implementing this strategy brings some advantages when working with a dataset collected from sensors spread across a vast territory, and when the spatial distribution of sensors is irregular. Consider Example 6.3 below.

Example 6.3 Figure 6.4 illustrates how one can implement a graph inference in practical setups. The Brazilian weather station dataset, employed to generate the graph on the left, contains the latitude and longitude coordinates of active weather stations, besides a monthly average temperature recorded in some of these stations, during the 1961–1990 period. From these data, we obtained a total of $N = 299$ nodes for the graph $\mathcal{G} = (\mathcal{V}, \mathcal{E})$, in which each of these vertices represents a weather station. Thus, a node v_n of the graph $\mathcal{G}$ and its signal value $x(v_n)$ are given, respectively, by the geographical coordinates and the average temperature of the associated weather station in a given month. As we have no explicit connection between weather stations, we are free to design the set $\mathcal{E}$ and choose the adjacency matrix $\mathbf{A}$, inferring the underlying graph structure. As the Brazilian weather stations are irregularly distributed, to prevent a graph with both dense and sparsely populated regions we adopted an approach that constructs the graph edges by connecting a vertex v_n to, at least, its eight closest – in the Haversine-distance sense – neighboring nodes, with edge weights defined as in Equation (6.6) for $\theta = 2 \cdot 10^3$. Thus, based on this procedure, we obtained the Brazilian graph structure/signal displayed in Figure 6.4. Similarly, considering weather stations distributed on the United States' territory, we used the temperature hourly records from the first 95 hours of January 2010 and, after discarding the stations with missing inputs, we obtained a total of $N = 205$ stations in the US mainland. By employing the same methodology as before, but now using seven nearest neighbors, we obtained the US graph structure displayed in Figure 6.4.

In Example 6.3, the graph structure was inferred from some physical side information (latitude and longitude coordinates of weather stations). There are some cases, however, that lack such side information and for which the GSs themselves are employed to infer the graph structure – for example, to infer the adjacency matrix from GSs' correlations; refer to [5–13] for further details.

6.2.2 Graph Fourier Transform

There are several different definitions for the so-called GFT. Two of them are most widely used [1]. They both developed independently throughout the last years and have been established as default mindsets when dealing with GSs. The first approach is based on the *algebraic signal processing* (ASP) theory [14–17] and employs the graph adjacency matrix $\mathbf{A}$ as elementary block by associating it with the *graph shift* operation, from where most of the further developments, covering aspects such as graph filtering, arise [18–21]. The second approach relies on the classical *spectral graph theory* (SGT) [22], a well-established research branch of graph theory concerned with the eigendecomposition of graph characteristic matrices, and uses the graph Laplacian matrix $\mathbf{L}$ for defining the basis of its signal space [23–27]. For further information on the differences between those approaches, the interested reader can refer to [1].

While the ASP-based approach suits better digraph-related problems with possibly complex-valued adjacency weights, the SGT-based approach is very well supported by a vast literature developed over many decades and will be

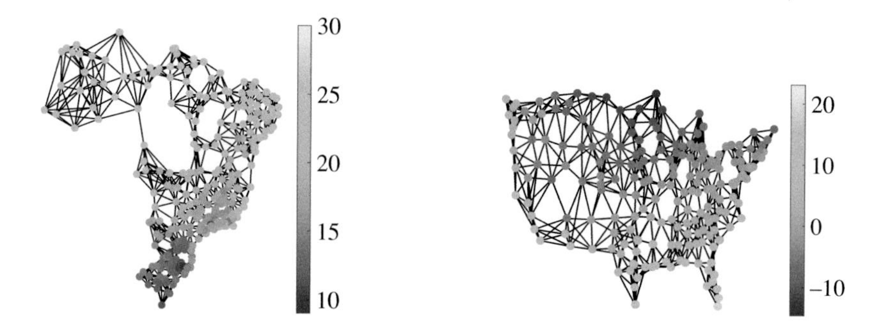

Figure 6.4 Brazilian and US graph signals: (left) July's 1961–1990 average temperatures (in °C) from 299 Brazilian weather stations and (right) hourly temperature (in °C) record on January 1, 2010 at 01:00 from 205 US weather stations.

adopted here for the sake of simplicity, since we will work with *undirected graphs with positive adjacency weights*, that is,

$$a_{mn} = a_{nm} \geq 0.$$

The spectral decomposition of the Laplacian matrix plays an important role in this approach and, therefore, we shall study this matrix in more detail.[3]

The Graph Laplacian: Connection with the Continuous Laplacian Operator

The Laplacian matrix, when seen as a linear transformation on the space of GSs, acts according to the following formula: If $\mathbf{x} \in \mathbb{C}^{\mathcal{G}}$ is a GS, then $\mathbf{Lx}$ is another GS with components given by

$$\begin{aligned} [\mathbf{Lx}]_n &= [(\mathbf{K} - \mathbf{A})\,\mathbf{x}]_n \\ &= k_{nn}x_n - \sum_{m\in\mathcal{N}} a_{nm}x_m \\ &= k_{nn}\left(x_n - \frac{\sum\limits_{m\in\mathcal{N}} a_{nm}x_m}{\sum\limits_{m\in\mathcal{N}} a_{nm}}\right). \end{aligned} \tag{6.7}$$

Each component is proportional to the degree k_{nn}, but the most interesting factor – the one that determines the polarity of $[\mathbf{Lx}]_n$, and also the only one that actually depends on the signal x itself – is the other multiplying factor. This second factor is the difference between the value of the signal x at vertex v_n and the average value of the signal at the neighboring vertices; this average is pondered by the weights of the edges that connect v_n to each of its respective neighbors.

In other words, each sample of the signal $\mathbf{Lx}$ is proportional to the difference between the original signal value at that vertex, and the average signal values at neighboring vertices, pondered by how related they are to the central vertex, as measured by the connection strength. This is similar to the behavior of the Laplacian operator in the continuous domain $\mathbb{R}^N$.

The continuous Laplacian operator in $\mathbb{R}^N$, usually denoted by $\triangle$, ∇^2, or $\nabla \cdot \nabla$, is given by:

$$\begin{gathered} \text{if } f : \mathbb{R}^N \to \mathcal{G} \text{ is twice-continuously differentiable,} \\ \text{then } \triangle f(\mathbf{t}) = \sum_n \frac{\partial^2 f}{\partial t_n^2}(\mathbf{t})\,, \forall \mathbf{t} = \begin{bmatrix} t_0 & \cdots & t_{N-1} \end{bmatrix}^{\mathrm{T}}. \end{gathered}$$

The Laplacian of a continuous function f also yields, at each point $\mathbf{t} \in \mathbb{R}^N$ of the Euclidean space, a measure of how much $f(\mathbf{t})$ differs from the mean value of f at the neighborhood of $\mathbf{t}$, as made precise by the following theorem:

[3] At this point, we focus on the SGT-based GSP. However, the adaptive filtering algorithms described in Section 6.3 also work in the context of ASP-based GSP. The problems in the end of this chapter will explore this aspect.

THEOREM 6.1 *Let $f : \mathbb{R}^N \to \mathbb{R}$ be twice-continuously differentiable. For each $r > 0$, let $S_r(\mathbf{t})$ be the surface of the sphere of radius r centered at $\mathbf{t} \in \mathbb{R}^N$, and let $m(S_r(\mathbf{t}))$ be its area. For a fixed $\mathbf{t}$, let $\nu(r)$ be the mean value of f on this surface, as given by the surface integral*

$$\nu(r) = \frac{1}{m(S_r(\mathbf{t}))} \int_{S_r(\mathbf{t})} f(\mathbf{s}) \, \mathrm{d}S(\mathbf{s}).$$

Then, as r approaches 0 from above (that is, $r \searrow 0$)

$$\nu(r) = f(\mathbf{t}) + \frac{1}{2} \frac{\triangle f(\mathbf{t})}{N} r^2 + o\left(r^2\right),$$

where $o\left(r^2\right)$ represents a function of r that converges to 0 faster than r^2 when $r \to 0$.

Alternatively, stated without the little-o notation:

$$\lim_{r \searrow 0} \frac{f(\mathbf{t}) - \nu(r)}{r^2} = -\frac{1}{2} \frac{\triangle f(\mathbf{t})}{N}.$$

The last formula in the theorem above states that, for small r, the difference between the value of f at some point $\mathbf{t} \in \mathbb{R}^N$ and the average value of f in a neighborhood of $\mathbf{t}$ given by the surface of the sphere of radius r is proportional to minus the Laplacian of f at $\mathbf{t}$:

$$f(\mathbf{t}) - \nu(r) \approx -\frac{r^2}{2N} \triangle f(\mathbf{t}).$$

Using this formula, we see that the continuous Laplacian operator for the Euclidean space is actually analogous to *minus* the Laplacian operator for graphs. If, for some GS $\mathbf{x} \in \mathcal{G}^{\mathcal{G}}$, the Laplacian $[\mathbf{Lx}]_n$ at vertex v_n is *positive*, this means that the value x_n of the signal at v_n is *larger*, on average, than the values at x at neighboring vertices. On the other hand, for a twice-continuously differentiable function in $\mathbb{R}^N$, if its Laplacian $\triangle f(\mathbf{t})$ is positive, then its value $f(\mathbf{t})$ is *lower*, on average, than the values of f in the neighborhood of $\mathbf{t}$.

The expression in Equation (6.7) serves to show that $\mathbf{L}$ is indeed a discrete version of the continuous Laplacian operator, in the sense that both of them *store*, at each point in the domain, a measure of how much the signal value at that point differs from the average signal value in the point's neighborhood. It is in this sense that we call $\mathbf{L}$ a *memory* building block in some particular structures of adaptive graph filters in Section 6.3.

The Graph Laplacian: Spectrum

From Equation (6.7), we note that, if the GS x is a constant signal, then $\mathbf{Lx} = \mathbf{0}$. In addition, since the averages computed in Equation (6.7) are all *local*, meaning that all differences are differences between values of x at connected vertices, if x is constant at each connected component of the graph, then $\mathbf{Lx} = \mathbf{0}$.

This is a recipe for finding eigenvectors of the Laplacian with eigenvalue 0. Suppose that $\mathcal{G}$ has C connected components; that is, the set of vertices $\mathcal{V}$ can be partitioned into C subsets: $\mathcal{V} = \mathcal{V}_0 \cup \mathcal{V}_1 \cup \cdots \cup \mathcal{V}_{C-1}$ in such a way that there are no edges from $\mathcal{V}_i$ to $\mathcal{V}_j$, for $i \neq j$. For each $i \in \{0, 1, \ldots, C-1\}$, we define the i-th indicator signal $\mathbf{x}^{(i)}$ as $x_j^{(i)} = 1$ whenever $v_j \in \mathcal{V}_i$, and $x_j^{(i)} = 0$ otherwise. Then all C indicator signals are orthogonal to one another, and all of them satisfy $\mathbf{L}\mathbf{x}^{(i)} = \mathbf{0}$. Conversely, if $\mathbf{x}$ is such that $\mathbf{L}\mathbf{x} = \mathbf{0}$, then it is possible to show, using mathematical induction on the size of the graph $\mathcal{G}$, that $\mathbf{x}$ is constant at each connected component of the graph.[4] Therefore, the null space of $\mathbf{L}$ has dimension C, and $\left\{\mathbf{x}^{(0)}, \ldots, \mathbf{x}^{(C-1)}\right\}$ is a basis for this null space.

We will usually assume that a graph has only one connected component, in which case its Laplacian matrix has exactly one eigenvalue equal to 0, and its respective eigenvector is a constant signal.

As for the other eigenvalues of $\mathbf{L}$, they are all positive, since the Laplacian is a semi-definite positive matrix. Indeed, for any $\mathbf{x} \in \mathbb{R}^N$, and using Equation (6.7), we have

$$\begin{aligned}
\mathbf{x}^{\mathrm{T}}\mathbf{L}\mathbf{x} &= \sum_n k_{nn} x_n^2 - \sum_n \sum_m a_{nm} x_n x_m \\
&= \sum_n \sum_m a_{nm} \left(x_n^2 - x_n x_m\right) \\
&= \frac{1}{2} \sum_n \sum_m a_{nm} \left(x_n^2 - x_n x_m\right) + \frac{1}{2} \sum_m \sum_n a_{mn} \left(x_m^2 - x_m x_n\right) \\
&= \frac{1}{2} \sum_n \sum_m a_{nm} \left(x_n^2 - 2x_n x_m + x_m^2\right) \\
&= \frac{1}{2} \sum_n \sum_m a_{nm} \left(x_n - x_m\right)^2 \geq 0 .
\end{aligned} \tag{6.8}$$

As $\mathbf{L}$ is symmetric, it has a complete set of eigenvectors and eigenvalues. The eigenvalues can be ordered as $0 = \lambda_0 < \lambda_1 \leq \cdots \leqslant \lambda_{N-1}$ for a connected graph. Let $\mathbf{u}_n \in \mathbb{R}^N$ be the orthonormal eigenvector of $\mathbf{L}$ corresponding to λ_n. Since $\mathbf{L}$ is a real symmetric matrix, then one can write

$$\mathbf{L} = \mathbf{U}\boldsymbol{\Lambda}\mathbf{U}^{\mathrm{T}} , \tag{6.9}$$

with $\boldsymbol{\Lambda} = \operatorname{diag}\{\lambda_0, \cdots, \lambda_{N-1}\}$ and $\mathbf{U}^{-1} = \mathbf{U}^{\mathrm{T}}$.

The GFT

One of the first problems to which Fourier analysis was applied was in solving partial differential equations, the simplest of which was the steady-state heat

[4] This is a manifestation of the finite character of graphs. In infinite (and connected) domains, a signal can have null Laplacian, but still be nonconstant. In the realm of continuous functions on the Euclidean space, for instance, such signals are called harmonic functions, and finding all of them given boundary conditions amounts to solving a partial differential equation called the Laplace equation.

equation, also called the Laplace equation: $\triangle f(\mathbf{t}) = 0$. As in any linear equation, it is instructive to search for eigenvalues and eigenvectors of the linear operators that arise in the problem. In the case for the one-dimensional Laplacian operator, for example, the eigenfunctions are the so-called complex exponentials

$$\triangle \left\{ \mathrm{e}^{\mathrm{j}2\pi\xi t} \right\} = -\left(2\pi\xi\right)^2 \mathrm{e}^{\mathrm{j}2\pi\xi t}. \tag{6.10}$$

Therefore, the eigenvalues, $-\left(2\pi\xi\right)^2$, of the Laplacian operator are related to the *frequencies* ξ. The Fourier transform on $\mathbb{R}$ is exactly the change of basis that diagonalizes the Laplacian operator $\triangle$, which yields, for each frequency ξ, the inner product between a signal f and the complex exponential $u_\xi(t) = \mathrm{e}^{\mathrm{j}2\pi\xi t}$:

$$\hat{f}(\xi) = \langle f, u_\xi \rangle = \int_{\mathbb{R}} f(t)\,\mathrm{e}^{-\mathrm{j}2\pi\xi t} \mathrm{d}t. \tag{6.11}$$

Although it is not straightforward to define the analog of complex exponentials for GSs, the Laplacian operator can be exploited to this end. The aforementioned analogy between $\mathbf{L}$ and $-\triangle$ will be used to transform traditional DSP tools into GSP tools.

The first thing to do in order to make this connection is to define a Fourier transform for signals on graphs. If $\mathbf{x} \in \mathbb{C}^{\mathcal{G}}$ is a GS, we mimic the definition for continuous time, and define the GFT of $\mathbf{x}$ to be the vector $\hat{\mathbf{x}} \in \mathbb{C}^N$ with components

$$\hat{x}(\lambda_n) = \hat{x}_n = \langle \mathbf{x}, \mathbf{u}_n \rangle = \mathbf{u}_n^{\mathrm{T}} \mathbf{x}. \tag{6.12}$$

Using this definition, the vector $\hat{\mathbf{x}}$ can be directly calculated as

$$\hat{\mathbf{x}} = \mathbf{U}^{\mathrm{T}} \mathbf{x} \quad \text{(analysis)}. \tag{6.13}$$

Thus, $\mathbf{U}^{\mathrm{T}}$ is called the GFT matrix. Since $\mathbf{U}$ is orthogonal, the inverse GFT is

$$\mathbf{x} = \sum_{n \in \mathcal{N}} \hat{x}_n \mathbf{u}_n = \mathbf{U}\hat{\mathbf{x}} \quad \text{(synthesis)}. \tag{6.14}$$

As a toy-example, consider the GS in Example 6.2, $\mathbf{x} = [10 \;\; 15 \;\; 20 \;\; 15 \;\; 20]^{\mathrm{T}}$, of the five-node graph structure, which we represent by the Laplacian matrix $\mathbf{L}_a$ in Equation (6.3). We compute the orthonormal eigenvectors matrix $\mathbf{U}$ and obtain the respective frequency-domain representation $\hat{\mathbf{x}}$ as

$$\hat{\mathbf{x}} = \begin{bmatrix} -0.4472 & 0.0000 & -0.4865 & 0.7505 & 0.0000 \\ -0.4472 & 0.1133 & -0.2979 & -0.4596 & 0.6980 \\ -0.4472 & 0.6980 & 0.5412 & 0.0843 & -0.1133 \\ -0.4472 & -0.1133 & -0.2979 & -0.4596 & -0.6980 \\ -0.4472 & -0.6980 & 0.5412 & 0.0843 & 0.1133 \end{bmatrix}^{\mathrm{T}} \begin{bmatrix} 10 \\ 15 \\ 20 \\ 15 \\ 20 \end{bmatrix} = \begin{bmatrix} -35.7771 \\ 0.0000 \\ 7.8445 \\ -2.9093 \\ 0.0000 \end{bmatrix}. \tag{6.15}$$

We observe that the GS $\mathbf{x}$ presents an interesting property: It can be represented by a reduced number of non-null frequency components. This useful "sparsity" property in an alternative domain can be exploited in order to provide a more compact representation of the original signal – see Section 6.2.4.

Frequency Interpretation

In the GFT construction, the graph Laplacian eigenvalues λ_n are analog to frequencies in traditional DSP, whereas the eigenvectors $\mathbf{u}_n$ are analog to the complex exponentials. From Equation (6.8), we know that

$$\frac{1}{2}\sum_i\sum_j a_{ij}\left(U_{ni}-U_{nj}\right)^2 = \mathbf{u}_n^{\mathrm{T}}\mathbf{L}\mathbf{u}_n = \mathbf{u}_n^{\mathrm{T}}\left(\lambda_n\mathbf{u}_n\right) = \lambda_n\|\mathbf{u}_n\|_2^2 = \lambda_n\,, \quad (6.16)$$

which means that the n-th eigenvalue corresponds to a weighted measure of variation of the GS $\mathbf{u}_n$. For instance, $\lambda_0 = 0$ corresponds to $\mathbf{u}_0 = \frac{1}{\sqrt{N}}\mathbf{1}_N$.

A way of assessing the frequency of classical sinusoids is to measure how often the signal changes polarity; that is, to measure how often the signal changes from positive samples to negative samples and vice versa. In a graph setting, we can simply count the number of edges connecting samples with different polarity, as illustrated in Example 6.4.

Example 6.4 Figure 6.5 shows the result of an experiment in which we measure the rate of change of polarity of various eigenvectors of the Laplacian operator for a 50-node randomly generated graph. Edge weights are assigned using the Gaussian kernel function and the Euclidean distance, and the underlying $\mathbf{A}$ is sparsified using a threshold for the distance between points. The edge weights are scaled to force $\lambda_{N-1} = 1$. The first plot shows the generated graph. Edge colors are drawn in grayscale to represent the weights. It can be seen that edge weight values are not so diverse. The rest of the plots show color-coded plots of the Laplacian eigenvector GSs. As mentioned above, the first eigenvector is always constant. All other eigenvectors, being orthogonal to the first, show both positive and negative entries, since the entries' sum must be 0. We say that an edge is a *zero-crossing edge* if it connects a positive sample to a negative one. In the eigenvector signal plots, we have drawn zero-crossing edges in red, solid lines, and nonzero-crossing edges in faint-gray dotted lines. It can be seen that the higher the eigenvalue, the more zero-crossing edges there are, corroborating the interpretation of the eigenvalues λ_n as frequencies.

Differences between GSP and DSP

There are important differences between traditional DSP and GSP. Indeed, the irregular nature of graphs can make it difficult to translate traditional DSP tools into the GSP framework. One important feature of Euclidean domains is that the algebraic structure of the real field $\mathbb{R}$ gives rise to *algebraic asymmetries* – that is, special numbers with unique properties around which the algebraic structure is formed.

The number $0 \in \mathbb{R}$ is one such a number, sometimes called the *origin* in the real number line. It serves special purpose in classical Fourier analysis

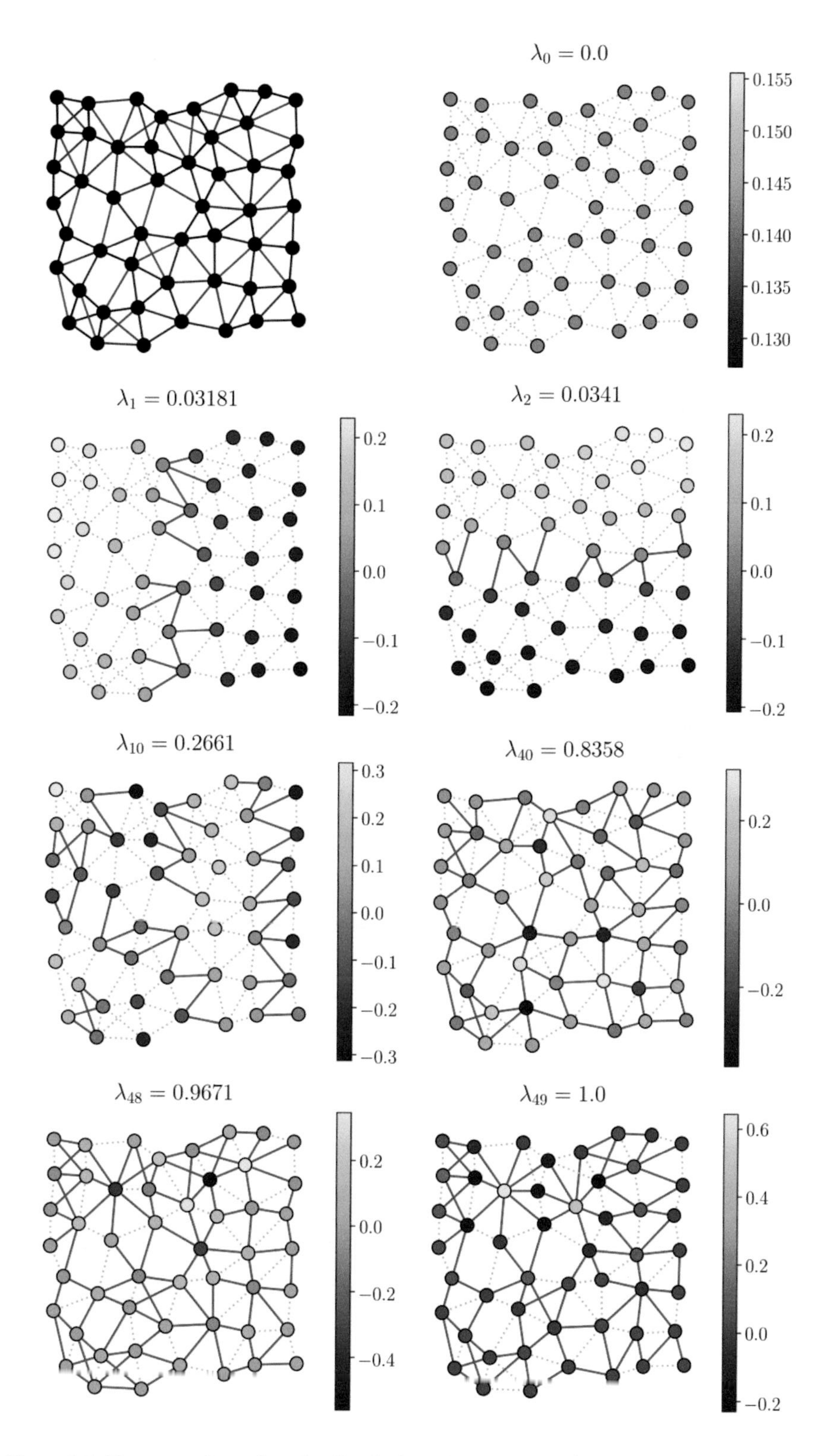

Figure 6.5 Zero-crossing edges for Laplacian eigenvectors of Example 6.4.

also: whenever a time-domain signal exhibits some kind of symmetry around 0, for that is an even or odd signal, it will also exhibit special properties in the frequency domain. For instance, if a signal is shifted away from 0 in the time domain, its frequency-domain representation becomes modulated. When we define a signal on the frequency domain and apply the inverse Fourier transform to derive the time-domain representation, zero also plays a special role. For instance, if we define a signal with frequency component equal to 1 for all frequencies – clearly a construction that is agnostic with respect to domain elements – the resulting time-domain signal is an impulse signal supported *exactly* on 0.

The examples we provided in the above paragraph rely on the fact that the very definition of the Fourier transform uses the origin of the time domain as a special element. Indeed, the orthonormal basis of eigenvectors of the continuous Laplacian is not unique – each eigenvector $u_\xi(t)$ could be multiplied by some constant $c_\xi \in \mathbb{C}$ on the unit circle ($|c_\xi| = 1$), and the resulting basis $\{c_\xi u_\xi(t)\}_{\xi \in \mathbb{R}}$ would still be an orthonormal eigenvector basis for the continuous Laplacian operator; given any such basis, the Fourier transform is defined by choosing these constants c_ξ in such a way that the value of any of the basis signals at the origin $t = 0$ is exactly 1. This makes $t = 0$ a somehow privileged point that is taken as reference for normalization, giving rise to many properties related to what happens around the time-domain origin.

When the domain is a graph, however, there is no algebraic structure giving rise to special domain elements. For an arbitrary graph $\mathcal{G}$, there is simply no systematic way to pick one "central" vertex. This is a fundamental limitation on the analogies between GSP and traditional DSP tools. For example, the problem of choosing one of the many orthonormal basis of eigenvectors of the graph Laplacian based on the values of such signals on one specific domain element is harder, because there is no central vertex we can use to inform our decision on the direction of each eigenvector. Indeed, for each eigenvector $c_n \mathbf{u}_n$, we need to choose between $c_n = +1$ or $c_n = -1$, since graph Laplacian eigenvectors are required to be real. The best we could do is choose one vertex v_m – arbitrarily, or perhaps based on the application – and choose the direction of each eigenvector $c_n \mathbf{u}_n$ in such a way that $c_n U_{mn} \geq 0$ for all $n \in \mathcal{N}$.

Example 6.5 shows the result of an experiment that explores this issue.

Example 6.5 Using the same graph of Example 6.4, choose the Laplacian eigenvectors $c_n \mathbf{u}_n$ so as to mimic the Euclidean-domain phenomenon of constructing an impulse signal out of in-phase complex exponentials.

Solution

Figure 6.6 details the solution to the problem.

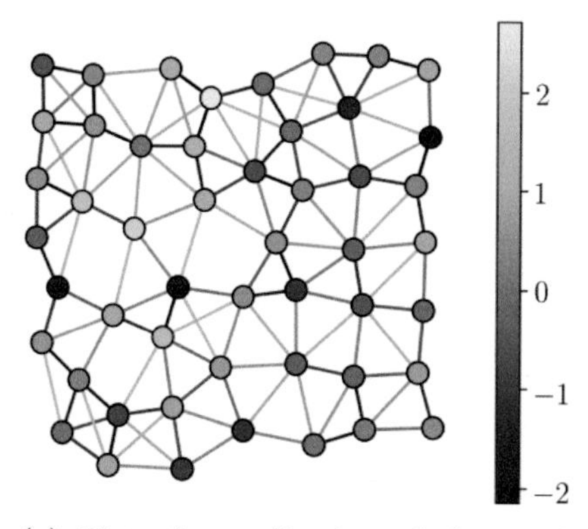

(a) Plot of a realization of the random graph signal $\mathbf{Uc} = \sum_n c_n \mathbf{u}_n$, where the coefficients c_n are independent and identically distributed, drawn randomly from the set $\{-1, +1\}$. This is a simulation of a sum of out-of-phase complex exponentials.

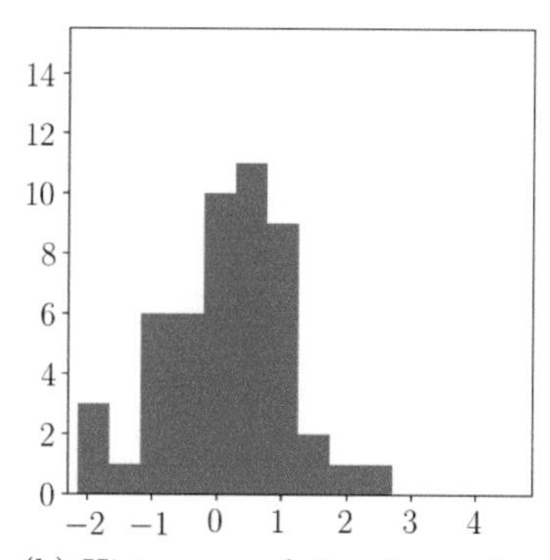

(b) Histogram of signal sample values for the graph signal on the left. The values are clustered around 0, showing no indication of a single outlier sample (stemming from a constructive sum $\sum_n c_n U_{mn}$) that could be interpreted as an impulse.

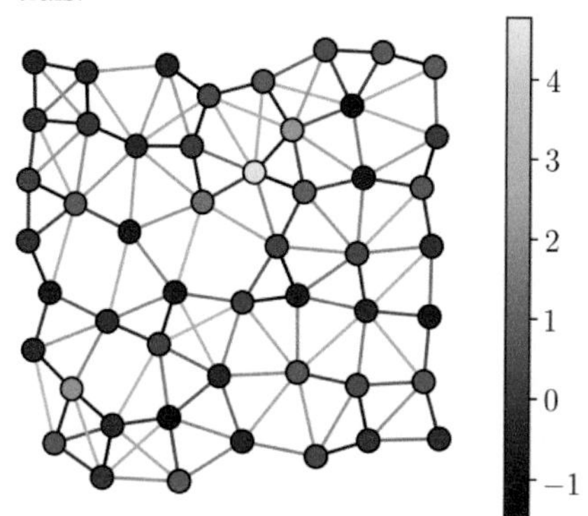

(c) Plot of the signal $\mathbf{Uc} = \sum_n c_n \mathbf{u}_n$, wherein the coefficients c_n are deterministically chosen to make each $c_n \mathbf{u}_n$ positive at one specific (arbitrarily determined) node m, colored in yellow in the figure. This simulates in-phase complex-exponentials, since $c_n = \text{sign}\{U_{mn}\}$, so that a constructive sum $\sum_n c_n U_{mn}$ is indeed obtained.

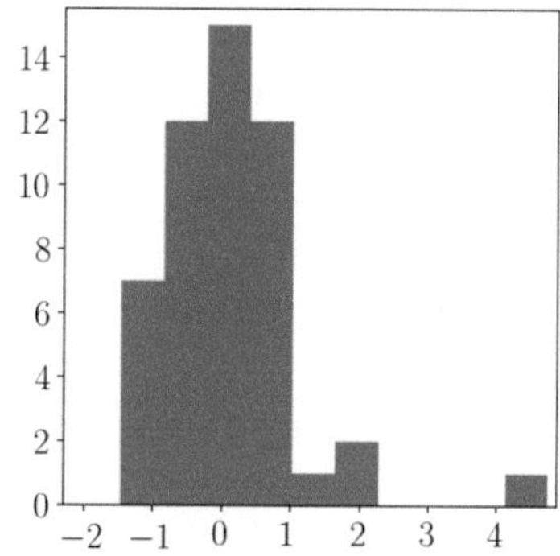

(d) Histogram of signal sample values for the graph signal on the left. The variance of the values around 0 is smaller than before, and this time there is clearly an outlier (a node m with signal value larger than 4), indicating that this signal could be interpreted as an approximation of an impulse signal. Here, the node m on which the impulse is supported is responsible for $\approx 46\%$ of the GS energy.

Figure 6.6 An attempt to construct an impulse signal.

Despite these striking differences between traditional DSP and GSP, as graph frequencies λ_n are real numbers, they are more amenable to algebraic manipulations and, therefore, can serve as a starting point for translating traditional DSP tools into the GSP framework. We will now explore a few more operations on GSs that can be borrowed from the traditional DSP toolset.

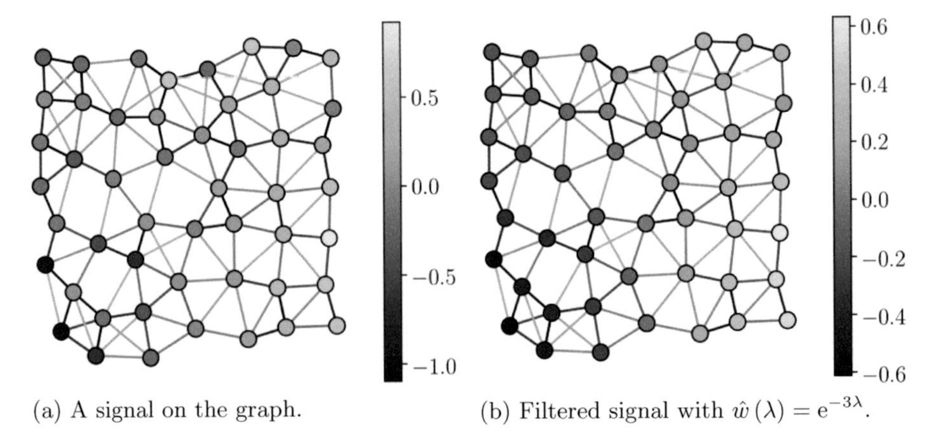

(a) A signal on the graph. (b) Filtered signal with $\hat{w}(\lambda) = e^{-3\lambda}$.

Figure 6.7 An example of frequency-domain filtering.

6.2.3 Graph Filters

Frequency-domain filtering consists in decomposing a signal into its Fourier components, and then choosing which components to keep, which to enhance, and which to discard. More specifically, we take a signal $f(t)$, multiply its Fourier transform $\hat{f}(\xi)$ by a filter function $\hat{h}(\xi)$, and then take the inverse Fourier transform of the resulting product. This is a simple enough specification of an operation that can be translated into the GSP framework without much effort.

We call any continuous function $\hat{w} : \mathbb{R}_+ \to \mathbb{C}$ a *graph filter*. Given any graph $\mathcal{G}$ and its Laplacian's eigendecomposition $\mathbf{L} = \mathbf{U}\boldsymbol{\Lambda}\mathbf{U}^{\mathrm{T}}$, we define the following linear transformation corresponding to the *graph filter* $\hat{w}$:[5]

$$\mathbf{W} = \hat{w}(\mathbf{L}) = \mathbf{U}\hat{w}(\boldsymbol{\Lambda})\mathbf{U}^{\mathrm{T}} = \mathbf{U}\begin{bmatrix} \hat{w}(\lambda_0) & & & \\ & \hat{w}(\lambda_1) & & \\ & & \ddots & \\ & & & \hat{w}(\lambda_{N-1}) \end{bmatrix}\mathbf{U}^{\mathrm{T}}. \tag{6.17}$$

The action of $\mathbf{W}$ on a GS $\mathbf{x}$ – yielding $\mathbf{W}^{\mathrm{H}}\mathbf{x}$ – is to apply successively the matrices $\mathbf{U}^{\mathrm{T}}, \hat{w}^*(\boldsymbol{\Lambda}), \mathbf{U}$ to $\mathbf{x}$. These linear transformations represent, respectively, a change of basis from x to the frequency-domain $\hat{x}(\lambda_n)$, a frequency-wise multiplication of each $\hat{x}(\lambda_n)$ by the filter gain $\hat{w}^*(\lambda_n)$, and a change of basis back to the graph domain. Notice that $\mathbf{W}^{\mathrm{H}}\mathbf{u}_n = \hat{w}^*(\lambda_n)\mathbf{u}_n$, meaning that the Fourier basis are invariant under graph filtering. Figure 6.7 illustrates an example of graph filtering.

[5] We shall focus on linear-in-parameters adaptive filters here. The interested reader can find further information on nonlinear adaptive filters, for instance, in [28–32]

If the graph filter $\hat{w}(\lambda)$ is a polynomial $w_0 + w_1\lambda + \cdots + w_Q\lambda^Q$, then the corresponding linear transformation $\mathbf{W}$ can be written as a polynomial in $\mathbf{L}$:

$$\begin{aligned}\mathbf{W} &= \mathbf{U}\left[\sum_q w_q \mathbf{\Lambda}^q\right]\mathbf{U}^{\mathrm{T}} \\ &= \sum_q w_q \mathbf{U}\mathbf{\Lambda}^q\mathbf{U}^{\mathrm{T}} \\ &= \sum_q w_q \mathbf{L}^q \,. \end{aligned} \tag{6.18}$$

As shown in Equation (6.7), the Laplacian matrix acts upon a GS as a *difference operator* that calculates the difference between the signal value on a node and the weighted average signal value of the corresponding neighboring nodes. Thus, we can interpret $\mathbf{L}$ as a graph *memory/energy storage* elementary building block and $\mathbf{W} = w_0\mathbf{I} + w_1\mathbf{L} + \cdots + w_Q\mathbf{L}^Q$ as description of a graph system in terms of this memory block. Sometimes, $\mathbf{L}$ is called a graph *shift* in the GSP literature, motivated by the similar role played by the adjacency matrix $\mathbf{A}$ in the developments based on ASP.

As a final remark, note that graph filters (as defined above) commute:

$$\begin{aligned}\mathbf{W}_a^{\mathrm{H}}\left(\mathbf{W}_b^{\mathrm{H}}\mathbf{x}\right) &= \mathbf{U}\hat{w}_a^*\left(\mathbf{\Lambda}\right)\mathbf{U}^{\mathrm{T}}\left(\mathbf{U}\hat{w}_b^*\left(\mathbf{\Lambda}\right)\mathbf{U}^{\mathrm{T}}\mathbf{x}\right) \\ &= \mathbf{U}\hat{w}_b^*\left(\mathbf{\Lambda}\right)\hat{w}_a^*\left(\mathbf{\Lambda}\right)\mathbf{U}^{\mathrm{T}}\mathbf{x} \\ &= \mathbf{U}\hat{w}_b^*\left(\mathbf{\Lambda}\right)\mathbf{U}^{\mathrm{T}}\left(\mathbf{U}\hat{w}_a^*\left(\mathbf{\Lambda}\right)\mathbf{U}^{\mathrm{T}}\mathbf{x}\right) \\ &= \mathbf{W}_b^{\mathrm{H}}\left(\mathbf{W}_a^{\mathrm{H}}\mathbf{x}\right), \forall \mathbf{x} \in \mathbb{C}^{\mathcal{G}} \,. \end{aligned} \tag{6.19}$$

6.2.4 Sampling and Reconstruction of Graph Signals

Motivated by several practical applications – see the discussions in Section 6.3.1 –, the problem of reconstructing GSs from their sampled noisy measurements was the inspiration for the first attempt to merge the well-established adaptive filtering field with the emerging area of GSP, culminating in the development of LMS-, NLMS-, and RLS-based algorithms for GS interpolation.

We can extend the idea of bandlimited signals to graph structures and say that a GS $\mathbf{x} \in \mathbb{C}^{\mathcal{G}}$ is *bandlimited* or *spectrally sparse (ssparse)* when its frequency representation $\hat{\mathbf{x}}$ given by the GFT in Equation (6.13) is sparse, as in the example in Equation (6.15). Taking the *support* or *frequency set* $\mathcal{F}$ as an index subset of $\mathcal{N}$ ($\mathcal{F} \subseteq \mathcal{N}$), a GS $\mathbf{x}$ is said $\mathcal{F}$-ssparse when the components $\hat{x}(\lambda_n)$, with indices $n \in \mathcal{N}\backslash\mathcal{F}$, are null, in which $\mathcal{N}\backslash\mathcal{F}$ denotes the difference between sets $\mathcal{N}$ and $\mathcal{F}$. For example, the support in Equation (6.15) is $\mathcal{F} = \{1, 3, 4\}$.

Moreover, if we collect the nonzero entries of $\hat{\mathbf{x}}$ to form the vector $\hat{\mathbf{x}}_{\mathcal{F}} \in \mathbb{C}^{|\mathcal{F}|}$ and the corresponding columns of $\mathbf{U}$ to form the reduced matrix $\mathbf{U}_{\mathcal{F}} \in \mathbb{R}^{N\times|\mathcal{F}|}$, we can then write

$$\mathbf{x} = \mathbf{U}_{\mathcal{F}}\hat{\mathbf{x}}_{\mathcal{F}} \,. \tag{6.20}$$

As in the case of bandlimited time signals that can be sampled and reconstructed with no loss of information, we describe now a similar result for GSs. Sampling is the operation of collecting only a limited number of values from the GS, whose reading positions are determined by the *sampling set* $\mathcal{S} \subseteq \mathcal{V}$. In this context, let $\mathbf{D}_{\mathcal{S}} \in \mathbb{R}^{N \times N}$ denote a diagonal matrix with diagonal entries $d_{nn}^{(\mathcal{S})} = 1$ if $v_n \in \mathcal{S}$ and $d_{nn}^{(\mathcal{S})} = 0$ otherwise. Thus, we can write the sampled vector $\mathbf{x}_{\mathcal{S}} \in \mathbb{C}^N$ as

$$\mathbf{x}_{\mathcal{S}} = \mathbf{D}_{\mathcal{S}} \mathbf{x} . \tag{6.21}$$

If we use an *interpolation matrix* $\mathbf{\Phi} \in \mathbb{R}^{N \times N}$ to perfectly recover any $\mathcal{F}$-ssparse GS $\mathbf{x}$ from its sampled version $\mathbf{x}_{\mathcal{S}}$, that is $\mathbf{x} = \mathbf{\Phi} \mathbf{D}_{\mathcal{S}} \mathbf{x}$ for all $\mathcal{F}$-ssparse $\mathbf{x}$, we have to compute

$$\mathbf{\Phi} = \mathbf{U}_{\mathcal{F}} \left(\mathbf{U}_{\mathcal{F}}^{\mathrm{T}} \mathbf{D}_{\mathcal{S}} \mathbf{U}_{\mathcal{F}} \right)^{-1} \mathbf{U}_{\mathcal{F}}^{\mathrm{T}} , \tag{6.22}$$

which requires matrix $\left(\mathbf{U}_{\mathcal{F}}^{\mathrm{T}} \mathbf{D}_{\mathcal{S}} \mathbf{U}_{\mathcal{F}} \right)$ to be full rank. Therefore, we conclude that perfect reconstruction of an $\mathcal{F}$-ssparse GS from its sampled version $\mathbf{x}_{\mathcal{S}}$ is possible as long as the chosen sampling set $\mathcal{S}$ guarantees that

$$\operatorname{rank}(\mathbf{D}_{\mathcal{S}} \mathbf{U}_{\mathcal{F}}) = |\mathcal{F}| . \tag{6.23}$$

As $\operatorname{rank}(\mathbf{D}_{\mathcal{S}}) = |\mathcal{S}|$, we conclude that a necessary condition for perfect recovery of a sampled GS is that $|\mathcal{S}| \geq |\mathcal{F}|$, that is, the number of samples retained must be at least the amount of nonzero frequency components of $\hat{\mathbf{x}}$. This condition is not sufficient, though, and for guaranteeing perfect reconstruction the sampling set $\mathcal{S}$ must be chosen in such a way that Equation (6.23) is satisfied. This fact clarifies the importance of an adequate choice for the sampling set $\mathcal{S}$ and its connection to the graph structure $\mathbf{U}_{\mathcal{F}}$.

Sampling Set Selection

So far we have considered recovering a bandlimited GS $\mathbf{x}$ from sampled values $\mathbf{x}_{\mathcal{S}}$ and any choice of sampling set $\mathcal{S}$ respecting condition Equation (6.23) results in a perfect recovery of the original GS when applying the interpolation matrix in Equation (6.22). However, in a practical situation where one acquires *noisy* data, we expect that the performance of the reconstruction method will depend on the sampling set.

Based on this assumption, a more adequate modeling for a practical GS is

$$\boldsymbol{d} = \boldsymbol{x} + \boldsymbol{\eta} , \tag{6.24}$$

where $\boldsymbol{d}$ is the noisy random GS,[6] $\boldsymbol{x} \in \mathbb{C}^N$ is the original bandlimited GS, and $\boldsymbol{\eta}$ is a zero-mean noise vector with covariance matrix $\mathbf{C} \in \mathbb{C}^{N \times N}$.

[6] Here, $\boldsymbol{x}$ denotes a random vector with realizations denoted as $\mathbf{x}$.

Algorithm 6.38 Greedy algorithm for sampling set selection

Initialization

N' (number of sampled nodes $\leq N$)

$\mathcal{S} \leftarrow \emptyset$

while $|\mathcal{S}| < N'$ **do**

$n' = \underset{n \in \mathcal{N}}{\text{argmax}} \ \lambda_{\min}(\mathbf{U}_{\mathcal{F}}^{\mathrm{T}} \mathbf{D}_{\mathcal{S} \cup \{n\}} \mathbf{U}_{\mathcal{F}})$

$\mathcal{S} \leftarrow \mathcal{S} + \{n'\}$

end

return $\mathcal{S}$

Considering the model in Equation (6.24), we can evaluate the quality of a GS estimate $\mathbf{y}$ through the *mean squared deviation* (MSD):

$$\text{MSD} = \mathbb{E}\left\{\|\boldsymbol{y} - \mathbf{x}\|_2^2\right\}. \tag{6.25}$$

It is straightforward to verify that an estimate $\mathbf{y}$ based on the linear interpolation procedure $\mathbf{\Phi D}_{\mathcal{S}}$ results in

$$\text{MSD}(\mathcal{S}) = \mathbb{E}\left\{\| \left(\mathbf{U}_{\mathcal{F}}^{\mathrm{T}} \mathbf{D}_{\mathcal{S}} \mathbf{U}_{\mathcal{F}}\right)^{-1} \mathbf{U}_{\mathcal{F}}^{\mathrm{T}} \mathbf{D}_{\mathcal{S}} \boldsymbol{\eta}\|_2^2\right\}, \tag{6.26}$$

where we recalled that $\mathbf{U}_{\mathcal{F}}^{\mathrm{T}} \mathbf{U}_{\mathcal{F}} = \mathbf{I}_{|\mathcal{F}|}$.

The MSD expression in Equation (6.26) presents an explicit dependency on $\mathbf{D}_{\mathcal{S}}$, which indicates that the sampling set $\mathcal{S}$ could be properly chosen to reduce the MSD. Nonetheless, the resulting optimization problem of choosing N' indices from $\mathcal{N} = \{0, 1, ..., N-1\}$ to form the set $\mathcal{S}$ is combinatorial in nature, hindering its applicability for large N. Hence, in order to obtain an interesting trade-off between reasonable reconstruction performance and time required for calculating the sampling set $\mathcal{S}$, a common approach is to employ a *greedy* algorithm for minimizing a specific *figure of merit* (FoM). A greedy algorithm basically reduces the overall computational complexity by searching for an optimal selection at each stage and expecting to find a near-optimal final value.

An example of a greedy scheme for sampling set selection is presented in Algorithm 6.38. It employs at each iteration a greedy search for the index $n \in \mathcal{N}$ to be added to the current set $\mathcal{S}$ in order to maximize the minimum eigenvalue of $\mathbf{U}_{\mathcal{F}}^{\mathrm{T}} \mathbf{D}_{\mathcal{S} \cup \{n\}} \mathbf{U}_{\mathcal{F}}$, denoted as $\lambda_{\min}(\mathbf{U}_{\mathcal{F}}^{\mathrm{T}} \mathbf{D}_{\mathcal{S} \cup \{n\}} \mathbf{U}_{\mathcal{F}})$.

Example 6.6 shows the result of an experiment that applies Algorithm 6.38. For more information on sampling and reconstruction of GSs, including probabilistic sampling, the reader may refer to [33–46].

Example 6.6 Calculate how many frequency components are necessary for representing the GSs depicted in Figure 6.4 with reasonable error.

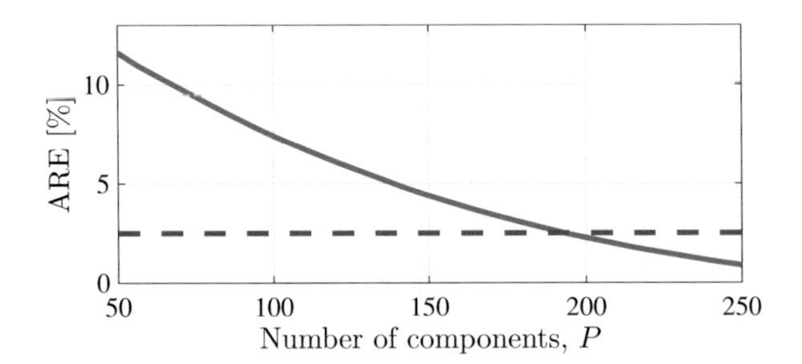

Figure 6.8 Average reconstruction error when the original signal is compressed using P components. The dashed curve shows the 2.5% threshold chosen for this example.

Solution

Let us start with the Brazilian GS. As the GS $\mathbf{x}$ obtained from temperature measurements across the country presents a smooth behavior (despite minor outlier points), we expect $\mathbf{x}$ to be approximately bandlimited. To solve the problem, we start solving a signal-compression problem, in which we evaluate the average reconstruction error (ARE) $\|\mathbf{x} - \mathbf{x}_P\|_2/\|\mathbf{x}_P\|_2$ for different estimates $\mathbf{x}_P$ using only the P-largest frequency components of the original bandlimited GS $\mathbf{x}$, taken as the signal from the July dataset depicted in Figure 6.4. Then, the signal $\mathbf{x}_P$ is obtained by sorting the absolute values of the frequency-domain signal $\hat{\mathbf{x}}$ obtained from Equation (6.13) and selecting the indices p of the P-largest components $|\hat{x}(\lambda_n)|$ to form the auxiliary set $\mathcal{F}_P$. Based on these indices $p \in \mathcal{F}_P \subseteq \mathcal{N}$, we pick the p-th eigenvector of $\mathbf{U}$ and the p-th frequency component of $\hat{\mathbf{x}}$ to define $\mathbf{U}_{\mathcal{F}_P}$ and $\hat{\mathbf{x}}_{\mathcal{F}_P}$. Then, the estimate $\mathbf{x}_P$ using P components is given by $\mathbf{U}_{\mathcal{F}_P}\hat{\mathbf{x}}_{\mathcal{F}_P}$.

Following this compression procedure, we compute the ARE percentage for different values of components P and display the results for the range $[42, 250]$ in Figure 6.8. Assuming that a deviation error of 2.5% is acceptable in the current application, we approximate the original GS by its $P = 200$ largest frequency components. From this assumption we can define the bandlimited set $\mathcal{F} = \mathcal{F}_P$, where $|\mathcal{F}| = P$.

Based on this approximately $\mathcal{F}$-ssparse signal $\mathbf{x}$, we need to take a practical project decision and select both the amount $|\mathcal{S}|$ and which vertices $v_n \in \mathcal{V}$ of the GS should be sampled. Increasing the number of samples in $\mathcal{S}$ always decreases the MSD in Equation (6.25). However, as we also want to reduce the amount of nodes to be measured, we consider that $|\mathcal{S}| = 210$ provides a reasonable trade-off and then find the sampling set $\mathcal{S}$ by using Algorithm 6.38, with $N' = |\mathcal{S}| = 210$. Then, at this point we obtain the reference bandlimited GS $\mathbf{x}$, the frequency set $\mathcal{F}$, the sampling set $\mathcal{S}$, and their respective matrices $\mathbf{U}_{\mathcal{F}}$ and $\mathbf{D}_{\mathcal{S}}$.

The same approach can be applied to the US GS depicted in Figure 6.4. In this case, if we assume an acceptable reconstruction error of 5%, we found that $|\mathcal{F}| = 125$ resulted in a reasonable approximation. We therefore used $|\mathcal{S}| = 130$.

6.3 Adaptive Graph Filtering

In this section, we introduce the two main families of adaptive graph filtering algorithms, viz.: the TML class of algorithms and the *interpolation* class of algorithms.[7] Both families rely on the GSP tools described in Section 6.2 and are comprised of graph counterparts of the traditional LMS, NLMS, and RLS algorithms, being respectively called graph LMS (GLMS), graph LNMS (GNLMS), and graph RLS (GRLS) algorithms.

6.3.1 Tapped Memory Line and Interpolation Approaches

As pointed out before, the main difficulty in translating traditional DSP techniques and algorithms into the GSP framework is to pose the technique in such a way that a generalization becomes apparent. This is true of filtering – which, as we have seen, is more easily handled in the frequency domain given the available order of the real eigenvalues of the Laplacian matrix – and even more so when dealing with adaptive filters, which rely heavily on the regularity of the signal domain. To this date, the path for designing adaptive filtering techniques for GSP is not crystal clear. Since the field of signal processing on graphs is still incipient, we verify that many works are still ad hoc in nature.

A common place of the majority of works in this field is to bypass some of the difficulties of dealing with the transformation of time domain to graph domain by working with both domains simultaneously. This approach allows us to consider a linear and uniform time dimension, while modeling the irregular dimension using graphs. In this way, the GSP theory can be applied to the *snapshots* of the resulting multivariable signal

$$\mathbf{x}(\cdot) : \mathbb{Z} \to \mathbb{C}^{\mathcal{G}}$$

since each sample in time – when thinking in terms of traditional adaptive filtering – has now become a whole GS $\mathbf{x}(k) \in \mathbb{C}^{\mathcal{G}}$, instead of just a scalar value.

Most readers acquainted with traditional adaptive filtering would expect that the basic block diagram for adaptive graph filtering would be the one depicted in Figure 6.9, in which $\mathbf{x}(k), \mathbf{y}(k), \mathbf{d}(k), and\ \mathbf{e}(k) \in \mathbb{C}^{\mathcal{G}}$, respectively represent the input, output, desired, and error GSs of the graph filter. From this basic structure, many applications analogous to the traditional system identification, signal enhancement, deconvolution/equalization, and signal prediction could then be derived. However, most efforts in the literature actually concentrate on a rather different structure, as the one depicted in Figure 6.10, wherein $\boldsymbol{\eta}(k) \in \mathbb{C}^{\mathcal{G}}$ denotes the noise measurement; of course, this block diagram is

[7] This taxonomy is actually new; the reader will not find such a classification in the research papers in the literature. Instead, most research papers concentrate on the interpolation class, but naming the underlying task as GS *estimation* or *reconstruction*.

motivated by some common practical interest in specific applications of GSs, as will be explained later on.

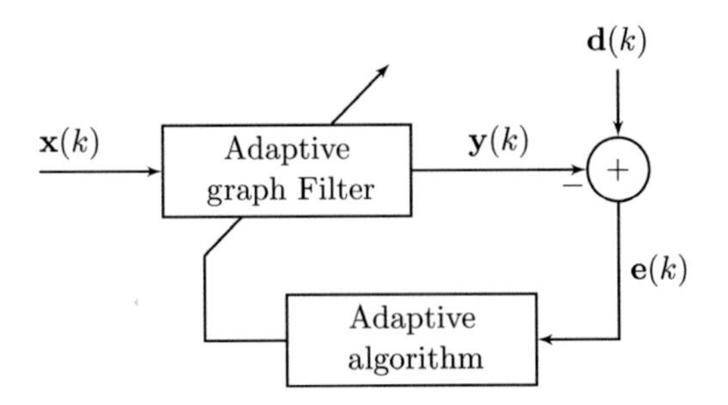

Figure 6.9 General block diagram for TML-based filtering.

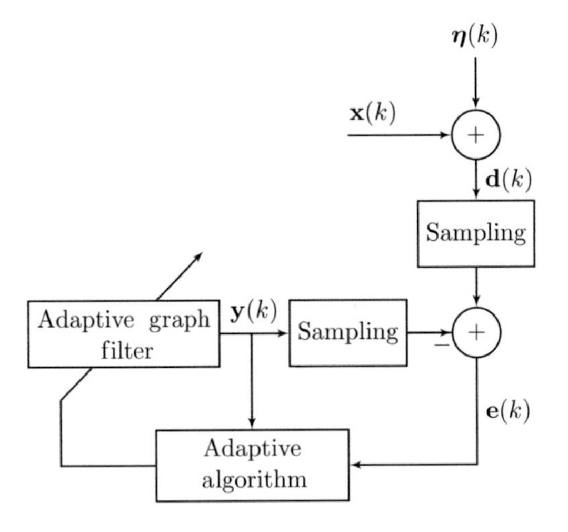

Figure 6.10 Block diagram for interpolation-based filtering.

In the following pages, we work out the details underlying the diagrams of Figures 6.9 and 6.10 and explain how these diagrams are related to the tapped-memory-line-based and interpolation-based classes of adaptive graph filtering algorithms.

TML

When dealing with discrete-time scalar signals, the most widely used structure for traditional adaptive filtering employs FIR filters with transversal realizations, called *tapped delay lines* (TDLs), as illustrated in Figure 6.11. In this case, the input vector $\begin{bmatrix} x(k) & \cdots & x(k-M) \end{bmatrix}^{\mathrm{T}}$ of the adaptive filter is defined

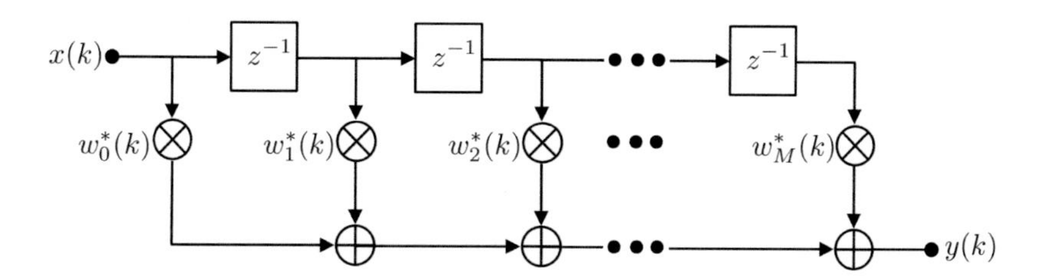

Figure 6.11 TDL structure for traditional adaptive filtering.

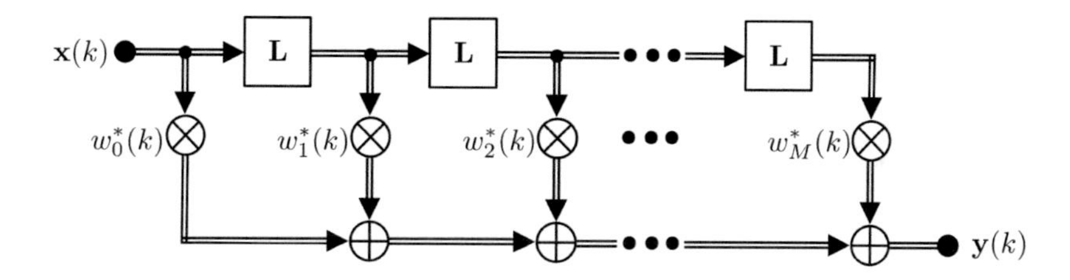

Figure 6.12 TML structure for adaptive graph filtering.

in such a way that the overall process corresponds to a "linear" convolution[8] between the streaming input signal $x(k)$ and the time-varying adaptive filter parameterized by

$$W(z,k) = \sum_{m=0}^{M} w_m(k)\left(z^{-1}\right)^m , \tag{6.27}$$

where M is the *filter order*. In other words, the input signal is "linearly" filtered by the adaptive system.

It is possible to generalize the TDL concept for GSs by defining an FIR adaptive graph filter, based on the polynomial filter definition in Equation (6.18), which is employed to implement a "linear" *graph* filtering over a snapshot $\mathbf{x}(k)$, as illustrated in Figure 6.12. The graph filter may change for each snapshot yielding the time-varying graph filter with *parameter matrix*

$$\mathbf{W}(k) = \sum_{m=0}^{M} w_m(k)\mathbf{L}^m . \tag{6.28}$$

We call this scheme TML approach, wherein the Laplacian matrix $\mathbf{L}$ plays the role of the *memory* building block.

As compared to the traditional TDL, the main modifications to form a TML structure are: (i) the input signal of the structure, at iteration k, is no longer

[8] To be more precise, this process is not linear because the parameter vector of the adaptive filter is computed using the input signal. Anyway, we consider this somehow loose usage of the term "linear" convolution helps grasp the concept better than otherwise using many other words just to describe more precisely the underlying processing.

a scalar sample, but rather a vector $\mathbf{x}(k) \in \mathbb{C}^{\mathcal{G}}$, which is a GS corresponding to a snapshot of the multivariable signal $\mathbf{x}(\cdot)$ – the use of double-line arrows in Figure 6.12 highlights this aspect; (ii) the delays are replaced by the memory building block of a graph filter, that is, the Laplacian matrix $\mathbf{L}$. The remaining modifications are compatible with these two main changes.

Given an input GS $\mathbf{x}(k) \in \mathbb{C}^{\mathcal{G}}$, the output GS $\mathbf{y}(k) \in \mathbb{C}^{\mathcal{G}}$ is computed as

$$\begin{aligned}
\mathbf{y}(k) &= \mathbf{W}^{\mathrm{H}}(k)\mathbf{x}(k) \\
&= \sum_{m=0}^{M} w_m^*(k)\mathbf{L}^m\mathbf{x}(k) \\
&= \underbrace{\begin{bmatrix}\mathbf{x}(k) & \mathbf{L}\mathbf{x}(k) & \cdots & \mathbf{L}^M\mathbf{x}(k)\end{bmatrix}}_{=\mathbf{X}^{\mathrm{T}}(k)\in\mathbb{C}^{N\times(M+1)}} \cdot \underbrace{\begin{bmatrix} w_0^*(k) \\ w_1^*(k) \\ \vdots \\ w_M^*(k)\end{bmatrix}}_{=\mathbf{w}^*(k)\in\mathbb{C}^{(M+1)}} \\
&= \mathbf{X}^{\mathrm{T}}(k)\mathbf{w}^*(k),
\end{aligned} \tag{6.29}$$

wherein $\mathbf{X}(k)$ is the *input matrix*, which plays the same role of the input vector in a standard TDL adaptive filtering scheme: the $(m+1)$-th row of $\mathbf{X}(k)$, given by $\mathbf{x}^{\mathrm{T}}(k)\mathbf{L}^{m+1} = \left[\mathbf{x}^{\mathrm{T}}(k)\mathbf{L}^m\right]\mathbf{L}$, is obtained from the m-th row, $\mathbf{x}^{\mathrm{T}}(k)\mathbf{L}^m(k)$, with $m \in \{0, \ldots, M-1\}$.

Given a desired GS $\mathbf{d}(k) \in \mathbb{C}^{\mathcal{G}}$, the (*a priori*) error GS signal $\mathbf{e}(k) \in \mathbb{C}^{\mathcal{G}}$ is

$$\begin{aligned}
\mathbf{e}(k) &= \mathbf{d}(k) - \mathbf{y}(k) \\
&= \mathbf{d}(k) - \mathbf{X}^{\mathrm{T}}(k)\mathbf{w}^*(k).
\end{aligned} \tag{6.30}$$

The TML framework is fully compliant with the general diagram in Figure 6.9, and it can be tailored to be used in several particular scenarios, including the analogous of system identification, signal enhancement, deconvolution/equalization, and signal prediction. Figure 6.13 exemplifies a general graph system identification setup. In this case, it is assumed that the desired GS is a noisy filtered version of the input GS, that is,

$$\begin{aligned}
\mathbf{d}(k) &= \mathbf{W}_{\mathrm{o}}^{\mathrm{H}}(k)\mathbf{x}(k) + \boldsymbol{\eta}(k) \\
&= \mathbf{X}^{\mathrm{T}}(k)\mathbf{w}_{\mathrm{o}}^*(k) + \boldsymbol{\eta}(k),
\end{aligned} \tag{6.31}$$

and the goal is to track the *actual* system parameter matrix

$$\mathbf{W}_{\mathrm{o}}(k) = \sum_{m=0}^{M} w_{\mathrm{o},m}(k)\mathbf{L}^m, \tag{6.32}$$

or equivalently, the actual parameter vector $\mathbf{w}_{\mathrm{o}}(k) = \begin{bmatrix} w_{\mathrm{o},0} & \cdots & w_{\mathrm{o},M}\end{bmatrix}^{\mathrm{T}}$, which is assumed to be unknown.

In the GSP context, this system identification setup might be interesting for coding GS patterns using $M+1$ coefficients, instead of N node attributes per

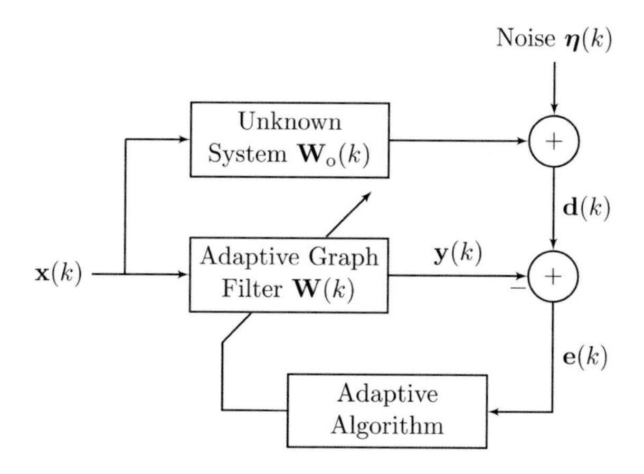

Figure 6.13 Block diagram of a graph system identification.

iteration, whenever $N \gg M+1$. It is therefore useful for signal modeling, where $\mathbf{x}(k)$ works as an excitation GS.

We can now derive TML-based adaptive graph filtering algorithms by using the above definitions. But, before doing so, let us introduce the interpolation class of algorithms illustrated in Figure 6.10.

Interpolation

Although the TML approach follows naturally from the definition of polynomial graph filters in Equation (6.18), the most widely used framework in the literature for developing adaptive graph filters tackles the problem of interpolating sampled GSs from a different perspective.

As mentioned before, graphs are particularly useful for modeling networks. Real-world networks can be huge, and attributes of a number of nodes may be unavailable, or deliberately undisclosed due to privacy matters [47]. Further, some practical networks change their set of nodes over time, so that new nodes may come to exist without enough time to sharing their node attributes with the central unit for processing. The interpolation problem, therefore, naturally arises in these and many other practical applications, including semi-supervised learning of categorical data, ranking problems, and matrix-completion problems [34]. All of these applications rely on predicting the property of some nodes (e.g., class, ranking, or function value) considering the knowledge of signal values at neighboring nodes, or in other words, *interpolating* those values. The quality of the interpolation depends on the smoothness of the GS, which can be assessed taking into account the similarity between nodes captured by link-weights in the graph – there is an underlying assumption that the graph model is accurate. For instance, in social networks, people tend to rate movies similar to their friends, and in financial networks, companies that trade with each other usually belong to the same category.

The appeal for using adaptive filtering ideas in the GS interpolation problem arises from the expected benefits of enabling online reconstruction and tracking time-varying GSs in noisy environments. Although traditional interpolation methods like kriging [48] can be used for handling signal inference in irregular domains with static reference values, adaptive algorithms seem more suitable for dealing with dynamic GSs due to their relative reduced complexity and benefit of online estimation.

An interpolation structure for adaptive graph filtering is illustrated in Figure 6.10. One can note that this structure differs a lot from the general structure of TML-based filtering depicted in Figure 6.9 or from the specific system identification setup in Figure 6.13, although some parallels can be traced back. Basically, the adaptive estimation of GSs deals with a scenario where one intends to recover a *bandlimited* – or approximately bandlimited in practice – GS $\mathbf{x}(k) \in \mathbb{C}^{\mathcal{G}}$, with frequency set $\mathcal{F} \subseteq \mathcal{N}$, from a reduced set $\mathcal{S}(k) \subseteq \mathcal{V}$ of noisy measurements drawn from $\mathbf{d}(k) = \mathbf{x}(k) + \boldsymbol{\eta}(k)$ – recall the definitions in Section 6.2.4. Since only the node signals indexed by $\mathcal{S}(k)$ are acquired at iteration k, the reference measurements at time instant k are in fact $\mathbf{D}_{\mathcal{S}(k)}\mathbf{d}(k)$.

We define the (*a priori*) error vector $\mathbf{e}(k) \in \mathbb{C}^{\mathcal{G}}$ as the difference between the measured reference GS $\mathbf{D}_{\mathcal{S}(k)}\mathbf{d}(k)$ and the respective positions of the current estimate $\mathbf{y}(k)$, that is,

$$\begin{aligned}\mathbf{e}(k) &= \mathbf{D}_{\mathcal{S}(k)}\mathbf{d}(k) - \mathbf{D}_{\mathcal{S}(k)}\mathbf{y}(k) \\ &= \mathbf{D}_{\mathcal{S}(k)}\left[\mathbf{d}(k) - \mathbf{U}_{\mathcal{F}}\hat{\mathbf{y}}_{\mathcal{F}}(k)\right],\end{aligned} \tag{6.33}$$

where $\hat{\mathbf{y}}_{\mathcal{F}}(k) \in \mathbb{C}^{|\mathcal{F}|}$ is the frequency-domain estimate of $\mathbf{y}(k)$, as given in Equation (6.20).

Unlike the TML framework, the adaptive filter is not explicitly parameterized in the interpolation-based scheme. In addition, the graph filtering algorithm employs only some *ancillary information* related to the GS properties, such as the knowledge of the sampling and frequency sets $\mathcal{S}(k)$ and $\mathcal{F}$, respectively. In other words, the filtering process does not act upon a given *input* GS and the only external driving force of the algorithm is dictated by the reference GS $\mathbf{D}_{\mathcal{S}(k)}\mathbf{d}(k)$.

We now have all the basic definitions to start the derivations of the adaptive algorithms, which rely on the following assumptions:

- A1 The graph structure is given via the knowledge of $\mathbf{L}$ and/or $\mathbf{U}$.
- A2 The graph structure is fixed.
- A3 For the interpolation-based algorithms, the frequency set $\mathcal{F}$ of the GS is fixed and known beforehand.
- A4 For the interpolation-based algorithms, the sampling set $\mathcal{S}(k)$ is given at iteration k.

Assumption A1 means that we are not inferring the underlying graph structure. Regarding assumption A2, it means we are not considering the applications where nodes appear/disappear as time goes by or when the link-weights are time varying. However, some of the results can be straightforwardly adapted to address these cases. As for assumption A3, it requires that some prior information about the data residing on the graph nodes must be available/inferred beforehand. Finally, assumption A4 needs further discussion. Many works in the literature *design* the sampling set along with the adaptive filter – see, for instance, [49–51]. In other words, the sampling process is an integral part of their methodology. This is motivated by some applications, like wireless sensor networks (WSNs), where it is reasonable to assume that recording a sample value from a sensor requires power, and therefore, we would like to minimize the amount of sensor readings at each time instant. Nevertheless, in many other applications, the decision on which nodes to sample relies on external factors that are beyond the designer control, such as the cost for obtaining the data acquired by a given node. For instance, consider that one wishes to collect data from a distributed sensor network but some of its autonomous nodes are difficult to reach. In these cases, selecting the sampling set $\mathcal{S}(k)$ is out of the scope of the adaptive system designer, and the assumption of having $\mathcal{S}(k)$ known *a priori* sounds reasonable. An even more appealing example is for pure interpolation processes: consider that we have a GS and we assume this signal corresponds to a sampled version of a signal defined over a (virtual) denser graph. The idea in this case is to find the signal values in new points of this denser graph. We use the same theory of GS estimation, but there is no sampling per se. This type of application might also find special interest in sensor networks.

From now on, we will present the adaptive graph filtering algorithms for each approach, starting with the TML-based schemes. Given the popularity of the interpolation-based algorithms, we will provide more information about this class of algorithms, especially concerning their computational complexity and steady-state error analyses.

6.3.2 The TML-Based GLMS Algorithm

The TML-Based graph counterpart of the LMS algorithm [52–55], called GLMS, aims to minimize the cost-function $\|\mathbf{d}(k) - \mathbf{X}^{\mathrm{T}}(k)\mathbf{w}^*\|_2^2$ via a steepest-descent approach, giving rise to the following update equation:

$$\mathbf{w}(k+1) = \mathbf{w}(k) + \mu\mathbf{X}(k)\mathbf{e}^*(k)\,, \tag{6.34}$$

in which $\mu > 0$ is the *convergence factor* of the algorithm.

The factor μ is a design parameter whose purpose is to control the trade-off between increasing the convergence speed and reducing the steady-state error. Considering the system identification setup in Figure 6.13, we can think of $\boldsymbol{w}(k)$ as an estimator for the actual system response $\mathbf{w}_{\mathrm{o}}(k)$.[9] We can therefore

[9] An *estimator* is a random variable whose realizations are called *estimates*. Once again, we are denoting multivariable random sequences by boldface italic letters, whereas their realizations are not in italic – for example, the random sequence $\boldsymbol{w}(k)$ and its realization $\mathbf{w}(k)$.

Algorithm 6.39 The TML-based GLMS algorithm

Initialization

$\mathbf{w}(0) = \mathbf{0}_{M+1}$

$\mu \in \left(0, \frac{2}{\lambda_{\max}(\mathbf{R})}\right)$

For $k \geq 0$ **do**

$\mathbf{e}(k) = \mathbf{d}(k) - \mathbf{X}^{\mathrm{T}}(k)\mathbf{w}^*(k)$

$\mathbf{w}(k+1) = \mathbf{w}(k) + \mu\mathbf{X}(k)\mathbf{e}^*(k)$

end

choose μ so as to induce an *asymptotically unbiased* estimator for a *fixed* system response $\mathbf{w}_{\mathrm{o}}$. Indeed, as $\boldsymbol{d}(k) = \boldsymbol{X}^{\mathrm{T}}(k)\mathbf{w}_{\mathrm{o}}^* + \boldsymbol{\eta}(k)$, then by defining

$$\Delta\boldsymbol{w}(k) = \boldsymbol{w}(k) - \mathbf{w}_{\mathrm{o}}, \tag{6.35}$$

we can rewrite Equation (6.34) as

$$\Delta\boldsymbol{w}(k+1) = \left[\mathbf{I}_{M+1} - \mu\boldsymbol{X}(k)\boldsymbol{X}^{\mathrm{H}}(k)\right]\Delta\boldsymbol{w}(k) + \mu\boldsymbol{X}(k)\boldsymbol{\eta}^*(k)\,. \tag{6.36}$$

If we assume that the multivariable random sequences $\boldsymbol{\eta}(k)$ and $\boldsymbol{x}(k)$ are independent and wide-sense stationary (WSS),[10] with $\mathbf{R} = \mathbb{E}\left\{\boldsymbol{X}(k)\boldsymbol{X}^{\mathrm{H}}(k)\right\}$ and $\mathbb{E}\{\boldsymbol{\eta}(k)\} = \mathbf{0}$, then we have

$$\begin{aligned}\mathbb{E}\{\Delta\boldsymbol{w}(k+1)\} &= \left[\mathbf{I} - \mu\mathbb{E}\left\{\boldsymbol{X}(k)\boldsymbol{X}^{\mathrm{H}}(k)\right\}\right]\mathbb{E}\{\Delta\boldsymbol{w}(k)\} + \mu\mathbb{E}\{\boldsymbol{X}(k)\}\,\mathbb{E}\{\boldsymbol{\eta}^*(k)\}\\ &= (\mathbf{I} - \mu\mathbf{R})^{k+1}\,\mathbb{E}\{\Delta\boldsymbol{w}(0)\}\,. \end{aligned} \tag{6.37}$$

Hence, we guarantee that $\mathbb{E}\{\Delta\boldsymbol{w}(k+1)\} \xrightarrow{k\to\infty} \mathbf{0}$ no matter the initial condition as long as

$$0 < \mu < \frac{2}{\lambda_{\max}(\mathbf{R})}\,. \tag{6.38}$$

The TML-based GLMS scheme is displayed in Algorithm 6.39.

6.3.3 The TML-Based GNLMS Algorithm

The TML-based graph counterpart of the NLMS algorithm, called GNLMS, can be obtained from the GLMS by optimizing the convergence factor so as to increase the convergence speed of the algorithm. The idea is to choose the convergence factor to minimize the energy of the *a posteriori* error

$$\boldsymbol{\varepsilon}(k) = \mathbf{d}(k) - \mathbf{X}^{\mathrm{T}}(k)\mathbf{w}^*(k+1)\,. \tag{6.39}$$

Unlike the standard NLMS algorithm that deals with scalar errors, in the GSP case we have a vector error, which gives us more degrees of freedom for this optimization. To take advantage of this added degrees of freedom, we

[10] A multivariable random sequence $\boldsymbol{x}(k)$ is wide-sense stationary (WSS) when $\mathbb{E}\{\boldsymbol{x}(k)\} = \bar{\mathbf{x}}$ and $\mathbb{E}\left\{\boldsymbol{x}(k)\boldsymbol{x}^{\mathrm{H}}(k+\kappa_0)\right\} = \mathbf{R}_x(\kappa_0)$.

generalize the convergence factor by using a convergence matrix $\mathbf{M}(k)$, which can be time varying. In the GLMS, we can identify $\mathbf{M}(k)$ as a fixed multiple of the identity matrix, that is, $\mathbf{M}(k) = \mu\mathbf{I}$. The dimension of the identity matrix depends on how we write the update equation:

- GLMS update (i): $\mathbf{w}(k+1) = \mathbf{w}(k) + \mathbf{X}(k)\mathbf{M}(k)\mathbf{e}^*(k)$, with $\mathbf{M}(k) = \mu\mathbf{I}_N$;
- GLMS update (ii): $\mathbf{w}(k+1) = \mathbf{w}(k) + \mathbf{M}(k)\mathbf{X}(k)\mathbf{e}^*(k)$, with $\mathbf{M}(k) = \mu\mathbf{I}_{M+1}$.

Thus, to derive the update equation of the GNLMS, we consider $\mathbf{M}(k)$ as a general convergence matrix and start from one of the forms of the GLMS update equation in order to minimize $\|\boldsymbol{\varepsilon}(k)\|_2^2$ with respect to $\mathbf{M}(k)$.

Let us start with the update equation

$$\mathbf{w}(k+1) = \mathbf{w}(k) + \mathbf{X}(k)\mathbf{M}(k)\mathbf{e}^*(k)\,. \tag{6.40}$$

The *a posteriori* error can be rewritten in function of the *a priori* error:

$$\boldsymbol{\varepsilon}(k) = \left[\mathbf{I} - \mathbf{X}^{\mathrm{T}}(k)\mathbf{X}^*(k)\mathbf{M}^*(k)\right]\mathbf{e}(k)\,. \tag{6.41}$$

Note that the matrix $\mathbf{M}(k)$ that minimizes $\|\boldsymbol{\varepsilon}(k)\|_2^2 \geq 0$ is

$$\mathbf{M}(k) = \left[\mathbf{X}^{\mathrm{H}}(k)\mathbf{X}(k)\right]^{-1}, \tag{6.42}$$

as long as the input matrix $\mathbf{X}(k)$ has full rank, with N linearly independent columns, thus requiring $M+1 \geq N$. In this case, the update equation, including a normalized convergence factor $\mu_{\mathrm{n}} > 0$, is

$$\mathbf{w}(k+1) = \mathbf{w}(k) + \mu_{\mathrm{n}}\mathbf{X}(k)\left[\mathbf{X}^{\mathrm{H}}(k)\mathbf{X}(k)\right]^{-1}\mathbf{e}^*(k)\,. \tag{6.43}$$

As the *a posteriori* error is zeroed, then the same update equation can be derived by solving the optimization problem

$$\begin{aligned} &\underset{\mathbf{w}}{\text{minimize}} && \|\mathbf{w} - \mathbf{w}(k)\|_2^2\,,\\ &\text{subject to} && \mathbf{d}(k) - \mathbf{X}^{\mathrm{T}}(k)\mathbf{w}^* = \mathbf{0}\,. \end{aligned} \tag{6.44}$$

Let us now consider the other possibility for the update equation

$$\mathbf{w}(k+1) = \mathbf{w}(k) + \mathbf{M}(k)\mathbf{X}(k)\mathbf{e}^*(k), \tag{6.45}$$

which yields

$$\boldsymbol{\varepsilon}(k) = \left[\mathbf{I} - \mathbf{X}^{\mathrm{T}}(k)\mathbf{M}^*(k)\mathbf{X}^*(k)\right]\mathbf{e}(k)\,, \tag{6.46}$$

so that

$$\begin{aligned}\|\boldsymbol{\varepsilon}(k)\|_2^2 =& \|\mathbf{e}(k)\|_2^2 + \mathbf{e}^{\mathrm{H}}(k)\mathbf{X}^{\mathrm{T}}(k)\left[-\mathbf{M}^*(k) - \mathbf{M}^{\mathrm{T}}(k)\right.\\ &\left.+\mathbf{M}^{\mathrm{T}}(k)\mathbf{X}^*(k)\mathbf{X}^{\mathrm{T}}(k)\mathbf{M}^*(k)\right]\mathbf{X}^*(k)\mathbf{e}(k).\end{aligned} \tag{6.47}$$

To select the matrix $\mathbf{M}(k)$ that minimizes $\|\boldsymbol{\varepsilon}(k)\|_2^2$, we take the derivative of the above equation with respect to $\mathbf{M}^{\mathrm{H}}(k)$ and set the result to $\mathbf{0}$, yielding

$$\left[-\mathbf{I} + \mathbf{X}^*(k)\mathbf{X}^{\mathrm{T}}(k)\mathbf{M}^*(k)\right]\mathbf{X}^*(k)\mathbf{e}(k)\mathbf{e}^{\mathrm{H}}(k)\mathbf{X}^{\mathrm{T}}(k) = \mathbf{0}, \tag{6.48}$$

so that

$$\mathbf{M}(k) = \left[\mathbf{X}(k)\mathbf{X}^{\mathrm{H}}(k)\right]^{-1}, \tag{6.49}$$

as long as the input matrix $\mathbf{X}(k)$ has full rank, with $M+1$ linearly independent rows, thus requiring $M+1 \leq N$. In this case, the update equation, including a normalized convergence factor $\mu_{\mathrm{n}} > 0$, is

$$\mathbf{w}(k+1) = \mathbf{w}(k) + \mu_{\mathrm{n}}\left[\mathbf{X}(k)\mathbf{X}^{\mathrm{H}}(k)\right]^{-1}\mathbf{X}(k)\mathbf{e}^{*}(k). \tag{6.50}$$

In summary, the GNLMS update equation is

$$\mathbf{w}(k+1) = \begin{cases} \mathbf{w}(k) + \mu_{\mathrm{n}}\mathbf{X}(k)\left[\mathbf{X}^{\mathrm{H}}(k)\mathbf{X}(k)\right]^{-1}\mathbf{e}^{*}(k) & \text{for } M+1 \geq N \\ \mathbf{w}(k) + \mu_{\mathrm{n}}\left[\mathbf{X}(k)\mathbf{X}^{\mathrm{H}}(k)\right]^{-1}\mathbf{X}(k)\mathbf{e}^{*}(k) & \text{for } M+1 \leq N \end{cases}. \tag{6.51}$$

The first case, when $M+1 \geq N$, tends to be much rarer than the second case in current GSP applications, in which the number of vertexes is quite large, like in big data applications.

The normalized factor μ_{n} can also be set up to achieve an asymptotically unbiased estimator in a system identification task. Indeed, considering the case $N \geq M+1$, we can write

$$\Delta\boldsymbol{w}(k+1) = (1-\mu_{\mathrm{n}})\,\Delta\boldsymbol{w}(k) + \mu_{\mathrm{n}}\left[\boldsymbol{X}(k)\boldsymbol{X}^{\mathrm{H}}(k)\right]^{-1}\boldsymbol{X}(k)\boldsymbol{\eta}^{*}(k), \tag{6.52}$$

thus yielding

$$\mathbb{E}\left\{\Delta\boldsymbol{w}(k+1)\right\} = (1-\mu_{\mathrm{n}})^{k+1}\,\mathbb{E}\left\{\Delta\boldsymbol{w}(0)\right\}. \tag{6.53}$$

The desired condition is met when

$$0 < \mu_{\mathrm{n}} < 2. \tag{6.54}$$

Unlike the GLMS, the GNLMS does not require the knowledge of second-order moments of the input GS to set up the convergence factor.

The TML-based GNLMS scheme is displayed in Algorithm 6.40 for the case $N \geq M+1$.

Algorithm 6.40 The TML-based GNLMS algorithm

Initialization

  $\mathbf{w}(0) = \mathbf{0}_{M+1}$

  $\mu_{\mathrm{n}} \in (0, 2)$

For $k \geq 0$ **do**

  $\mathbf{e}(k) = \mathbf{d}(k) - \mathbf{X}^{\mathrm{T}}(k)\mathbf{w}^{*}(k)$

  $\mathbf{w}(k+1) = \mathbf{w}(k) + \mu_{\mathrm{n}}\left[\mathbf{X}(k)\mathbf{X}^{\mathrm{H}}(k)\right]^{-1}\mathbf{X}(k)\mathbf{e}^{*}(k)$

end

6.3.4 The TML-Based GRLS Algorithm

The TML-based graph counterpart of the RLS algorithm, called *graph RLS* (GRLS), can be obtained by solving the unconstrained problem

$$\underset{\mathbf{w}}{\text{minimize}} \quad \sum_{i=0}^{k} \beta^{k-i} \|\mathbf{d}(i) - \mathbf{X}^{\mathrm{T}}(i)\mathbf{w}^*\|_2^2 \,, \tag{6.55}$$

whose solution is

$$\mathbf{w}(k+1) = \mathbf{R}^{-1}(k)\mathbf{p}(k)\,, \tag{6.56}$$

in which

$$\begin{aligned}\mathbf{R}(k) &= \sum_{i=0}^{k} \beta^{k-i}\mathbf{X}(i)\mathbf{X}^{\mathrm{H}}(i) \\ &= \beta\mathbf{R}(k-1) + \mathbf{X}(k)\mathbf{X}^{\mathrm{H}}(k)\,, \end{aligned} \tag{6.57}$$

$$\begin{aligned}\mathbf{p}(k) &= \sum_{i=0}^{k} \beta^{k-i}\mathbf{X}(i)\mathbf{d}^*(i) \\ &= \beta\mathbf{p}(k-1) + \mathbf{X}(k)\mathbf{d}^*(k)\,, \end{aligned} \tag{6.58}$$

with β being denominated the *forgetting factor.*

The GRLS unconstrained problem is convex only when $\beta > 0$. Moreover, for WSS inputs, we have

$$\mathbb{E}\left\{\boldsymbol{R}(k)\right\} = \frac{1-\beta^{k+1}}{1-\beta}\mathbf{R}\,, \tag{6.59}$$

which converges when $\beta < 1$. In fact, when $\beta = 1$, one has $\mathbb{E}\left\{\boldsymbol{R}(k)\right\} = (k+1)\mathbf{R}$ and $\mathbb{E}\left\{\boldsymbol{p}(k)\right\} = (k+1)\mathbb{E}\left\{\boldsymbol{X}(i)\boldsymbol{d}^*(i)\right\} = (k+1)\mathbf{p}$, so that Equation (6.56) can still be stably computed with proper normalization when $\beta = 1$ in steady state. Therefore, we have to choose the forgetting factor in the interval

$$0 < \beta \leq 1. \tag{6.60}$$

Moreover, we can rewrite Equation (6.56) as

$$\begin{aligned}\mathbf{R}(k)\mathbf{w}(k+1) &= \beta\mathbf{p}(k-1) + \mathbf{X}(k)\mathbf{d}^*(k) \\ &= \beta\mathbf{R}(k-1)\mathbf{w}(k) + \mathbf{X}(k)\mathbf{d}^*(k) \\ &= \left[\mathbf{R}(k) - \mathbf{X}(k)\mathbf{X}^{\mathrm{H}}(k)\right]\mathbf{w}(k) + \mathbf{X}(k)\mathbf{d}^*(k)\,, \end{aligned} \tag{6.61}$$

yielding the following alternative update equation:

$$\mathbf{w}(k+1) = \mathbf{w}(k) + \mathbf{R}^{-1}(k)\mathbf{X}(k)\mathbf{e}^*(k)\,. \tag{6.62}$$

The TML-based GRLS scheme is displayed in Algorithm 6.41 for the case $N \geq M+1$.

Algorithm 6.41 The TML-based GRLS algorithm

Initialization

$\mathbf{w}(0) = \mathbf{0}_{M+1}$
$\beta \in (0, 1]$
$\mathbf{R}(-1) = \mathbf{0}_{(M+1)\times(M+1)}$

For $k \geq 0$ **do**

$\mathbf{e}(k) = \mathbf{d}(k) - \mathbf{X}^{\mathrm{T}}(k)\mathbf{w}^*(k)$
$\mathbf{R}(k) = \beta\mathbf{R}(k-1) + \mathbf{X}(k)\mathbf{X}^{\mathrm{H}}(k)$
$\mathbf{w}(k+1) = \mathbf{w}(k) + \mathbf{R}^{-1}(k)\mathbf{X}(k)\mathbf{e}^*(k)$

end

Example 6.7 Considering a second-order system identification scenario (see the general block diagram depicted in Figure 6.13), compare the mean squared error (MSE) and MSD performances attained by the TML-based GLMS, GNLMS, and GRLS algorithms for 2000 randomly generated graphs with $N = 50$ nodes.

Solution

As the problem does not completely specify the graphs to be used, we generated a different binary Erdös–Renyi random undirected graph with edge probability 0.2 in each simulation run (total of 2000 runs). The link weights were then generated following a uniform distribution in the interval $[0, 1]$, and the resulting Laplacian matrix was normalized to have maximum eigenvalue $\lambda_{N-1} = 1$. In addition, at each run, we consider a realization of the actual system parameter $\mathbf{w}_{\mathrm{o}} \in \mathbb{R}^3$ according to a zero-mean Gaussian distribution with covariance matrix $\mathbf{I}_3$. The time-invariant output GS $\mathbf{y} \in \mathbb{R}^{50}$ was generated at each run by filtering a realization of a random vector $\mathbf{x} \in \mathbb{R}^{50}$, drawn from a zero-mean Gaussian distribution with uncorrelated samples; the variance of each entry was randomly chosen in the interval $[0.5, 1.5]$. The measurement noise $\boldsymbol{\eta} \in \mathbb{R}^{50}$ was generated in a similar way of the input $\mathbf{x}$, but with the variance of each entry randomly picked within the interval $[0.05, 0.15]$. We then empirically set the algorithms' parameters ($\mu = 10^{-4}, \mu_{\mathrm{n}} = 2.5 \cdot 10^{-2}, \beta = 0.95$) so as to achieve similar steady-state MSE performance (around 8 dB for the entire GS or -9 dB per node). Figure 6.14 shows that the GRLS algorithm converges much faster than the others in terms of MSE (on the left), followed by the GNLMS algorithm, and the slowest GLMS algorithm, which eventually – after thousands of iterations – achieves the same MSE performance. Regarding the normalized MSD (on the right), the algorithms have distinct steady-state and convergence characteristics. It is worth pointing out the poor MSD performance of the GLMS algorithm in this specific scenario, even though its steady-state MSE is as good as the GNLMSs and the GRLSs.

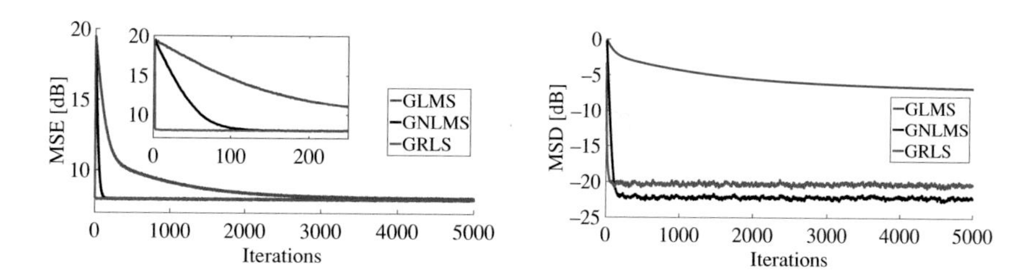

Figure 6.14 Learning curves of the TML-based algorithms: (left) MSE and (right) normalized MSD.

Regarding the learning curves presented in Example 6.7, it is worth mentioning that different performance characteristics could be obtained if we consider different matrices – for example the adjacency matrix – for the definition of the polynomial graph filter. This aspect will be further explored by the problems in the end of this chapter.

6.3.5 Generalization of the TML Approach

The TML-based adaptive algorithms do not capture the temporal dimension of the multivariable signal $\mathbf{x}(\cdot)$. Indeed, the basic TML structure in Figure 6.12 considers only a GS $\mathbf{x}(k)$ (snapshot of the multivariable signal at time instant k) and the resulting filter $\mathbf{w}(k)$ acts upon such a snapshot to yield the output $\mathbf{y}(k) = \mathbf{X}^{\mathrm{T}}(k)\mathbf{w}^{*}(k)$. The input matrix $\mathbf{X}(k)$ depends only on $\mathbf{x}(k)$ and, therefore, disregards the temporal dynamics of the signals at each node.

If we go one step back and compare Figures 6.11 and 6.12, we see that their main difference is the replacement of the line of delays with the line of Laplacian matrices. As for the taps themselves, they are kept unchanged in the TML structure, when compared to the TDL one. But we can do something different: We keep "unchanged" the line of delays and replace the taps by graph filters, as illustrated in Figure 6.15. In a way, this can be thought as a *dual* modification – with respect to the primal modification that yielded the TML approach – over the basic TDL structure.

Each graph filter $\mathbf{W}_m(k)$ in Figure 6.15 can be written as

$$\mathbf{W}_m(k) = \sum_{q=0}^{Q_m} w_{m,q}(k)\mathbf{L}^q , \tag{6.63}$$

wherein Q_m is the order of the m-th graph filter and $w_{m,q}(k) \in \mathbb{C}$ is the filter's q-th coefficient. Hence, the basic TML structure in Figure 6.12 can be used to represent the m-th graph filter in Figure 6.15, with input $\mathbf{x}(k-m)$, for $m \in \{0, \ldots, M\}$. This eventually implies that the basic TML structure can be recovered from this mixed TDL-TML structure when $M = 0$.

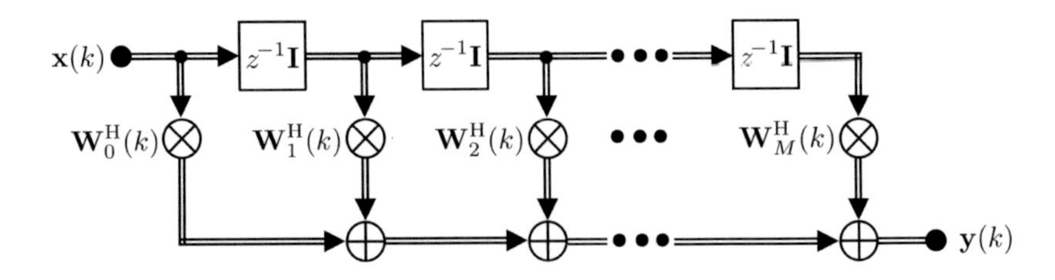

Figure 6.15 Mixed TDL-TML structure for adaptive graph filtering.

It is possible to derive LMS-based and RLS-based algorithms for the structure in Figure 6.15; see [56] for a related approach. But we leave these generalizations as exercises in the end of this chapter, and we will discuss another topic: the interpolation-based adaptive graph filters, whose basic block diagram is illustrated in Figure 6.10.

6.3.6 The Interpolation-Based GLMS Algorithm

The interpolation-based graph counterpart of the LMS algorithm [57–62], also called GLMS, aims to solve the problem

$$\underset{\hat{\mathbf{y}}_{\mathcal{F}}}{\text{minimize}}\ \mathbb{E}\left\{ \|\mathbf{D}_{\mathcal{S}(k)}\left[\boldsymbol{d}(k) - \mathbf{U}_{\mathcal{F}}\hat{\mathbf{y}}_{\mathcal{F}}\right]\|_2^2 \right\} \tag{6.64}$$

via a stochastic-gradient approach, giving rise to the frequency-domain update equation

$$\hat{\mathbf{y}}_{\mathcal{F}}(k+1) = \hat{\mathbf{y}}_{\mathcal{F}}(k) + \mu\mathbf{U}_{\mathcal{F}}^{\mathrm{T}}\mathbf{e}(k)\,, \tag{6.65}$$

or, in the vertex domain:

$$\mathbf{y}(k+1) = \mathbf{y}(k) + \mu\mathbf{U}_{\mathcal{F}}\mathbf{U}_{\mathcal{F}}^{\mathrm{T}}\mathbf{e}(k)\,. \tag{6.66}$$

As done for the TML-based algorithms, we can choose the convergence factor $\mu > 0$ so as to achieve an asymptotically unbiased estimator. By assumption, the GS is bandlimited to the support $\mathcal{F}$; so we define

$$\Delta\hat{\boldsymbol{y}}_{\mathcal{F}}(k) = \hat{\boldsymbol{y}}_{\mathcal{F}}(k) - \hat{\mathbf{x}}_{\mathcal{F}}(k)\,, \tag{6.67}$$

to quantify the estimation error in the graph-frequency domain. For the asymptotic analysis, we assume that the signal $\hat{\mathbf{x}}_{\mathcal{F}}(k) = \hat{\mathbf{x}}_{\mathcal{F}}$ is fixed. Thus, one can rewrite the frequency-domain version of Equation (6.66) as follows:

$$\begin{aligned}\Delta\hat{\boldsymbol{y}}_{\mathcal{F}}(k+1) &= \Delta\hat{\boldsymbol{y}}_{\mathcal{F}}(k) + \mu\mathbf{U}_{\mathcal{F}}^{\mathrm{T}}\mathbf{D}_{\mathcal{S}}(\mathbf{U}_{\mathcal{F}}\hat{\mathbf{x}}_{\mathcal{F}} + \boldsymbol{\eta}(k) - \mathbf{U}_{\mathcal{F}}\hat{\boldsymbol{y}}_{\mathcal{F}}(k))\\ &= \left(\mathbf{I} - \mu\mathbf{U}_{\mathcal{F}}^{\mathrm{T}}\mathbf{D}_{\mathcal{S}}\mathbf{U}_{\mathcal{F}}\right)\Delta\hat{\boldsymbol{y}}_{\mathcal{F}}(k) + \mu\mathbf{U}_{\mathcal{F}}^{\mathrm{T}}\mathbf{D}_{\mathcal{S}}\boldsymbol{\eta}(k)\end{aligned} \tag{6.68}$$

yielding

$$\mathbb{E}\left\{\Delta\hat{\boldsymbol{y}}_{\mathcal{F}}(k+1)\right\} = \left(\mathbf{I} - \mu\mathbf{U}_{\mathcal{F}}^{\mathrm{T}}\mathbf{D}_{\mathcal{S}}\mathbf{U}_{\mathcal{F}}\right)^{k+1}\mathbb{E}\left\{\Delta\hat{\boldsymbol{y}}_{\mathcal{F}}(0)\right\} \tag{6.69}$$

Algorithm 6.42 The interpolation-based GLMS algorithm

Initialization

$\hat{\mathbf{y}}_{\mathcal{F}}(0) = \mathbf{0}_{|\mathcal{F}|}$

$\mu \in \left(0, \frac{2}{\lambda_{\max}\left(\mathbf{U}_{\mathcal{F}}^{\mathrm{T}}\mathbf{D}_{\mathcal{S}}\mathbf{U}_{\mathcal{F}}\right)}\right)$

$\mathcal{S}(0)$

$\mathbf{B}_{\mathrm{L}} = \mathbf{U}_{\mathcal{F}}\mathbf{U}_{\mathcal{F}}^{\mathrm{T}}$

For $k \geq 0$ **do**

$\mathbf{e}(k) = \mathbf{D}_{\mathcal{S}(k)}\left[\mathbf{d}(k) - \mathbf{y}(k)\right]$

$\mathbf{y}(k+1) = \mathbf{y}(k) + \mu\mathbf{B}_{\mathrm{L}}\mathbf{e}(k)$

If necessary, update the sampling set $\mathcal{S}(k)$; otherwise $\mathcal{S}(k+1) = \mathcal{S}(k)$

end

so that

$$0 < \mu < \frac{2}{\lambda_{\max}\left(\mathbf{U}_{\mathcal{F}}^{\mathrm{T}}\mathbf{D}_{\mathcal{S}}\mathbf{U}_{\mathcal{F}}\right)}. \tag{6.70}$$

This simple LMS-based procedure for online estimation of GSs is displayed in Algorithm 6.42.

6.3.7 The Interpolation-Based GNLMS Algorithm

As done for the TML GNLMS, the interpolation-based graph counterpart of the NLMS algorithm [63], also called GNLMS, can be derived starting from the GLMS and optimizing the convergence matrix. Again, the goal is to choose the convergence matrix $\mathbf{M}(k)$ to minimize the energy of the *a posteriori* error

$$\begin{aligned}\varepsilon(k) &= \mathbf{D}_{\mathcal{S}(k)}\left[\mathbf{d}(k) - \mathbf{y}(k+1)\right] \\ &= \mathbf{D}_{\mathcal{S}(k)}\left[\mathbf{d}(k) - \mathbf{U}_{\mathcal{F}}\hat{\mathbf{y}}_{\mathcal{F}}(k+1)\right].\end{aligned} \tag{6.71}$$

From Equation (6.66), the frequency-domain GLMS update equation becomes

$$\hat{\mathbf{y}}_{\mathcal{F}}(k+1) = \hat{\mathbf{y}}_{\mathcal{F}}(k) + \mu\mathbf{U}_{\mathcal{F}}^{\mathrm{T}}\mathbf{D}_{\mathcal{S}(k)}\mathbf{e}(k). \tag{6.72}$$

Once again, we replace $\mu\mathbf{I}$ with the convergence matrix $\mathbf{M}(k)$. This time, however, we consider only one form of update equation:

$$\hat{\mathbf{y}}_{\mathcal{F}}(k+1) = \hat{\mathbf{y}}_{\mathcal{F}}(k) + \mathbf{M}(k)\mathbf{U}_{\mathcal{F}}^{\mathrm{T}}\mathbf{D}_{\mathcal{S}(k)}\left[\mathbf{d}(k) - \mathbf{U}_{\mathcal{F}}\hat{\mathbf{y}}_{\mathcal{F}}(k)\right] \tag{6.73}$$

since $|\mathcal{F}| \leq N$ always.

It is possible to show that the convergence matrix $\mathbf{M}(k)$ that minimizes $\|\varepsilon(k)\|_2^2$ is

$$\mathbf{M}(k) = \left(\mathbf{U}_{\mathcal{F}}^{\mathrm{T}}\mathbf{D}_{\mathcal{S}(k)}\mathbf{U}_{\mathcal{F}}\right)^{-1}. \tag{6.74}$$

Hence, the frequency-domain update expression, including a convergence factor $\mu_{\rm n} > 0$, is

$$\hat{\mathbf{y}}_{\mathcal{F}}(k+1) = \hat{\mathbf{y}}_{\mathcal{F}}(k) + \mu_{\rm n}(\mathbf{U}_{\mathcal{F}}^{\rm T}\mathbf{D}_{\mathcal{S}(k)}\mathbf{U}_{\mathcal{F}})^{-1}\mathbf{U}_{\mathcal{F}}^{\rm T}\mathbf{e}(k) \tag{6.75}$$

or in the vertex domain:

$$\mathbf{y}(k+1) = \mathbf{y}(k) + \mu_{\rm n}\mathbf{U}_{\mathcal{F}}(\mathbf{U}_{\mathcal{F}}^{\rm T}\mathbf{D}_{\mathcal{S}(k)}\mathbf{U}_{\mathcal{F}})^{-1}\mathbf{U}_{\mathcal{F}}^{\rm T}\mathbf{e}(k)\,. \tag{6.76}$$

Regarding the choice of $\mu_{\rm n}$ to guarantee asymptotic unbiasedness, based on the definition of $\Delta\hat{\boldsymbol{y}}_{\mathcal{F}}(k)$ in Equation (6.67) and the frequency-domain version of Equation (6.76), if we define the ancillary matrix $\mathbf{M}_{\rm N} \in \mathbb{R}^{|\mathcal{F}|\times N}$ as

$$\mathbf{M}_{\rm N}(k) = (\mathbf{U}_{\mathcal{F}}^{\rm T}\mathbf{D}_{\mathcal{S}(k)}\mathbf{U}_{\mathcal{F}})^{-1}\mathbf{U}_{\mathcal{F}}^{\rm T}\mathbf{D}_{\mathcal{S}(k)} \tag{6.77}$$

it follows that

$$\Delta\hat{\boldsymbol{y}}_{\mathcal{F}}(k+1) = (1-\mu_{\rm n})\Delta\hat{\boldsymbol{y}}_{\mathcal{F}}(k) + \mu_{\rm n}\mathbf{M}_{\rm N}(k)\boldsymbol{\eta}(k)\,. \tag{6.78}$$

By taking the expected value on both sides of Equation (6.78), we find the recursive expression

$$\mathbb{E}\{\Delta\hat{\boldsymbol{y}}_{\mathcal{F}}(k+1)\} = (1-\mu_{\rm n})\,\mathbb{E}\{\Delta\hat{\boldsymbol{y}}_{\mathcal{F}}(k)\}\,, \tag{6.79}$$

which allows one to write

$$\mathbb{E}\{\Delta\hat{\boldsymbol{y}}_{\mathcal{F}}(k)\} = (1-\mu_{\rm n})^{k}\,\mathbb{E}\{\Delta\hat{\boldsymbol{y}}_{\mathcal{F}}(0)\}\,, \tag{6.80}$$

and, as a result,

$$0 < \mu_{\rm n} < 2\,. \tag{6.81}$$

Algorithm 6.43 summarizes the interpolation-based GNLMS algorithm. When the sampling set is fixed, that is, $\mathcal{S}(k) = \mathcal{S}$, then $\mathbf{B}_{\rm N} = \mathbf{U}_{\mathcal{F}}(\mathbf{U}_{\mathcal{F}}^{\rm T}\mathbf{D}_{\mathcal{S}}\mathbf{U}_{\mathcal{F}})^{-1}\mathbf{U}_{\mathcal{F}}^{\rm T}$ can be computed only once in the initialization stage, thus relieving the computational burden.

Algorithm 6.43 The interpolation-based GNLMS algorithm

Initialization

$\hat{\mathbf{y}}_{\mathcal{F}}(0) = \mathbf{0}_{|\mathcal{F}|}$
$\mu_{\rm n} \in (0, 2)$
$\mathcal{S}(0)$

For $k \geq 0$ **do**

$\mathbf{e}(k) = \mathbf{D}_{\mathcal{S}(k)}\,[\mathbf{d}(k) - \mathbf{y}(k)]$
$\mathbf{B}_{\rm N}(k) = \mathbf{U}_{\mathcal{F}}(\mathbf{U}_{\mathcal{F}}^{\rm T}\mathbf{D}_{\mathcal{S}(k)}\mathbf{U}_{\mathcal{F}})^{-1}\mathbf{U}_{\mathcal{F}}^{\rm T}$ (if the sampling set is fixed, then compute $\mathbf{B}_{\rm N}$ in the initialization instead)
$\mathbf{y}(k+1) = \mathbf{y}(k) + \mu_{\rm n}\mathbf{B}_{\rm N}(k)\mathbf{e}(k)$
If necessary, update the sampling set $\mathcal{S}(k)$; otherwise $\mathcal{S}(k+1) = \mathcal{S}(k)$

end

6.3.8 The Interpolation-Based GRLS Algorithm

The interpolation-based graph counterpart of the RLS algorithm [64], also called GRLS, can be obtained by solving the weighted least-squares (WLS) unconstrained problem

$$\underset{\hat{\mathbf{y}}_{\mathcal{F}}}{\text{minimize}} \sum_{i=1}^{k} \beta^{k-l} \|\mathbf{D}_{\mathcal{S}(k)}(\mathbf{d}(i) - \mathbf{U}_{\mathcal{F}}\hat{\mathbf{y}}_{\mathcal{F}})\|^2_{\mathbf{C}^{-1}(k)} + \beta^k \|\hat{\mathbf{y}}_{\mathcal{F}}\|^2_{\mathbf{\Pi}}, \tag{6.82}$$

where the forgetting factor β is in the range $0 \ll \beta \leq 1$, and the regularization matrix $\mathbf{\Pi}$, which is usually taken as $\mathbf{\Pi} = \delta \mathbf{I}_{|\mathcal{F}|}$ with a small $\delta > 0$, is included to account for ill-conditioning in the first iterations. The matrix $\mathbf{C}(k) = \mathbb{E}\left\{\boldsymbol{\eta}(k)\boldsymbol{\eta}^{\mathrm{H}}(k)\right\}$ is the noise covariance matrix.

An online method for evaluating the WLS solution of Equation (6.82) employs the ancillary variables $\mathbf{R}(k) \in \mathbb{C}^{|\mathcal{F}| \times |\mathcal{F}|}$ and $\mathbf{p}(k) \in \mathbb{C}^{|\mathcal{F}|}$ defined recursively as

$$\begin{aligned} \mathbf{R}(k) &= \beta \mathbf{R}(k-1) + \mathbf{U}^{\mathrm{T}}_{\mathcal{F}} \mathbf{D}_{\mathcal{S}(k)} \mathbf{C}^{-1}(k) \mathbf{D}_{\mathcal{S}(k)} \mathbf{U}_{\mathcal{F}}\,, \\ \mathbf{p}(k) &= \beta \mathbf{p}(k-1) + \mathbf{U}^{\mathrm{T}}_{\mathcal{F}} \mathbf{D}_{\mathcal{S}(k)} \mathbf{C}^{-1}(k) \mathbf{D}_{\mathcal{S}(k)} \mathbf{d}(k)\,. \end{aligned} \tag{6.83}$$

From these variables, the solution $\hat{\mathbf{y}}_{\mathcal{F}}(k+1)$ of Equation (6.82) satisfies $\mathbf{R}(k)$ $\hat{\mathbf{y}}_{\mathcal{F}}(k+1) = \mathbf{p}(k)$. We therefore find the GS estimate $\mathbf{y}(k+1)$

$$\mathbf{y}(k+1) = \mathbf{U}_{\mathcal{F}} \mathbf{R}^{-1}(k) \mathbf{p}(k). \tag{6.84}$$

Moreover, it is possible to show that the update Equations (6.83) and (6.84) are equivalent to

$$\begin{aligned} \mathbf{R}(k) &= \beta \mathbf{R}(k-1) + \mathbf{U}^{\mathrm{T}}_{\mathcal{F}} \mathbf{D}_{\mathcal{S}(k)} \mathbf{C}^{-1}(k) \mathbf{D}_{\mathcal{S}(k)} \mathbf{U}_{\mathcal{F}}\,, \\ \mathbf{y}(k+1) &= \mathbf{y}(k) + \mathbf{U}_{\mathcal{F}} \mathbf{R}^{-1}(k) \mathbf{U}^{\mathrm{T}}_{\mathcal{F}} \mathbf{D}_{\mathcal{S}(k)} \mathbf{C}^{-1}(k) \mathbf{e}(k)\,. \end{aligned} \tag{6.85}$$

Algorithm 6.44 presents the GRLS using its compact representation in Equation (6.85) and the initialization $\mathbf{R}(-1) = \mathbf{\Pi}$.

As a final remark, by considering $\mathbf{R}(-1) = \mathbf{\Pi}$, we rewrite $\mathbf{R}(k)$ in Equation (6.85) as

$$\mathbf{R}(k) = \beta^k \mathbf{\Pi} + (\mathbf{U}^{\mathrm{T}}_{\mathcal{F}} \mathbf{D}_{\mathcal{S}(k)} \mathbf{C}^{-1}(k) \mathbf{D}_{\mathcal{S}(k)} \mathbf{U}_{\mathcal{F}}) \frac{(1-\beta^k)}{(1-\beta)}\,. \tag{6.86}$$

Particularly, as $k \to \infty$, one has that $\beta^k \to 0$. Thus, by defining

$$\mathbf{M}_{\mathrm{R}} = (\mathbf{U}^{\mathrm{T}}_{\mathcal{F}} \mathbf{D}_{\mathcal{S}} \mathbf{C}^{-1} \mathbf{D}_{\mathcal{S}} \mathbf{U}_{\mathcal{F}})^{-1} \mathbf{U}^{\mathrm{T}}_{\mathcal{F}} \mathbf{D}_{\mathcal{S}}\,, \tag{6.87}$$

we verify that, when k increases, the GRLS update tends to the simple expression

$$\mathbf{y}(k+1) = \mathbf{y}(k) + (1-\beta) \mathbf{U}_{\mathcal{F}} \mathbf{M}_{\mathrm{R}} \mathbf{C}^{-1}(k) \mathbf{e}(k)\,, \tag{6.88}$$

which will be further explored for providing steady-state error metrics.

Algorithm 6.44 The interpolation-based GRLS algorithm

Initialization

$\hat{\mathbf{y}}_{\mathcal{F}}(0) = \mathbf{0}_{|\mathcal{F}|}$

$\beta \in (0,1)$

$\mathcal{S}(0)$

$\mathbf{R}(-1) = \mathbf{\Pi}$

For $k \geq 0$ **do**

$\mathbf{e}(k) = \mathbf{D}_{\mathcal{S}(k)}\left[\mathbf{d}(k) - \mathbf{y}(k)\right]$

$\mathbf{B}_{\mathrm{R}}(k) = \mathbf{U}_{\mathcal{F}}^{\mathrm{T}}\mathbf{D}_{\mathcal{S}(k)}$

$\mathbf{R}(k) = \beta\mathbf{R}(k-1) + \mathbf{B}_{\mathrm{R}}(k)\mathbf{C}^{-1}(k)\mathbf{B}_{\mathrm{R}}^{\mathrm{T}}(k)$

$\mathbf{y}(k+1) = \mathbf{y}(k) + \mathbf{U}_{\mathcal{F}}\mathbf{R}^{-1}(k)\mathbf{B}_{\mathrm{R}}(k)\mathbf{C}^{-1}(k)\mathbf{e}(k)$

If necessary, update the sampling set $\mathcal{S}(k)$; otherwise $\mathcal{S}(k+1) = \mathcal{S}(k)$

end

6.3.9 Analysis of the Interpolation Algorithms

Before starting the analyses of the interpolation-based algorithms, we introduce the FoMs for performance comparisons. The MSE and MSD are respectively given by

$$\mathrm{MSE}(k) = \mathbb{E}\left\{\|\boldsymbol{e}(k)\|_2^2\right\} \text{ and } \mathrm{MSD}(k) = \mathbb{E}\left\{\|\Delta\boldsymbol{y}(k)\|_2^2\right\}, \tag{6.89}$$

where $\Delta\boldsymbol{y}(k) = \boldsymbol{y}(k) - \mathbf{x}(k)$ is the difference between the current estimator $\boldsymbol{y}(k)$ and the original noiseless GS $\mathbf{x}(k)$. In particular, for bandlimited GSs, if we use the compact representation in Equation (6.20) and the property $\mathbf{U}_{\mathcal{F}}^{\mathrm{T}}\mathbf{U}_{\mathcal{F}} = \mathbf{I}_{|\mathcal{F}|}$, from Equation (6.89) we find that the MSD is equivalent to

$$\mathrm{MSD}(k) = \mathbb{E}\left\{\|\Delta\hat{\boldsymbol{y}}_{\mathcal{F}}(k)\|_2^2\right\}. \tag{6.90}$$

A disadvantage of using the scalar metric $\mathrm{MSE}(k)$ in Equation (6.89) is that it potentially masks the occurrence of large error entries in Equation (6.33). Then, we define an alternative FoM for estimating each component of the error vector $\mathbf{e}(k)$. This more general FoM relies on statistics of $e_n(k)$, the n-th component of $\mathbf{e}(k)$ in Equation (6.33), and provides a more accurate insight of the algorithm overall performance. Note that, from Equations (6.24), (6.33), and (6.67), one has

$$e_n(k) = d_{nn}\left[\eta_n(k) - \bar{\mathbf{u}}_{\mathcal{F}_n}^{\mathrm{T}}\Delta\hat{\boldsymbol{y}}_{\mathcal{F}}(k)\right], \tag{6.91}$$

where $d_{nn} \in \{0,1\}$, $\eta_n(k)$ is the n-th entry of the noise $\boldsymbol{\eta}(k)$, and $\bar{\mathbf{u}}_{\mathcal{F}_n}^{\mathrm{T}}$ is the n-th row of $\mathbf{U}_{\mathcal{F}}$.

If we consider an asymptotically unbiased estimator $\hat{\boldsymbol{y}}_{\mathcal{F}}(k)$ such that $\mathbb{E}[\Delta\hat{\boldsymbol{y}}_{\mathcal{F}}(k)]$ converges to the null vector in steady state, which holds true for the graph adaptive algorithms presented here, and by recalling that the noise vector $\boldsymbol{\eta}(k)$ has zero mean, then we have $\mathbb{E}[e_n(k)] \to 0$ as k grows to infinity.

By taking the expected value of the squared expression in Equation (6.91), and under the same independence assumptions we made before, we can compute the steady-state error variance

$$\sigma_{e_n}^2 = \lim_{k\to\infty} d_{nn} \left[\mathbb{E}\{|\eta_n(k)|^2\} + \bar{\mathbf{u}}_{\mathcal{F}_n}^{\mathrm{T}} \mathbb{E}[\Delta\hat{\boldsymbol{y}}_{\mathcal{F}}(k)\Delta\hat{\boldsymbol{y}}_{\mathcal{F}}^{\mathrm{H}}(k)]\bar{\mathbf{u}}_{\mathcal{F}_n} \right], \tag{6.92}$$

thus yielding, from Equation (6.89), the following steady-state MSE:

$$\mathrm{MSE} = \lim_{k\to\infty} \mathrm{MSE}(k) = \sum_{n\in\mathcal{N}} \sigma_{e_n}^2 . \tag{6.93}$$

Similarly, based on the MSD in Equation (6.90), one can also define

$$\mathrm{MSD} = \lim_{k\to\infty} \mathrm{MSD}(k) = \lim_{k\to\infty} \mathrm{tr}\{ \mathbb{E}[\Delta\hat{\boldsymbol{y}}_{\mathcal{F}}(k)\Delta\hat{\boldsymbol{y}}_{\mathcal{F}}^{\mathrm{H}}(k)] \}, \tag{6.94}$$

in which the matrix trace $\mathrm{tr}\{\cdot\}$ operator is employed to show the dependency of this FoM on the steady-state matrix $\mathbb{E}[\Delta\hat{\boldsymbol{y}}_{\mathcal{F}}[\infty]\Delta\hat{\boldsymbol{y}}_{\mathcal{F}}^{\mathrm{H}}[\infty]]$. This matrix plays a central role in the forthcoming analyses, being the first one to be computed for each algorithm in order to evaluate expressions (6.92), (6.93), and (6.94).

GLMS Steady-State FoMs

From Equation (6.68), we can write

$$\begin{aligned}\mathbb{E}\left\{\Delta\hat{\boldsymbol{y}}_{\mathcal{F}}(k+1)\Delta\hat{\boldsymbol{y}}_{\mathcal{F}}^{\mathrm{H}}(k+1)\right\} &= \mu^2\mathbf{U}_{\mathcal{F}}^{\mathrm{T}}\mathbf{D}_{\mathcal{S}(k)}\mathbf{C}\mathbf{D}_{\mathcal{S}(k)}\mathbf{U}_{\mathcal{F}}+\\ &+(\mathbf{I}-\mu\mathbf{U}_{\mathcal{F}}^{\mathrm{T}}\mathbf{D}_{\mathcal{S}(k)}\mathbf{U}_{\mathcal{F}})\mathbb{E}\left\{\Delta\hat{\boldsymbol{y}}_{\mathcal{F}}(k)\Delta\hat{\boldsymbol{y}}_{\mathcal{F}}^{\mathrm{H}}(k)\right\}(\mathbf{I}-\mu\mathbf{U}_{\mathcal{F}}^{\mathrm{T}}\mathbf{D}_{\mathcal{S}(k)}\mathbf{U}_{\mathcal{F}}).\end{aligned} \tag{6.95}$$

If μ is in the range that guarantees stability, then $\mathbb{E}\left\{\Delta\hat{\mathbf{y}}_{\mathcal{F}}(k)\Delta\hat{\mathbf{y}}_{\mathcal{F}}^{\mathrm{H}}(k)\right\}$ converges to $\mathbf{S}_{\mathrm{L}} \in \mathbb{R}^{|\mathcal{F}|\times|\mathcal{F}|}$ when $k\to\infty$. Thus, by defining $\mathbf{P},\mathbf{Q} \in \mathbb{R}^{|\mathcal{F}|\times|\mathcal{F}|}$ such that

$$\mathbf{P} = \mathbf{U}_{\mathcal{F}}^{\mathrm{T}}\mathbf{D}_{\mathcal{S}(k)}\mathbf{U}_{\mathcal{F}} \quad \text{and} \quad \mathbf{Q} = \mathbf{U}_{\mathcal{F}}^{\mathrm{T}}\mathbf{D}_{\mathcal{S}(k)}\mathbf{C}\mathbf{D}_{\mathcal{S}(k)}\mathbf{U}_{\mathcal{F}} , \tag{6.96}$$

we verify that the stationary expression (6.95) can be written as

$$\begin{aligned}\mathbf{S}_{\mathrm{L}} &= \mu^2\mathbf{Q} + (\mathbf{I}-\mu\mathbf{P})\mathbf{S}_{\mathrm{L}}(\mathbf{I}-\mu\mathbf{P})\\ &= \mu^2\mathbf{Q} + \mathbf{S}_{\mathrm{L}} - \mu\mathbf{P}\mathbf{S}_{\mathrm{L}} - \mu\mathbf{S}_{\mathrm{L}}\mathbf{P} + \mu^2\mathbf{P}\mathbf{S}_{\mathrm{L}}\mathbf{P},\end{aligned} \tag{6.97}$$

yielding

$$\mathbf{P}\mathbf{S}_{\mathrm{L}} + \mathbf{S}_{\mathrm{L}}\mathbf{P} - \mu\mathbf{P}\mathbf{S}_{\mathrm{L}}\mathbf{P} = \mu\mathbf{Q}, \tag{6.98}$$

which is equivalent to

$$[(\mathbf{I}\otimes\mathbf{P}) + (\mathbf{P}\otimes\mathbf{I}) - \mu(\mathbf{P}\otimes\mathbf{P})]\,\mathrm{vec}(\mathbf{S}_{\mathrm{L}}) = \mu\,\mathrm{vec}(\mathbf{Q}), \tag{6.99}$$

where $\otimes$ indicates the Kronecker product and $\mathrm{vec}(\mathbf{S}_{\mathrm{L}})$ represents the vectorization of $\mathbf{S}_{\mathrm{L}}$, performed by stacking its columns into a single column vector. When the left hand side matrix of Equation (6.99) has full rank, $\mathrm{vec}(\mathbf{S}_{\mathrm{L}})$ is obtained by solving

$$\mathrm{vec}(\mathbf{S}_{\mathrm{L}}) = \mu\,[(\mathbf{I}\otimes\mathbf{P})+(\mathbf{P}\otimes\mathbf{I})-\mu(\mathbf{P}\otimes\mathbf{P})]^{-1}\mathrm{vec}(\mathbf{Q}). \tag{6.100}$$

After recovering matrix $\mathbf{S}_\mathrm{L}$ from its vectorized version, the n-th variance $\sigma^2_{e_n}$ for the GSP LMS algorithm is computed by replacing $\mathbf{S}_\mathrm{L}$ in Equation (6.92), yielding

$$\sigma^2_{e_n} = d_{nn}\left(\sigma^2_\eta + \bar{\mathbf{u}}^\mathrm{T}_{\mathcal{F}_n}\mathbf{S}_\mathrm{L}\bar{\mathbf{u}}_{\mathcal{F}_n}\right), \tag{6.101}$$

and from Equation (6.93)

$$\mathrm{MSE} = \sum_{n\in\mathcal{N}} d_{nn}\left(\sigma^2_\eta + \bar{\mathbf{u}}^\mathrm{T}_{\mathcal{F}_n}\mathbf{S}_\mathrm{L}\bar{\mathbf{u}}_{\mathcal{F}_n}\right). \tag{6.102}$$

Moreover, an additional result that comes straightforwardly from the knowledge of $\mathbf{S}_\mathrm{L}$ is the MSD in Equation (6.94), so that

$$\mathrm{MSD} = \mathrm{tr}\{\mathbf{S}_\mathrm{L}\}. \tag{6.103}$$

GNLMS Steady-State FoMs

By defining $\mathbf{S}_\mathrm{N}(k) = \mathbb{E}[\Delta\hat{\boldsymbol{y}}_\mathcal{F}(k)\Delta\hat{\boldsymbol{y}}^\mathrm{H}_\mathcal{F}(k)]$, one has from Equation (6.78) that

$$\mathbf{S}_\mathrm{N}(k+1) = (1-\mu_\mathrm{n})^2\mathbf{S}_\mathrm{N}(k) + \mu_\mathrm{n}^2\mathbf{M}_\mathrm{N}\mathbf{C}\mathbf{M}_\mathrm{N}^\mathrm{T}, \tag{6.104}$$

which is a difference equation that converges to a solution as long as $|1-\mu_\mathrm{n}| < 1$, that is, the condition in Equation (6.81) holds true. In this case, stability is guaranteed and $\mathbb{E}[\Delta\hat{\boldsymbol{y}}_\mathcal{F}(k)\Delta\hat{\boldsymbol{y}}^\mathrm{H}_\mathcal{F}(k)]$ approaches $\mathbf{S}_\mathrm{N} \in \mathbb{R}^{|\mathcal{F}|\times|\mathcal{F}|}$ as $k \to \infty$, given as

$$\mathbf{S}_\mathrm{N} = \mathbf{S}_\mathrm{N}[\infty] = \frac{\mu_\mathrm{n}}{2-\mu_\mathrm{n}}\mathbf{M}_\mathrm{N}\mathbf{C}\mathbf{M}_\mathrm{N}^\mathrm{T}. \tag{6.105}$$

From Equations (6.92) and (6.105), the steady-state value for $\sigma^2_{e_n}$ is given by

$$\sigma^2_{e_n} = d_{nn}\left[\sigma^2_\eta + \frac{\mu_\mathrm{n}}{2-\mu_\mathrm{n}}\bar{\mathbf{u}}^\mathrm{T}_{\mathcal{F}_n}\mathbf{M}_\mathrm{N}\mathbf{C}\mathbf{M}_\mathrm{N}^\mathrm{T}\bar{\mathbf{u}}_{\mathcal{F}_n}\right]. \tag{6.106}$$

Moreover, according to Equation (6.93), we find the MSE for the GNLMS by simply summing $\sigma^2_{e_n}$ for all $n \in \mathcal{N}$, which results in

$$\mathrm{MSE} = \sum_{n=1}^{N} d_{nn}\left[\sigma^2_\eta + \frac{\mu_\mathrm{n}}{2-\mu_\mathrm{n}}\bar{\mathbf{u}}^\mathrm{T}_{\mathcal{F}_n}\mathbf{M}_\mathrm{N}\mathbf{C}\mathbf{M}_\mathrm{N}^\mathrm{T}\bar{\mathbf{u}}_{\mathcal{F}_n}\right]. \tag{6.107}$$

Finally, based on Equations (6.94) and (6.105), the MSD is

$$\mathrm{MSD} = \frac{\mu_\mathrm{n}}{2-\mu_\mathrm{n}}\mathrm{tr}\left\{\mathbf{M}_\mathrm{N}\mathbf{C}\mathbf{M}_\mathrm{N}^\mathrm{T}\right\}. \tag{6.108}$$

GRLS Steady-State FoMs

Similarly, the GRLS analysis starts by rewriting Equation (6.88) as

$$\Delta\hat{\boldsymbol{y}}_\mathcal{F}(k+1) = \beta\Delta\hat{\boldsymbol{y}}_\mathcal{F}(k) + (1-\beta)\mathbf{M}_\mathrm{R}\mathbf{C}^{-1}\boldsymbol{\eta}(k). \tag{6.109}$$

If we evaluate the outer product of Equation (6.109) and take its expected value we find

$$\begin{aligned}\mathbb{E}[\Delta\hat{\boldsymbol{y}}_{\mathcal{F}}(k+1)\Delta\hat{\boldsymbol{y}}_{\mathcal{F}}^{\mathrm{H}}(k+1)] &= \beta^2\mathbb{E}[\Delta\hat{\boldsymbol{y}}_{\mathcal{F}}(k)\Delta\hat{\boldsymbol{y}}_{\mathcal{F}}^{\mathrm{H}}(k)]+ \\ &+ (1-\beta)^2\mathbf{M}_{\mathrm{R}}\mathbf{C}^{-1}\mathbf{D}_{\mathcal{S}(k)}\mathbf{C}\mathbf{D}_{\mathcal{S}(k)}\mathbf{C}^{-1}\mathbf{M}_{\mathrm{R}}^{\mathrm{T}}.\end{aligned} \tag{6.110}$$

Then, by considering the convergence of $\mathbb{E}[\Delta\hat{\mathbf{y}}_{\mathcal{F}}(k)\Delta\hat{\mathbf{y}}_{\mathcal{F}}^{\mathrm{H}}(k)]$ to $\mathbf{S}_{\mathrm{R}} \in \mathbb{R}^{|\mathcal{F}|\times|\mathcal{F}|}$ as k grows to infinity, one gets

$$\mathbf{S}_{\mathrm{R}} = \frac{1-\beta}{1+\beta}\cdot\mathbf{M}_{\mathrm{R}}\mathbf{C}^{-1}\mathbf{D}_{\mathcal{S}}\mathbf{C}\mathbf{D}_{\mathcal{S}}\mathbf{C}^{-1}\mathbf{M}_{\mathrm{R}}^{\mathrm{T}}, \tag{6.111}$$

allowing us to write the corresponding stationary FoMs for the GRLS algorithm

$$\sigma_{e_n}^2 = d_{nn}\left(\sigma_\eta^2 + \bar{\mathbf{u}}_{\mathcal{F}_n}^{\mathrm{T}}\mathbf{S}_{\mathrm{R}}\bar{\mathbf{u}}_{\mathcal{F}_n}\right), \tag{6.112}$$

$$\mathrm{MSE} = \sum_{n=1}^{N} d_{nn}\left(\sigma_\eta^2 + \bar{\mathbf{u}}_{\mathcal{F}_n}^{\mathrm{T}}\mathbf{S}_{\mathrm{R}}\bar{\mathbf{u}}_{\mathcal{F}_n}\right), \tag{6.113}$$

$$\mathrm{MSD} = \mathrm{tr}\{\mathbf{S}_{\mathrm{R}}\}. \tag{6.114}$$

Computational Complexity

In order to provide a fair comparison of the computational complexity among the GLMS, GNLMS, and GRLS algorithms, we estimate the amount of numerical operations required to evaluate the estimate $\mathbf{y}(k+1)$ at each iteration. As all algorithms present some common steps, we focus on the differences among them, which consist basically in how the update of $\mathbf{y}(k+1)$ is performed.

For the sake of comparison, we consider that the product between a $T_1 \times T_2$ matrix and a $T_2 \times T_3$ matrix results in $T_1T_2T_3$ multiplications and $T_1T_3(T_2-1)$ sums, or a total of $T_1T_3(2T_2-1)$ operations. In particular, by taking two vectors $\mathbf{b}, \mathbf{c} \in \mathbb{C}^N$, with one vector having at most $|\mathcal{S}|$ nonzero elements, as the sampled error vector $\mathbf{e}(k)$ in Equation (6.33), we assume that an inner product operation $\mathbf{b}^{\mathrm{H}}\mathbf{c}$ can be efficiently computed in $|\mathcal{S}|$ operations. Then, the cost for computing $\mathbf{U}_{\mathcal{F}}^{\mathrm{T}}\mathbf{e}(k)$ is $|\mathcal{F}|(2|\mathcal{S}|-1)$ operations. Similarly, due to the reduced complexity required by the sampling operation $\mathbf{D}_{\mathcal{S}(k)}\mathbf{b}$, which is estimated as one operation, it is straightforward to conclude that $\mathbf{D}_{\mathcal{S}(k)}\mathbf{U}_{\mathcal{F}}$ demands $|\mathcal{F}|$ operations, while $\mathbf{U}_{\mathcal{F}}^{\mathrm{T}}\mathbf{D}_{\mathcal{S}(k)}\mathbf{U}_{\mathcal{F}}$ accounts for $|\mathcal{F}|^2(2|\mathcal{S}|-1)$ operations. Likewise, when evaluating Equation (6.85), the product of $\mathbf{U}_{\mathcal{F}}^{\mathrm{T}}\mathbf{D}_{\mathcal{S}(k)}\mathbf{C}^{-1}(k)$ by $\mathbf{e}(k)$ requires $|\mathcal{F}|(2|\mathcal{S}|-1)$ operations. In addition, it is assumed that the inversion of a $|\mathcal{F}|\times|\mathcal{F}|$ matrix like $\mathbf{U}_{\mathcal{F}}^{\mathrm{T}}\mathbf{D}_{\mathcal{S}(k)}\mathbf{U}_{\mathcal{F}}$ adds $\frac{1}{3}|\mathcal{F}|^3$ operations per iteration via Cholesky factorization. Thus, the computational complexity, in terms of operations, for obtaining the estimate update when using the interpolation-based GLMS, GNLMS, and GRLS adaptive algorithms is summarized in Table 6.2.

One can observe in Table 6.2 that the computational complexity required for evaluating the GLMS update in Equation (6.66) is significantly smaller in comparison to the other adaptive algorithms because it is obtained by simple matrix-vector products. On the other hand, the computation of the GNLMS and GRLS expressions in Equations (6.76) and (6.85), respectively, is

Table 6.2 Interpolation algorithms' complexity for computing $\mathbf{y}(k+1)$

Algorithm	$\mathbf{y}(k+1)$	Operations/iter.
GLMS	(6.66)	$\lvert\mathcal{F}\rvert\left[2(\lvert\mathcal{S}\rvert+N)-1\right]+N$
GNLMS	(6.76)	$\frac{1}{3}\lvert\mathcal{F}\rvert^3+\lvert\mathcal{F}\rvert^2(2\lvert\mathcal{S}\rvert+1)+2\lvert\mathcal{F}\rvert\,(N+\lvert\mathcal{S}\rvert-1)+N$
GRLS	(6.85)	$\frac{1}{3}\lvert\mathcal{F}\rvert^3+\lvert\mathcal{F}\rvert^2(2\lvert\mathcal{S}\rvert+3)+2\lvert\mathcal{F}\rvert\,(N\lvert\mathcal{S}\rvert+\lvert\mathcal{S}\rvert-1)+\lvert\mathcal{F}\rvert N$

more complexi since it requires matrix–matrix products and the inversion of a $|\mathcal{F}| \times |\mathcal{F}|$ matrix; yet, GNLMS algorithm requires less operations than the GRLS algorithm. Additionally, it is worth mentioning that GNLMS and GRLS algorithms are more complex than the GLMS, when the GS is properly represented by a few frequency components ($|\mathcal{F}| \ll N$), the order of complexity of both algorithms is also dominated by a linear term in N.

A scenario of practical interest occurs when the sampling set $\mathcal{S}(k)$ happens to be static and known *a priori*, as in the case of *strict* interpolation problems. In these cases, the evaluation of the GNLMS update in Equation (6.76) is considerably simplified by rewriting the constant matrix $(\mathbf{U}_{\mathcal{F}}^{\mathrm{T}}\mathbf{D}_{\mathcal{S}(k)}\mathbf{U}_{\mathcal{F}})^{-1}$ as $\mathbf{T}\mathbf{T}^{\mathrm{T}}$, where $\mathbf{T} \in \mathbb{R}^{|\mathcal{F}|\times|\mathcal{F}|}$ is a lower triangular matrix obtained using the Cholesky decomposition. Then, after defining the ancillary matrix $\mathbf{B}_{\mathrm{N}} \in \mathbb{R}^{N\times|\mathcal{F}|}$ as $\mathbf{B}_{\mathrm{N}} = \mathbf{U}_{\mathcal{F}}\mathbf{T}$, one verifies that the GNLMS update can be implemented in static sampling set scenarios, where $\mathcal{S}(k) = \mathcal{S}$ is known *a priori*, according to the expression

$$\mathbf{y}(k+1) = \mathbf{y}(k) + \mu_{\mathrm{n}}\mathbf{B}_{\mathrm{N}}\mathbf{B}_{\mathrm{N}}^{\mathrm{T}}\mathbf{e}(k)\,. \tag{6.115}$$

Therefore, by considering that matrix $\mathbf{B}_{\mathrm{N}}$ in Equation (6.115) is a pre-evaluated structure stored for efficient algorithm implementation, one easily concludes from comparing Equation (6.66) to Equation (6.115) that the complexity required for the NLMS algorithm resembles that of the LMS method in the static sampling set scenario, being both given by $|\mathcal{F}|\left[2(|\mathcal{S}|+N)-1\right]+N$ operations per iteration.

Example 6.8 The objective of this example is to showcase the performance of the interpolation-based algorithms in terms of convergence speed and computational complexity. Consider the Brazilian weather-station dataset (see Figure 6.4) of Examples 6.3 and 6.6. In addition, consider different measurement noise scenarios, according to the covariance matrix

$$\mathbf{C} = \mathrm{diag}\left\{\sigma_{\eta_a}^2\,\mathbf{1} + \sigma_{\eta_b}^2\,\mathbf{r}_\eta\right\}\,, \tag{6.116}$$

where $\mathbf{1} \in \mathbb{R}^N$ is a vector with all components equal to 1, $\mathbf{r}_\eta \in \mathbb{R}^N$ is a realization of a random vector whose entries follow a uniform distribution between $[0,1]$, and $\sigma_{\eta_a}^2, \sigma_{\eta_b}^2 \in \mathbb{R}_+$ are variances that scale the elements of $\mathbf{1}$ and $\mathbf{r}_\eta$, respectively. Define the noise signal $\boldsymbol{\eta}(k)$ used for each simulation as zero-mean Gaussian according to three scenarios:

(i) $\sigma_{\eta_a}^2 = 0.001$ and $\sigma_{\eta_b}^2 = 0.000$;
(ii) $\sigma_{\eta_a}^2 = 0.010$ and $\sigma_{\eta_b}^2 = 0.000$;
(iii) $\sigma_{\eta_a}^2 = 0.005$ and $\sigma_{\eta_b}^2 = 0.010$.

Run the GLMS, GNLMS, and GRLS algorithms under the aforementioned conditions and assess their performances in terms of MSD(k), steady-state MSD, number of iterations until convergence, and average time for computing the estimate $\mathbf{y}(k+1)$.

Solution

We performed numerical simulations based on the noise scenarios (i)–(iii); for the GNLMS, two different implementations were considered: the general one in Equation (6.76) and the particular one in Equation (6.115). For each noise scenario, the respective convergence/forgetting factors were adjusted in order to provide a similar MSD. At each simulation run, we evaluated 5000 iterations, where we scaled the reference GS $\mathbf{x}$ by a 1.2 factor at $k = 2500$ to observe the algorithms' tracking abilities. We considered that: the algorithm has converged after reaching $1.025 \cdot$ MSD for the first time, the steady-state FoMs are computed using the last 1000 iterations of each run, and the update time uses the "*tic/toc*" MATLAB functions to provide the reader with a rough idea of how long it takes to compute $\mathbf{y}(k+1)$ for each algorithm in our simulations. Based on the average values of a 1000-run ensemble, we obtained the numerical results presented in Table 6.3. For notation simplicity, we assigned a label to each entry of this table — for example, (L.I) denotes a simulation result for the GLMS algorithm considering the noise setup (i) and convergence factor $\mu = 0.28$, whereas (N.III') denotes a simulation result for the simplified (see Equation (6.115)) GNLMS algorithm considering the noise setup (iii) and normalized convergence factor $\mu_{\rm n} = 0.208$.

We observe that the GNLMS algorithm converges considerably (more than 10 times) faster than the GLMS algorithm, but slightly (about twice) slower than the RLS algorithm. This convergence speed comparison is made clear by the MSD(k) plots in Figure 6.16. Note that the GNLMS algorithm behaves like the GRLS for large k. Another conclusion from Table 6.3 is that the computation complexity for performing the GLMS update is noticeably smaller than the one required for computing the general implementation of the GNLMS in Equation (6.76), which is still about 2.9 times faster than the GRLS approach.[11] In particular, one verifies in Table 6.3 that the GNLMS approach in Equation (6.115) provides the same fast converging characteristic of its general implementation while being as complex as the GLMS algorithm; however, it requires $\mathcal{S}(k)$ to be static and *a priori* known. Additionally, it is worth mentioning that the large difference in update time among the adaptive algorithms occurs because the current simulation scenario does not rely on the condition $|\mathcal{F}| \ll N$ ($|\mathcal{F}| \approx 0.7N$), where the dominant term in Table 6.2 becomes N.

[11] Although these complexity results cannot be directly compared to the theoretical complexity estimation, they corroborate it in terms of order of magnitude.

Table 6.3 MSD, iterations until convergence, and time for computing $\mathbf{y}(k+1)$

Entry	Simulation setup				Simulation results		
label	Alg.	Factor	$\mathbf{C}$	$\mathbf{y}(k+1)$	MSD	Converg.	Upd. time
(L.I)	GLMS	0.28	(i)	(6.66)	0.0313	1662 iter.	40 μs
(N.I)	GNLMS	0.07	(i)	(6.76)	0.0308	131 iter.	2474 μs
(N.I')	GNLMS	0.07	(i)	(6.115)	0.0308	125 iter.	41 μs
(R.I)	GRLS	0.93	(i)	(6.85)	0.0309	57 iter.	7041 μs
(L.II)	GLMS	0.28	(iii)	(6.66)	0.3111	1416 iter.	41 μs
(N.II)	GNLMS	0.07	(iii)	(6.76)	0.3055	114 iter.	2446 μs
(N.II')	GNLMS	0.07	(iii)	(6.115)	0.3058	111 iter.	41 μs
(R.II)	GRLS	0.93	(iii)	(6.85)	0.3041	57 iter.	7014 μs
(L.III)	GLMS	0.721	(iii)	(6.66)	0.9987	488 iter.	40 μs
(N.III)	GNLMS	0.208	(iii)	(6.76)	0.9765	34 iter.	2457 μs
(N.III')	GNLMS	0.208	(iii)	(6.115)	0.9779	33 iter.	41 μs
(R.III)	GRLS	0.792	(iii)	(6.85)	0.9729	18 iter.	7034 μs

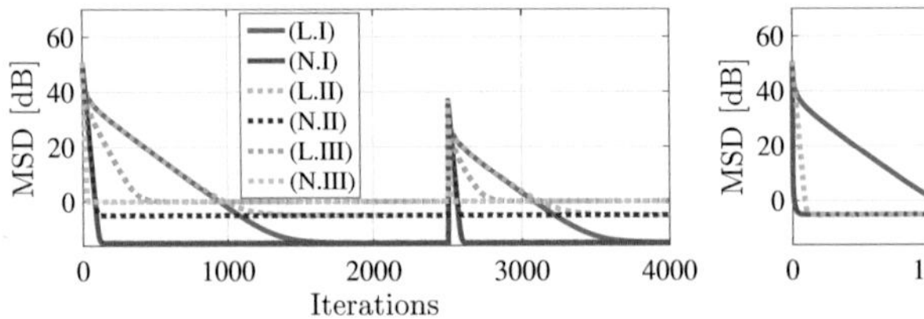

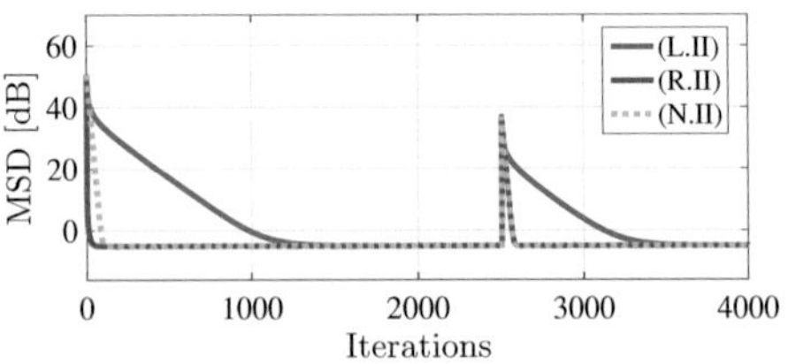

Figure 6.16 MSD behavior: (left) the GLMS and GNLMS algorithms for all noise scenarios and (right) the GLMS, GNLMS, and GRLS algorithms for the noise scenario (ii). See the entry labels in Table 6.3 for the legend definitions.

Table 6.4 Theoretical and experimental MSE and MSD, and their respective REs, for the GLMS, GNLMS, and GRLS algorithms with different factors μ_{n} and β, respectively. Ensemble of 1000 runs and noise scenario (ii)

Alg.	Factor	MSE			MSD		
		Theory	Simul.	RE[%]	Theory	Simul.	RE[%]
GLMS	0.20	2.2600	2.2599	0.004	0.2160	0.2159	0.046
	0.50	2.5717	2.5711	0.023	0.6179	0.6179	0.000
	1.00	3.4617	3.4638	-0.061	1.6803	1.6815	-0.071
GNLMS	0.05	2.1513	2.1512	0.005	0.2217	0.2212	0.226
	0.10	2.2053	2.2052	0.005	0.4550	0.4549	0.022
	0.25	2.3857	2.3855	0.008	1.2350	1.2358	-0.065
	0.50	2.7667	2.7665	0.007	2.8817	2.8822	-0.017
GRLS	0.95	2.1513	2.1516	-0.014	0.2217	0.2217	0.000
	0.90	2.2053	2.2050	0.014	0.4550	0.4551	-0.022
	0.75	2.3857	2.3857	0.000	1.2350	1.2362	-0.097

Example 6.9 The objective of this example is to showcase the accuracy of the steady-state MSE and MSD predicted for the interpolation-based algorithms. Consider the same conditions of Example 6.8 and the noise scenario (ii). Run the GLMS, GNLMS, and GRLS algorithms under the aforementioned conditions, assess the MSE and MSD steady-state performances, and compare with the theoretical predictions.

Solution

We performed numerical simulations based on the above scenario. We employed different convergence/forgetting factors (to assess the theoretical predictions in diverse conditions). At each simulation run we evaluated 12000, 3000, and 1500 iterations for the GLMS, GNLMS, and GRLS algorithms, respectively, where we considered the last 1000 iterations of each run to be part of the steady state. By taking an average of the MSE(k) and MSD(k) measurements at steady state for an ensemble of 1000 runs, we obtained the experimental results presented in Table 6.4. These results are compared to the MSE and MSD theoretical predictions presented in: Equations (6.102) and (6.103) for the GLMS algorithm; Equations (6.106), (6.93), and (6.108) for the GNLMS algorithm; and Equations (6.113) and (6.114) for the GRLS algorithm. Additionally, in Table 6.4, we also include a relative error (RE) metric computed as

$$\text{Relative error (RE)} = \frac{\text{Theory value} - \text{Simul. result}}{\text{Simul. result}}. \tag{6.117}$$

From Table 6.4 we verify that the MSE and MSD predictions provided for the GLMS, GNLMS, and GRLS algorithms are accurate across all different simulation scenarios. In particular, all results present an RE smaller than 0.25% with respect to their theoretical estimates.

Example 6.10 The objective of this example is to showcase the tracking capabilities of the interpolation-based algorithms. Consider now the US weather-station dataset – see Figure 6.4 – of Examples 6.3 and 6.6. Run the GLMS, GNLMS, and GRLS algorithms and plot the interpolated signal in a nonsampled (unobserved) node.

Solution

We performed numerical simulations based on the above scenario. The estimation of a randomly picked unobserved node, which have not been sampled according to the index set $\mathcal{S}$ – see Example 6.6 – is presented in Figure 6.17. Based on this result, we observe that the behavior of the GNLMS resembles the GRLS algorithm (apart from a slight difference in the first iterations), while

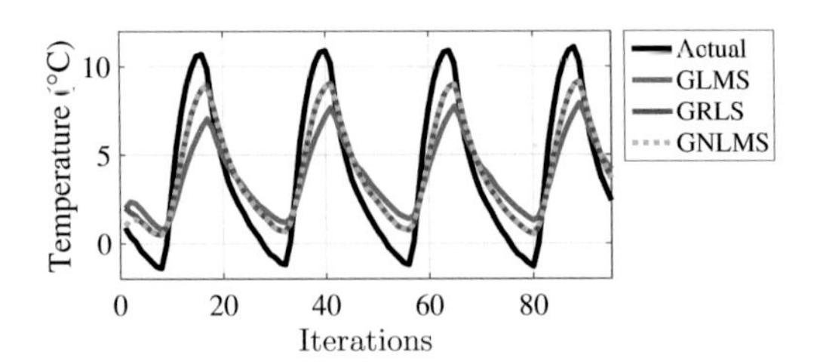

Figure 6.17 Actual temperature signal and adaptive estimates across time at a randomly picked unobserved US weather station.

presenting a better performance in estimating most unobserved nodes in comparison to the GLMS method.

6.4 Conclusion

This chapter dealt with adaptive graph filtering algorithms. The first part of the chapter described the basic GSP tools, including the very definitions of graphs and GSs as well as GFT linear graph filtering, and compact representations of bandlimited GSs along with proper methods for sampling and reconstruction of these signals. The second part of the chapter described the two main families of adaptive graph filtering algorithms, namely: the TML class of algorithms and the *interpolation* class of algorithms. Both families comprised graph counterparts of the traditional LMS, NLMS, and RLS algorithms, being respectively called GLMS, GNLMS, and GRLS algorithms.

All the adaptive solutions were shown in algorithmic formats and are expected to be used without difficulties in many distinct scenarios.

Problems

6.1 Considering the estimation error of the TML-based GLMS algorithm in Equation (6.36), show that the covariance error matrix $\mathbf{S}(k+1) = \mathbb{E}\left\{\Delta\boldsymbol{w}(k+1)\Delta\boldsymbol{w}^{\mathrm{H}}(k+1)\right\}$ satisfies the relation:

$$\begin{aligned}\operatorname{vec}\{\mathbf{S}(k+1)\} = \big[&\mathbf{I} - \mu(\mathbf{R}^{\mathrm{T}} \otimes \mathbf{I}) - \mu(\mathbf{I} \otimes \mathbf{R}) \\ &+ \mu^2 \mathbb{E}\left\{\left[(\boldsymbol{X}^*(k)\boldsymbol{X}^{\mathrm{T}}(k)) \otimes (\boldsymbol{X}(k)\boldsymbol{X}^{\mathrm{H}}(k))\right]\right\}\big] \operatorname{vec}\{\mathbf{S}(k)\} \\ &+ \mu^2\sigma_\eta^2 \operatorname{vec}\{\mathbf{R}\}.\end{aligned}$$

6.2 Assume that μ is sufficiently small to guarantee stability of the covariance error matrix of the TML-based GLMS algorithm, and propose a closed-form formula approximation for the steady-state MSE.

6.3 Considering the estimation error of the TML-based GLMS algorithm in Equation (6.52), show that the covariance error matrix $\mathbf{S}(k+1) = \mathbb{E}\left\{\Delta \boldsymbol{w}(k+1)\right.$ $\left.\Delta \boldsymbol{w}^{\mathrm{H}}(k+1)\right\}$ satisfies the relation:

$$\mathbf{S}(k+1) = (1-\mu_{\mathrm{n}})^2\,\mathbf{S}(k) + \mu_{\mathrm{n}}^2\sigma_{\eta}^2\mathbb{E}\left\{\left[\boldsymbol{X}(k)\boldsymbol{X}^{\mathrm{H}}(k)\right]^{-1}\right\}.$$

6.4 Assume that μ_{n} guarantees stability of the covariance error matrix of the TML-based GNLMS algorithm, and propose a closed-form formula approximation for the steady-state MSE.

6.5 Show that $\mathbb{E}\left\{\Delta \boldsymbol{w}(k+1)\right\} = \mathbf{0}$ for the TML-based GRLS algorithm.

6.6 Solve the problem described in Example 6.7 using the adjacency matrix as the memory building block of the TML-based adaptive graph filters.

6.7 Derive a TML-TDL–based GLMS algorithm for the structure depicted in Figure 6.15.

6.8 Derive a TML-TDL–based GNLMS algorithm for the structure depicted in Figure 6.15.

6.9 Derive a TML-TDL–based GRLS algorithm for the structure depicted in Figure 6.15.

6.10 Carry out the details to derive the optimal convergence matrix in Equation (6.74) employed by the interpolation-based GNLMS algorithm.

6.11 Carry out the details to derive the alternative update equation for the interpolation-based GRLS algorithm in Equation (6.85).

References

[1] A. Ortega, P. Frossard, J. Kovačević, J. M. F. Moura, and P. Vandergheynst, Graph signal processing: Overview, challenges, and applications, Proceedings of the IEEE **106**, pp. 808–828 (2018).

[2] L. A. S. Moreira, A. L. L. Ramos, M. L. R. de Campos, J. A. Apolinário Jr., and F. G. Serrenho, A graph signal processing approach to direction of arrival estimation, in Proceedings of the 27th European Signal Processing Conference (EUSIPCO), A Coruña, Spain, September 2019, pp. 1–5.

[3] A. Sandryhaila and J. M. F. Moura, Big data analysis with signal processing on graphs: Representation and processing of massive data sets with irregular structure, IEEE Signal Processing Magazine **31**, pp. 80–90 (2014).

[4] Y. Chen, S. Kar, and J. M. F. Moura, The internet of things: Secure distributed inference, IEEE Signal Processing Magazine **35**, pp. 64–75 (2018).

[5] J. Friedman, T. Hastie, and R. Tibshirani, Sparse inverse covariance estimation with the graphical LASSO, Biostatistics **9**, pp. 432–441 (2008).

[6] S. I. Daitch, J. A. Kelner, and D. A. Spielman, Fitting a graph to vector data, in Proceedings of the 26th Annual International Conference on Machine Learning (ICML), Montreal, Canada, June 2009, pp. 201–208.

[7] X. Dong, D. Thanou, P. Frossard, and P. Vandergheynst, Learning Laplacian matrix in smooth graph signal representations, IEEE Transactions on Signal Processing **64**, pp. 6160–6173 December (2016).

[8] S. Segarra, G. Mateos, A. G. Marques, and A. Ribeiro, Blind identification of graph filters, IEEE Transactions on Signal Processing **65**, pp. 1146–1159 (2017).

[9] J. Mei and J. M. F. Moura, Signal processing on graphs: Causal modeling of unstructured data, IEEE Transactions on Signal Processing **65**, pp. 2077–2092 (2017).

[10] D. Thanou, X. Dong, D. Kressner, and P. Frossard, Learning heat diffusion graphs, IEEE Transactions on Signal and Information Processing over Networks **3**, pp. 484–499 (2017).

[11] H. E. Egilmez, E. Pavez, and A. Ortega, Graph learning from data under Laplacian and structural constraints, IEEE Journal of Selected Topics in Signal Processing **11**, pp. 825–841 (2017).

[12] S. Segarra, A. G. Marques, G. Mateos, and A. Ribeiro, Network topology inference from spectral templates, IEEE Transactions on Signal and Information Processing over Networks **3**, pp. 467–483 (2017).

[13] B. Pasdeloup, V. Gripon, G. Mercier, D. Pastor, and M. G. Rabbat, Characterization and inference of graph diffusion processes from observations of stationary signals, IEEE Transactions on Signal and Information Processing over Networks **4**, pp. 481–496 (2018).

[14] M. Püschel and J .M. F. Moura, The algebraic approach to the discrete cosine and sine transforms and their fast algorithms, SIAM Journal on Computing **32**, pp. 1280–1316 (2003).

[15] M. Püschel and J .M. F. Moura, Algebraic signal processing theory: Cooley-Tukey type algorithms for DCTs and DSTs, IEEE Transactions on Signal Processing **56**, pp. 1502–1521 (2008).

[16] M. Püschel and J .M. F. Moura, Algebraic signal processing theory: Foundation and 1-D time, IEEE Transactions on Signal Processing **56**, pp. 3572–3585 (2008).

[17] M. Püschel and J .M. F. Moura, Algebraic signal processing theory: 1-D space, IEEE Transactions on Signal Processing **56**, pp. 3586–3599 (2008).

[18] A. Sandryhaila and J. M. F. Moura, Discrete signal processing on graphs, IEEE Transactions on Signal Processing **61**, pp. 1644–1656 (2013).

[19] A. Sandryhaila and J. M. F. Moura, Discrete signal processing on graphs: Frequency analysis, IEEE Transactions on Signal Processing **62**, pp. 3042–3050 (2014).

[20] B. Girault, P. Gonçalves, and É. Fleury, Translation on graphs: An isometric shift operator, IEEE Signal Processing Letters **22**, pp. 2416–2420 (2015).

[21] A. Gavili and X. P. Zhang, On the shift operator, graph frequency, and optimal filtering in graph signal processing, IEEE Transactions on Signal Processing **65**, pp. 6303–6318 (2017).

[22] F. R. K. Chung, *Spectral Graph Theory* (AMS, Providence, 1997).

[23] M. Belkin and P. Niyogi, Laplacian eigenmaps for dimensionality reduction and data representation, Neural Computation **15**, pp. 1373–1396 (2003).

[24] R. S. Wagner, R. G. Baraniuk, S. Du, D. B. Johnson, and A. Cohen, An architecture for distributed wavelet analysis and processing in sensor networks, in Proceedings of the 5th International Conference on Information Processing in Sensor Networks (IPSN), Nashville, USA, April 2006, pp. 243–250.

[25] X. Zhu and M. Rabbat, Approximating signals supported on graphs, in Proceedings of the IEEE International Conference on Acoustics, Speech and Signal Processing (ICASSP), Kyoto, Japan, March 2012, pp. 3921–3924.

[26] D. L. Donoho and C. Grimes, Hessian eigenmaps: Locally linear embedding techniques for high-dimensional data, Proceedings of the National Academy of Sciences of the United States of America **100**, pp. 5591–5596 (2013).

[27] D. I. Shuman, S. K. Narang, P. Frossard, A. Ortega, and P. Vandergheynst, The emerging field of signal processing on graphs: Extending high-dimensional data analysis to networks and other irregular domains, IEEE Signal Processing Magazine **30**, pp. 83–98 (2013).

[28] D. Romero, M. Ma, and G. B. Giannakis, Kernel-based reconstruction of graph signals, IEEE Transactions on Signal Processing **65**, pp. 764–778 (2017).

[29] D. Romero, V. N. Ioannidis, and G. B. Giannakis, Kernel-based reconstruction of space-time functions on dynamic graphs, IEEE Journal of Selected Topics in Signal Processing **11**, pp. 856–869 (2017).

[30] V. N. Ioannidis, D. Romero, and G. B. Giannakis, Inference of spatio-temporal functions over graphs via multikernel kriged Kalman filtering, IEEE Transactions on Signal Processing **66**, pp. 3228–3239 (2018).

[31] V. R. M. Elias, V. C. Gogineni, W. A. Martins, and S. Werner, Kernel regression on graphs in random Fourier features space, in Proceedings of the IEEE International Conference on Acoustics, Speech and Signal Processing (ICASSP), Toronto, Canada, June 2021, pp. 5235–5239.

[32] V. R. M. Elias, V. C. Gogineni, W. A. Martins, and S. Werner, Adaptive graph filters in reproducing kernel Hilbert spaces: Design and performance analysis, IEEE Transactions on Signal and Information Processing over Networks, **77**, pp. 62–74 (2021).

[33] I. Pesenson, Sampling in Paley-Wiener spaces on combinatorial graphs, Transactions of the American Mathematical Society **360**, pp. 5603–5627 (2008).

[34] S. K. Narang, A. Gadde, and A. Ortega, Signal processing techniques for interpolation in graph structured data, in Proceedings of the IEEE International Conference on Acoustics, Speech and Signal Processing (ICASSP), Vancouver, Canada, May 2013, pp. 5445–5449.

[35] A. Anis, A. Gadde, and A. Ortega, Towards a sampling theorem for signals on arbitrary graphs, in Proceedings of the IEEE International Conference on Acoustics, Speech and Signal Processing (ICASSP), Florence, Italy, May 2014, pp. 3864–3868.

[36] H. Shomorony and A. S. Avestimehr, Sampling large data on graphs, in Proceedings of the IEEE Global Conference on Signal and Information Processing (GlobalSIP), Atlanta, USA, December 2014, pp. 933–936.

[37] A. Gadde and A. Ortega, A probabilistic interpretation of sampling theory of graph signals, in Proceedings of the IEEE International Conference on Acoustics, Speech and Signal Processing (ICASSP), Brisbane, Australia, April 2015, pp. 3257–3261.

[38] S. Chen, A. Sandryhaila, and J. Kovačević, Sampling theory for graph signals, in Proceedings of the IEEE International Conference on Acoustics, Speech and Signal Processing (ICASSP), Brisbane, Australia, April 2015, pp. 3392–3396.

[39] X. Wang, P. Liu, and Y. Gu, Local-set-based graph signal reconstruction, IEEE Transactions on Signal Processing **63**, pp. 2432–2444 (2015).

[40] S. Chen, A. Sandryhaila, J. M. F. Moura, and J. Kovačević, Signal recovery on graphs: Variation minimization, IEEE Transactions on Signal Processing **63**, pp. 4609-4624 (2015).

[41] S. Chen, R. Varma, A. Sandryhaila, and J. Kovačević, Discrete signal processing on graphs: Sampling theory, IEEE Transactions on Signal Processing **63**, pp. 6510–6523 (2015).

[42] A. Anis, A. Gadde, and A. Ortega, Efficient sampling set selection for bandlimited graph signals using graph spectral proxies, IEEE Transactions on Signal Processing **64**, pp. 3775–3789 (2016).

[43] A. G. Marques, S. Segarra, G. Leus, and A. Ribeiro, Sampling of graph signals with successive local aggregations, IEEE Transactions on Signal Processing **64**, pp. 1832–1843 (2016).

[44] S. Chen, R. Varma, A. Singh, and J. Kovačević, Signal recovery on graphs: Fundamental limits of sampling strategies, IEEE Transactions on Signal and Information Processing over Networks **2**, pp. 539–554 (2016).

[45] L. F. O. Chamon and A. Ribeiro, Greedy sampling of graph signals, IEEE Transactions on Signal Processing **66**, pp. 34–47 (2018).

[46] G. Puy, N. Tremblay, R. Gribonval, and P. Vandergheynst, Random sampling of bandlimited signals on graphs, Applied and Computational Harmonic Analysis **44**, pp. 446–475 (2018).

[47] Y. Shen, G. Leus, and G. B. Giannakis, Online graph-adaptive learning with scalability and privacy, IEEE Transactions on Signal Processing **67**, pp. 2471–2483 (2019).

[48] N. Cressie, The origins of Kriging, Mathematical Geology **22**, pp. 239–252 (1990).

[49] P. D. Lorenzo, P. Banelli, and S. Barbarossa, Optimal sampling strategies for adaptive learning of graph signals, in Proceedings of the 25th European Signal Processing Conference (EUSIPCO), Cos, Greece, September 2017, pp. 1684–1688.

[50] P. D. Lorenzo, P. Banelli, E. Isufi, S. Barbarossa, and G. Leus, Adaptive graph signal processing: Algorithms and optimal sampling strategies, IEEE Transactions on Signal Processing **66**, pp. 3584–3598 (2018).

[51] P. Lorenzo, S. Barbarossa, and P. Banelli, Chapter 9 – Sampling and recovery of graph signals, in: P. M. Djurić, C. Richard (Eds.), Cooperative and Graph Signal Processing, Academic Press, 2018, pp. 261–282.

[52] R. Nassif, C. Richard, J. Chen, and A. H. Sayed, A graph diffusion LMS strategy for adaptive graph signal processing, in Proceedings of the 51st Asilomar Conference on Signals, Systems, and Computers, Pacific Grove, USA, November 2017, pp. 1973–1976.

[53] R. Nassif, C. Richard, J. Chen, and A. H. Sayed, Distributed diffusion adaptation over graph signals, in Proceedings of the IEEE International Conference on Acoustics, Speech and Signal Processing (ICASSP), Calgary, Canada, April 2018, pp. 4129–4133.

[54] F. Hua, R. Nassif, C. Richard, H. Wang, and A. H. Sayed, A preconditioned graph diffusion LMS for adaptive graph signal processing, in Proceedings of the 26th

European Signal Processing Conference (EUSIPCO), Rome, Italy, September 2018, pp. 111–115.

[55] F. Hua, R. Nassif, C. Richard, H. Wang, and A. H. Sayed, Decentralized clustering for node-variant graph filtering with graph diffusion LMS, in Proceedings of the 52nd Asilomar Conference on Signals, Systems, and Computers, Pacific Grove, USA, October 2018, pp. 1418–1422.

[56] M. Coutino, E. Isufi, and G. Leus, Distributed edge-variant graph filters, in Proceedings of the IEEE 7th International Workshop on Computational Advances in Multi-Sensor Adaptive Processing (CAMSAP), Curaçao, Dutch Antilles, December 2017, pp. 1–5.

[57] X. Wang, M. Wang, and Y. Gu, A distributed tracking algorithm for reconstruction of graph signals, IEEE Journal of Selected Topics in Signal Processing **9**, pp. 728–740 (2015).

[58] P. D. Lorenzo, S. Barbarossa, P. Banelli, and S. Sardellitti, Adaptive least mean squares estimation of graph signals, IEEE Transactions on Signal and Information Processing over Networks **2**, pp. 555–568 (2016).

[59] P. D. Lorenzo, P. Banelli, S. Barbarossa, and S. Sardellitti, Distributed adaptive learning of graph signals, IEEE Transactions on Signal Processing **65**, pp. 4193–4208 (2017).

[60] K. Qiu, X. Mao, X. Shen, X. Wang, T. Li, and Y. Gu, Time-varying graph signal reconstruction, IEEE Journal of Selected Topics in Signal Processing **11**, pp. 870–883 (2017).

[61] P. D. Lorenzo and E. Ceci, Online recovery of time-varying signals defined over dynamic graphs, in Proceedings of the 26th European Signal Processing Conference (EUSIPCO), Rome, Italy, September 2018, pp. 131–135.

[62] M. J. M. Spelta and W. A. Martins, Online temperature estimation using graph signals, in Proceedings of the XXXVI Simpósio Brasileiro de Telecomunicações e Processamento de Sinais (SBrT), Campina Grande, Brazil, September 2018, pp. 154–158.

[63] M. J. M. Spelta and W. A. Martins, Normalized LMS algorithm and data-selective strategies for adaptive graph signal estimation, Signal Processing **167**, 107326 (2020).

[64] P. D. Lorenzo, E. Isufi, P. Banelli, S. Barbarossa, and G. Leus, Distributed recursive least squares strategies for adaptive reconstruction of graph signals, in Proceedings of the 25th European Signal Processing Conference (EUSIPCO), Cos, Greece, September 2017, pp. 2289–2293.

Index

Printed in the United States
by Baker & Taylor Publisher Services